Modular Origami Airplanes

ISBN 978-1-7378499-0-2

Acknowledgements

I am deeply indebted to my wonderful spouse Merida for her editing skills and constant support..

I am also indebted to my wonderful neighbors, Dan and Meg Beech, for allowing me to use their photographic studio equipment.

I am also indebted to the wonderful Origami designer Ekaterina Lukasheva for critical comments on the manuscript.

Lastly, I am indebted to Sergey Pasterev for the photo of mountains used on the cover.

Table of Contents

What makes paper airplanes fly?
The exact same forces that make insects, birds, and airliners fly. While those forces act somewhat differently on those different flying things, they are all the same.

To discuss what makes things fly we have to talk about two long dead bright guys. The first was likely an alien. I mean an outer space alien, not some dude from another country. Newton was amazing, and laid the fundament of our understanding of the universe. If extraterrestrials ever came to Earth, he was certainly one of them.

One of Newton's postulates was that for every action there is an equal and opposite reaction. Thus if an air molecule strikes an object in flight, the object will react. The reaction of the object creates a force called **lift**.

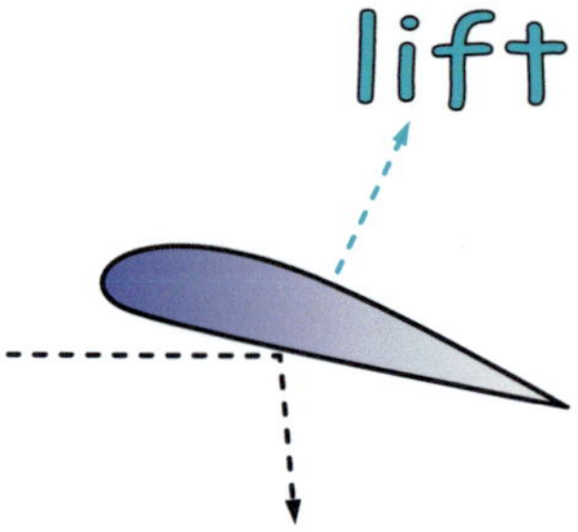

This is an easily observable phenomenon. Hold you hand at an upward angle when moving rapidly in a car or if you're cavorting on your unicycle. You'll feel it pushed upwards by the passing air. This applies to all bodies in motion. But not all bodies in motion fly (actually, they will if you throw them hard enough…).

An engineer named Bernoulli discovered the other force we need to consider. While measuring the speed of the river Seine he discovered that as liquids speed up the pressure they exert goes down. That's what happens at the surface of an airplane wing. Most wings have curvature on the upper surface (called camber) and none on the bottom surface. The air has to speed up on the upper surface, so it exerts less pressure. The net upward force is lift.

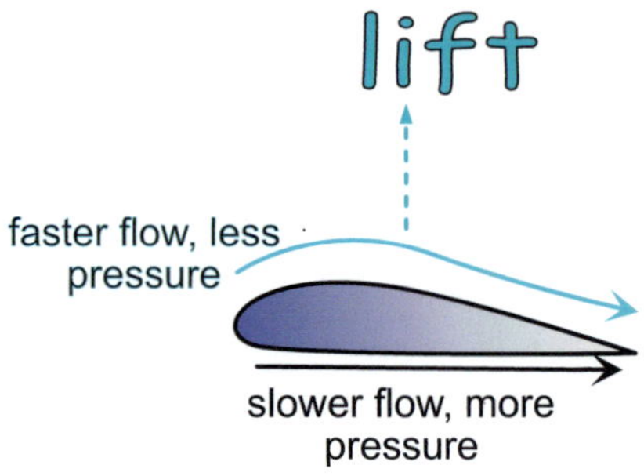

Nothing comes without a price. Lift creates drag, the enemy of flight. Lifting forces all produce drag, the more the lift the greater the drag. Fast airplanes usually have short thin wings; slower airplanes have thicker longer ones. The best flying paper airplanes have the lowest drag possible. Here this collection suffers somewhat. By their nature modular airplanes are draggy, and I doubt any of these will set any world records. That said, they will fly across the room just fine and look good doing it.

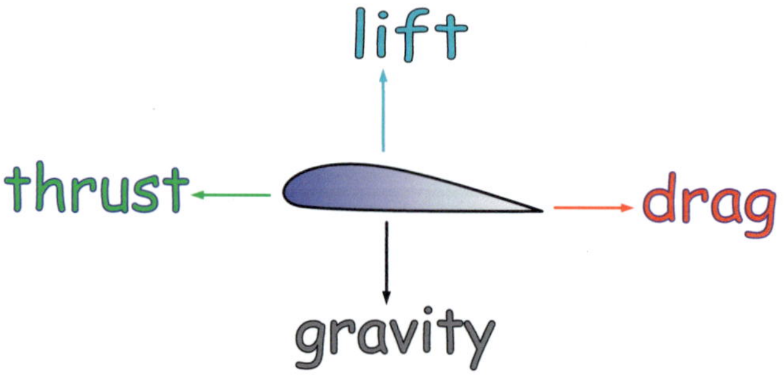

Why doesn't my paper airplane fly?
Sadly, this can happen. Modular aircraft can be a bit tricky, though I promise everything I've diagrammed in this book has flown well. Sometimes it takes just a bit of adjustment to get them n the right track.

My airplane is unstable in flight

Newly made airplanes can barrel roll and do all sorts of awful ungraceful things, and yes, it has happened to yours truly. The first thing to check is the construction of the airplane. Are the creases tight, and are the sets of wings at the same *angle of incidence* (angle relative to the fuselage)?

If everything is ship shape then a way to fix the situation is **dihedral**. This is the angle of the wings relative to the ground. A little dihedral can stabilize any airplane, indeed many real airplanes are built with it.

no dihedral

In a roll, lifting forces are the same on both wings. There is no force to counteract the roll.

positive dihedral

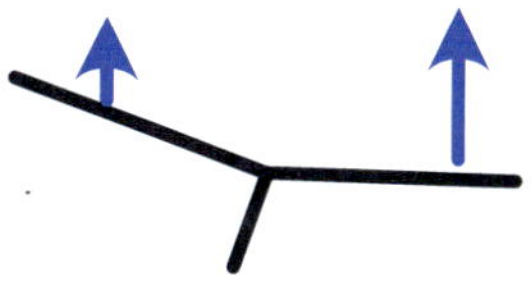

In a roll, the lifting force on the downward wing is greater, as the upward wing is at an angle and can't generate as much lift. The imbalanced lift forces generate a force counter to the direction of roll.

My airplane doesn't seem to have any lift, and just drops to the ground

A number of things can rob an airplane of lift. One problem with the Modular Aircraft is many are fairly nose heavy. A good way to treat a nose heavy airplane is to give it some up elevator. Elevator is what we use in real airplanes to point the nose upwards, it is usually at the back of the airplane and deflects wind up. Upward wind pushes the tail down and the nose up.

The most forceful elevator is a jet tail, it basically pushes up all the air flowing through the fuselage. Many of the aircraft are diagrammed with tails, as I found in testing that they flew better that way. And of course airplanes with jet tails look snazzy.

To fold a jet tail you need to *reverse fold* the end of the fuselage up between layers. It will improve the flight of some, on others it'll be too much. If the tail is too much, unfold it. The airplane's flight problem might even be fixed with the tail folded and then unfolded. I've seen it happen.

The Lift generated by the tail is spoiled when the elevator is raised. This causes the tail to drop and the nose to rise.

How to fold a tail:

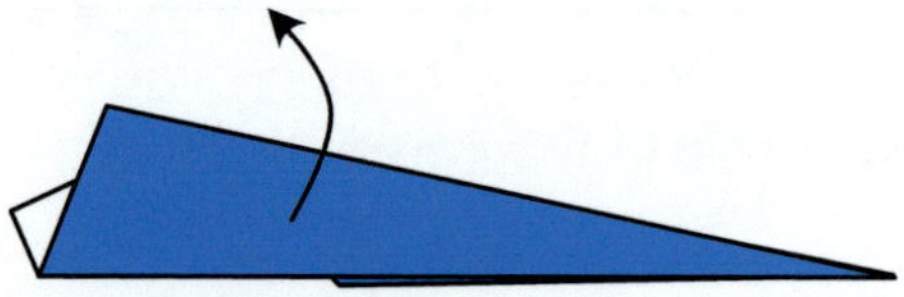

1. We'll start with a traditional paper airplane. Hopefully you've seen this before. Unfold the wings.

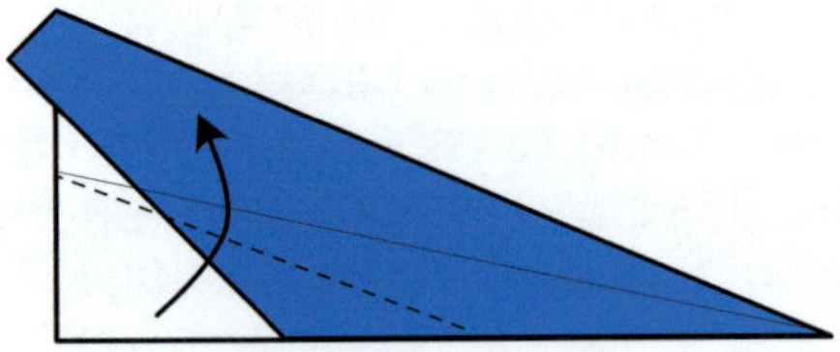

2. Valley fold up where you want your tail to go.

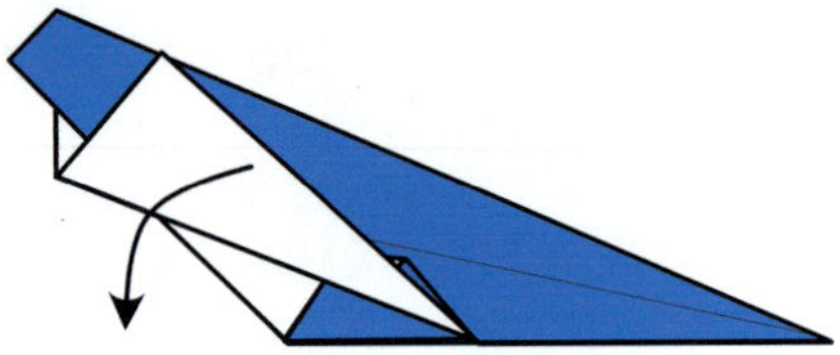

3. Unfold what will be the tail.

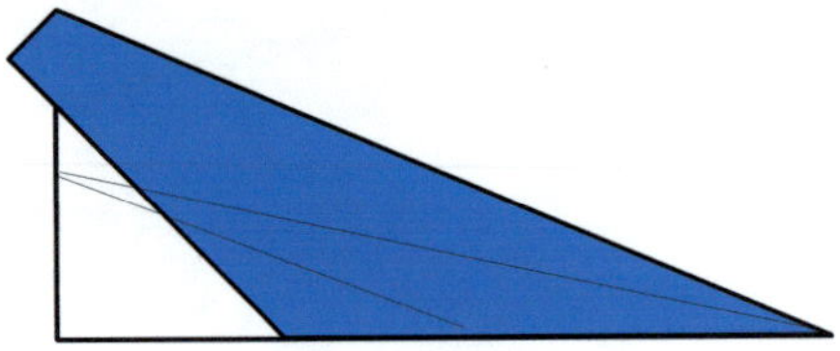

4. Reverse fold the tail between the layers of the fuselage.

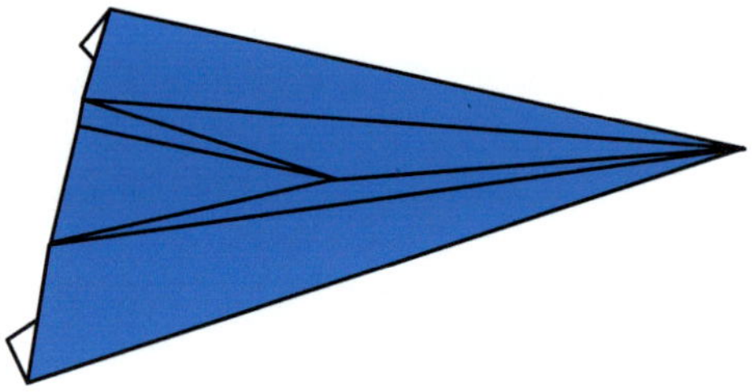

5. Reverse fold in progress.

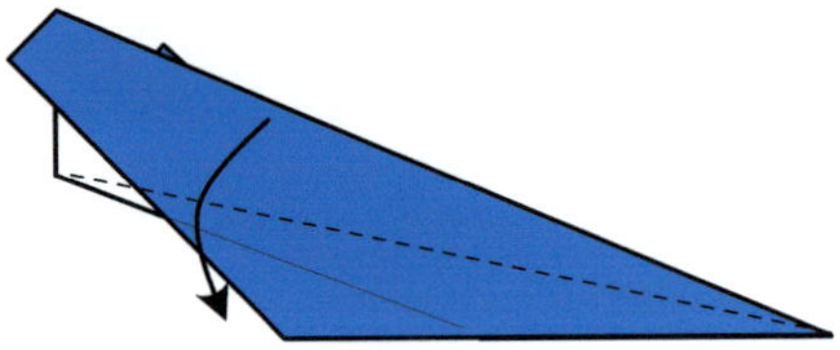

6. Once done you can fold the wings back down and you'll

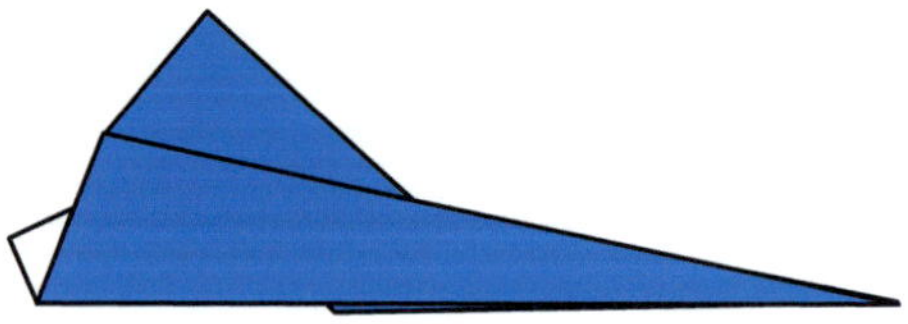

Paper

All of the aircraft in this collection are made from square pieces of paper, a bit unusual for origami airplanes. I started doing this for the extra challenge, as squares as shorter, so there's less paper to distribute weight and build structures. But a square can be folded in half rectangularly or diagonally, giving added design flexibility that proved crucial for this collection.

I designed all these airplanes using Kami, the Japanese origami paper found in book stores and on the internet. I would recommend you fold them from kami, as it's light, crisp, and quite attractive. If you're going to spend a lot of time folding airplanes you might as well make them look good. These days kami can be found in many, it not most book and craft stores and is also available from numerous internet outlets. If you use the traditional Kami that is colored on one side you'll get a color change, you may or may not like that. Origami papers colored on both sides are widely available on the internet.

You'll see a number of papers employed in the photographs in the book. I've used foil paper in a number of models. Foil doesn't help airplanes fly better, though it was handy to get better photographs. If you use thicker paper the airplanes will likely fly, though not as well. More weight carried by the same wing area means less flight time. The one problem you might run into with thicker papers is where the models join together, as it can get quite thick in there with multiple layers of paper all compressed together.

How to use this book?

I wrote this to allow you to make complex aerodynamic shapes from what I hope is simple and straightforward Origami. That said, some of these folds are simpler than others. The models in the beginning of the book, both of the front pieces and the back pieces, are simpler, while those toward the end tend to use more challenging Origami.

I begin with the ten Front Pieces. You begin by folding the join unit on Page 11, and then fold it into a Front Piece. You can then go find a Back Piece to make an airplane. I show the attachment of each Back Piece to at least one Front Piece. For a few I show multiple Front Pieces because there is a complication in the attachment for a Front Piece. The landmarks for folding wings is shown either in the folding of the Front Pieces, where they're attached to the Back Piece, or both.

I give lots of instructions on the landmarks to fold wings and how to position the sets of wings. Some of the combinations have their own special instructions. I have tried my best to empower you to make nicely flying paper airplanes.

How many airplanes are there in this book?

Most of the Front Pieces and most of the Back Pieces make fairly decent airplanes in their own right if the front pieces are folded without the join module, call it 15. So 15 airplanes for starters. And yes, we're only just getting started.

For dimers, that is airplanes made from a Front and Back piece, there are 10x10 possibilities, or 100. Not all the dimers flew, I count 91 that did. I should stress not all the dimers flew *for me*, you might have better luck.

Of trimers, or three piece airplanes, there are 10x10x10 possible combinations, or 1000. I didn't try all thousand, I have a life. I did put together 10 that flew well. So 116 airplanes diagrammed in some fashion in this collection.

I diagrammed the folding for all the Front Pieces and all the Back Pieces. I have shown the attachment strategy for each back piece. I only specifically diagrammed one front piece being attached for most, doing more would become repetitive. I do show plan views for all the combinations that made flyable aircraft.

Are there possibilities I didn't show in this book? Goodness yes! Most of the Front pieces can act as back pieces, hence there are 100 Front piece dimer possibilities. And I showed ten out of a possible thousand three piece airplanes. Lots of potential discovery for the aspiring paper airplane artist.

How to throw these paper airplanes

Any paper airplane should be thrown from its center of gravity. For those in this collection the center of gravity will be at the point of greatest thickness, where the two pieces join each other and is where the airplanes should be grasped for throwing. Angle the airplane upwards and throw. Some of the airplanes will respond well to a full power throw, some not as much. Have fun finding what works best for yours!

Symbols

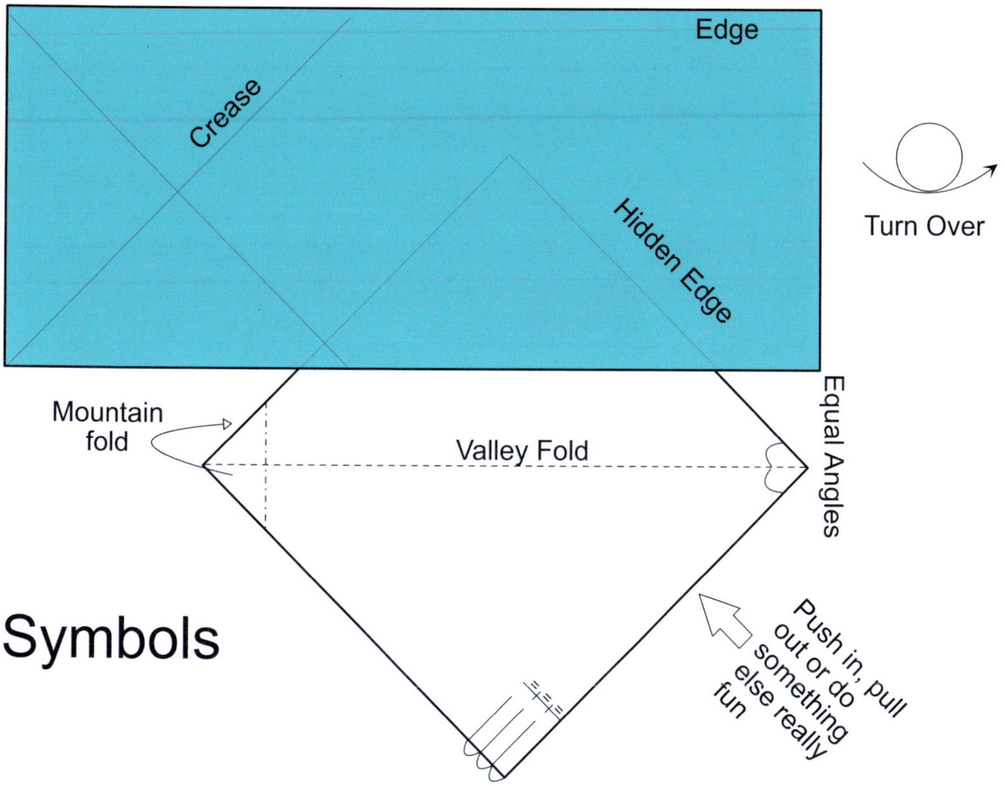

Valley fold

Fold such that the crease points away from you (and the flap towards you).

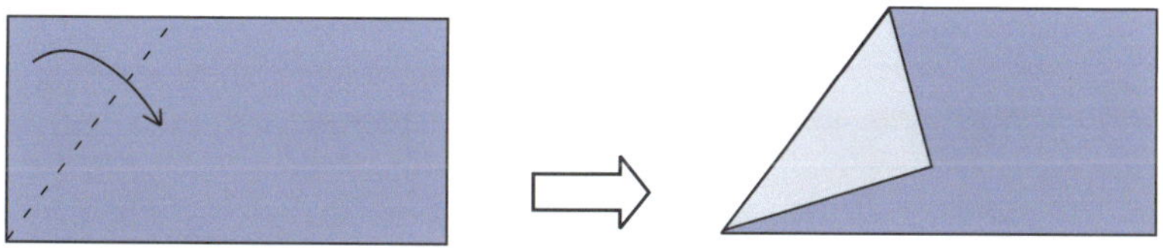

Mountain Fold

Fold so that the crease points toward you (and the flap away from you.

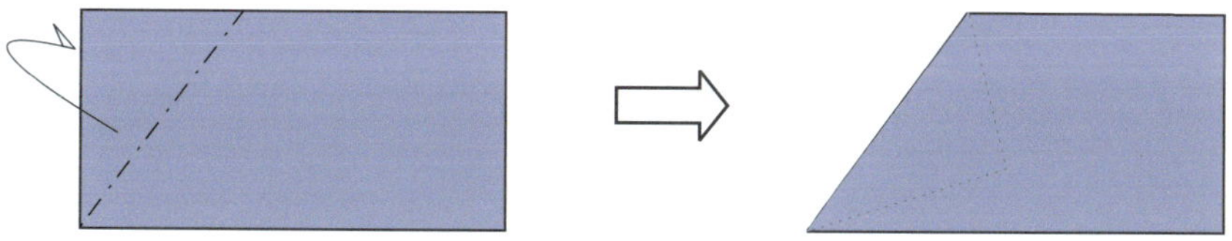

Inside Reverse Fold

Fold so that a point falls between layers.

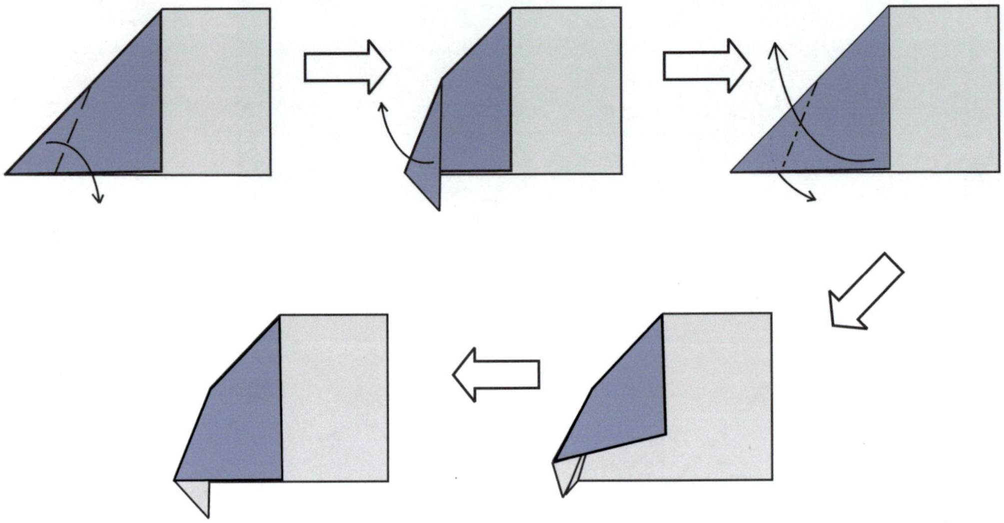

Outside Reverse Fold

Fold so that a point wraps around the outside of two layers.

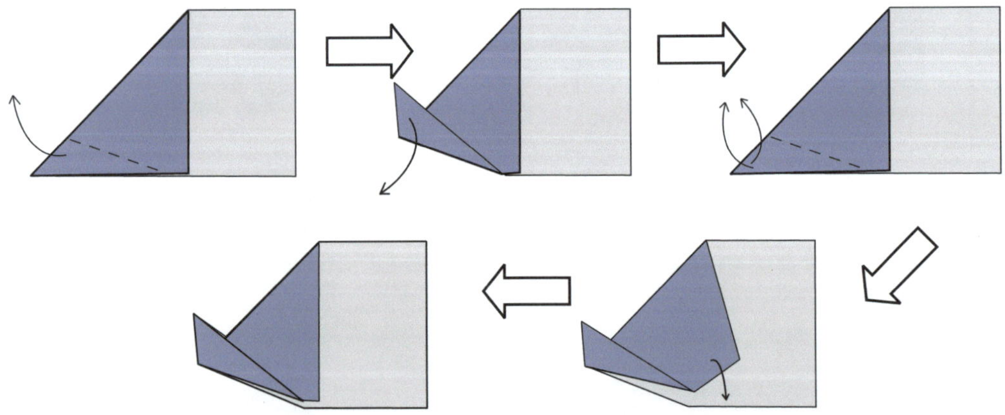

Rabbit Ear

Pull out a hidden layer and fold it flat.

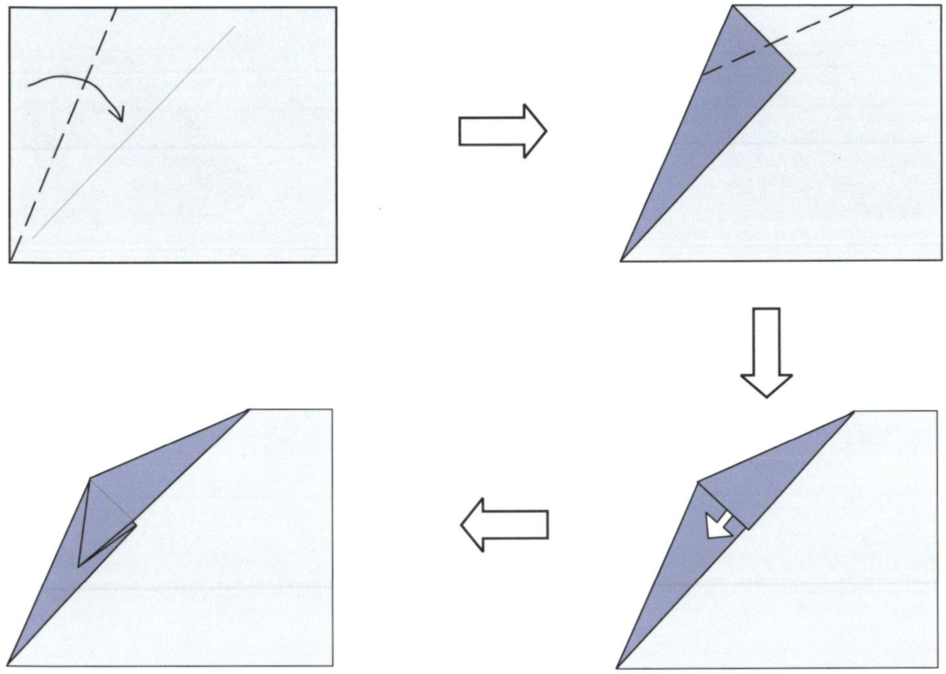

Squash fold

Separate two adjacent layers and flatten the pocket that forms between them.

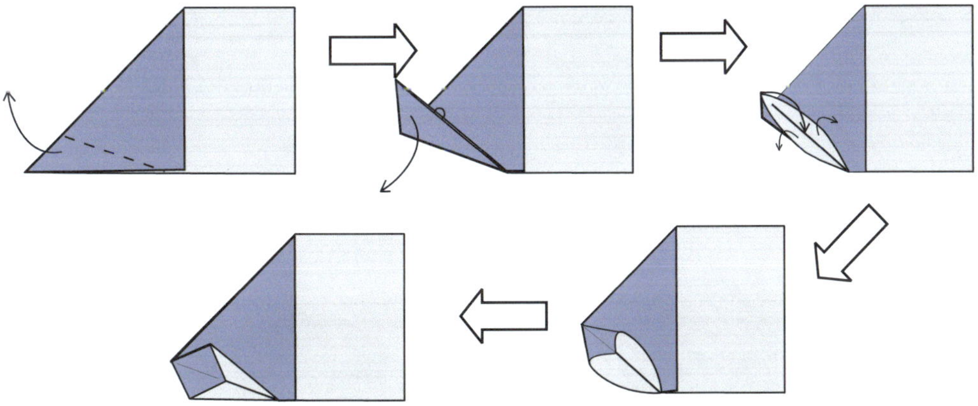

Petal fold

Separate two layers while flattening the pockets that form on either side.

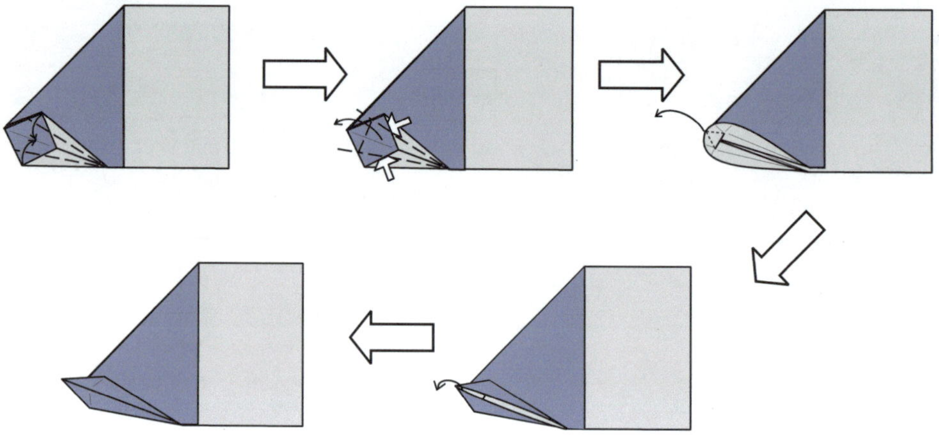

Sink Fold

A multilayered point gets pushed to the inside. The point is partially unfolded and mountain folded at the desired position.

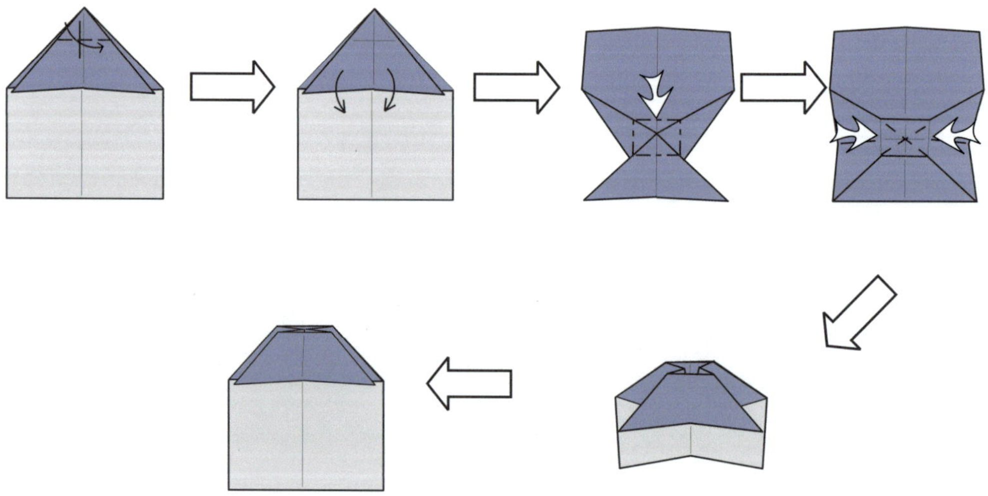

Front Piece Base

This is shown color side up, but can be started just as well with the white side up.

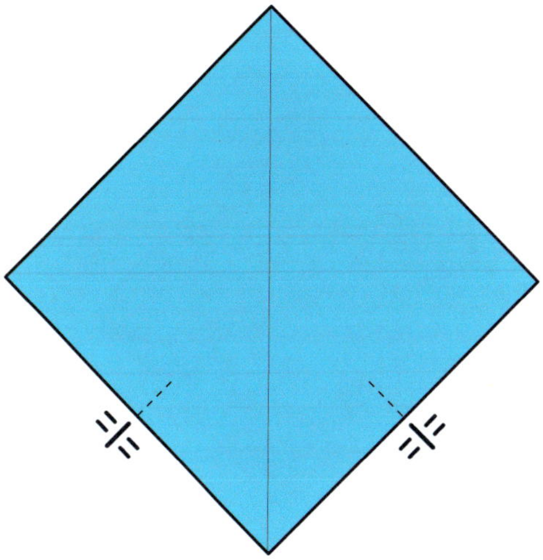

1. Make a pinch halfway down each side.

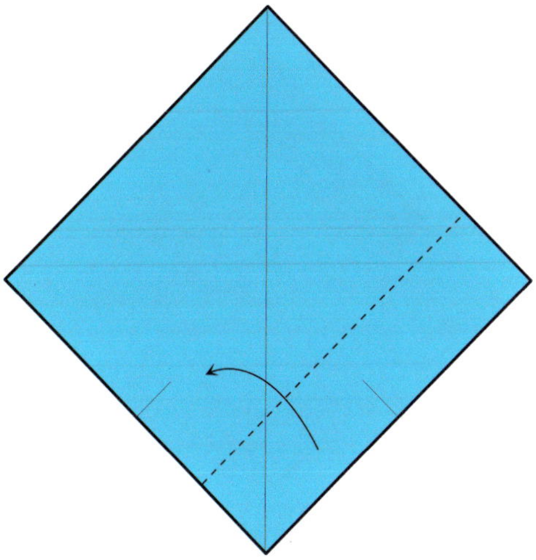

2. Valley fold the edge to meet the pinch.

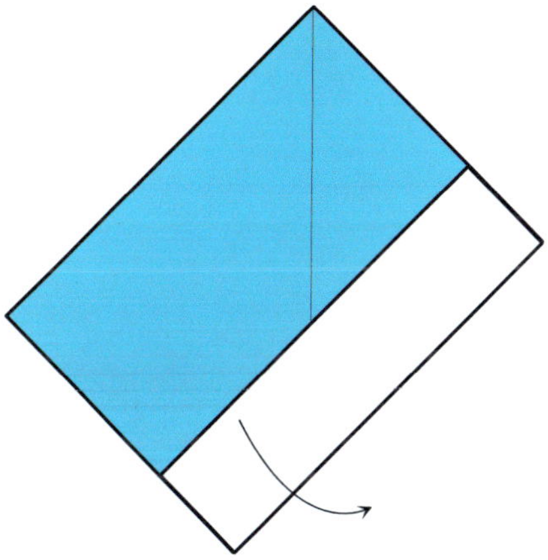

3. Unfold.

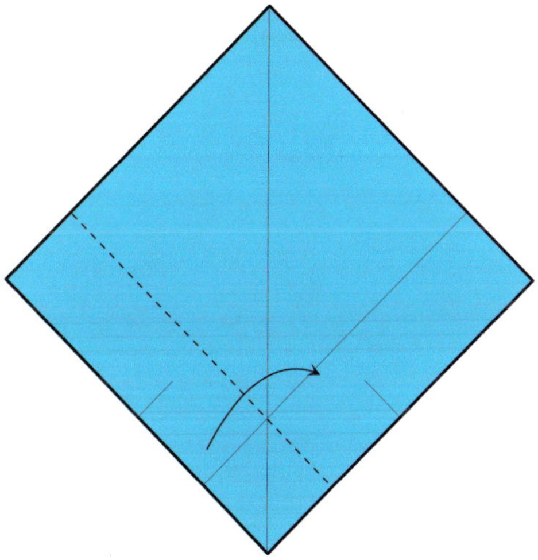

4. Repeat on the other side.

Front Piece Base Page 2

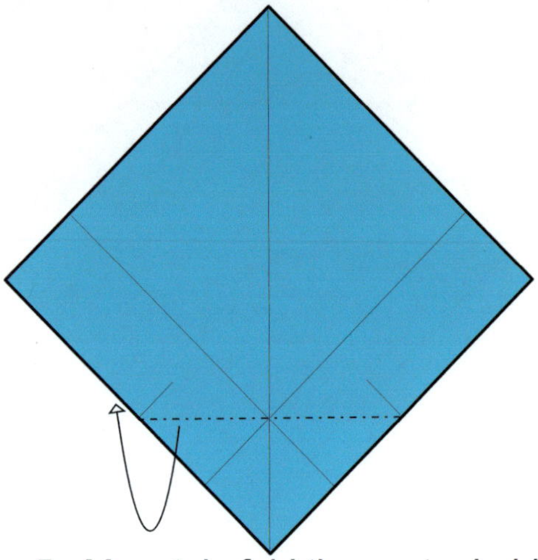

5. Mountain fold the center behind where the diagonal creases meet. Unfold.

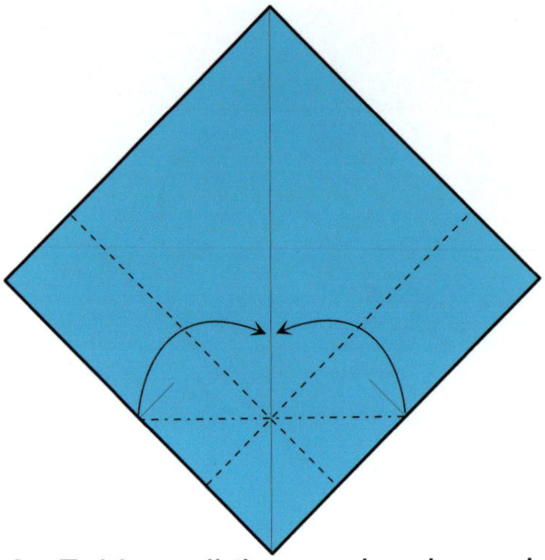

6. Fold on all the previously made creases simultaneously.

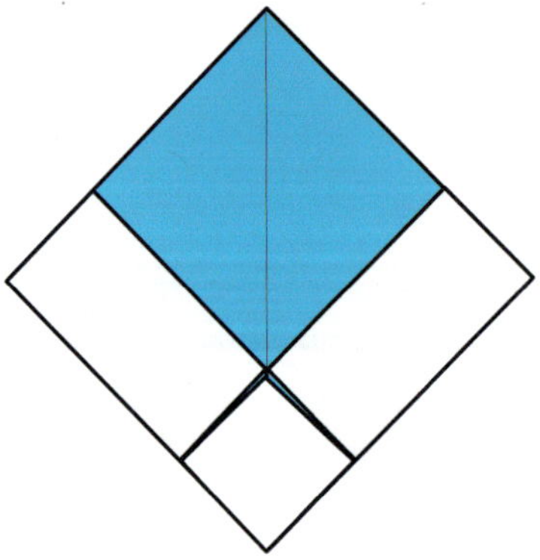

7. All of the front pieces will start with this base.

Front Piece Alpha

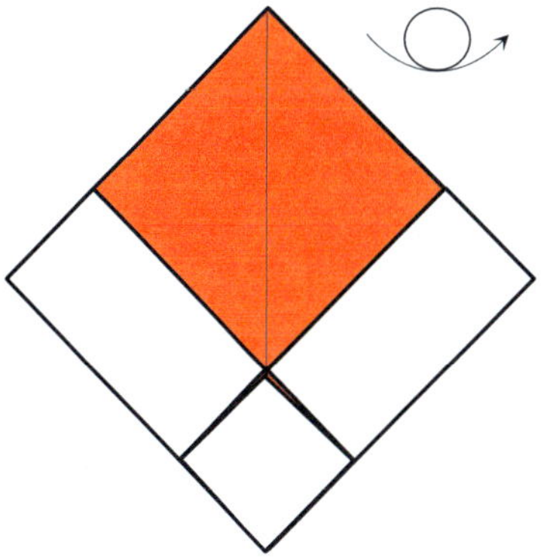

1. Begin with the Front Piece Base. Turn over.

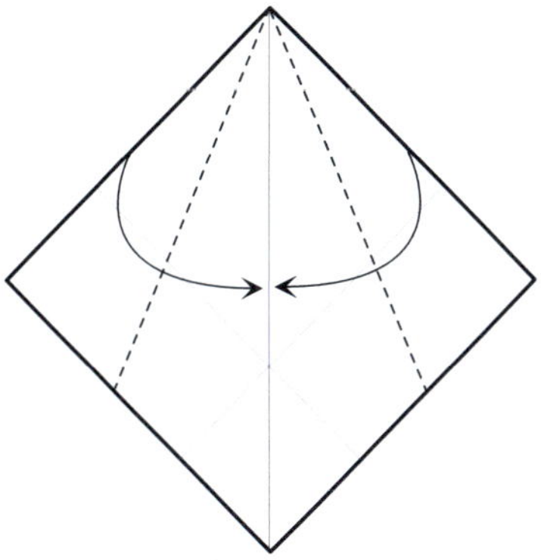

2. Valley fold both edges into the center.

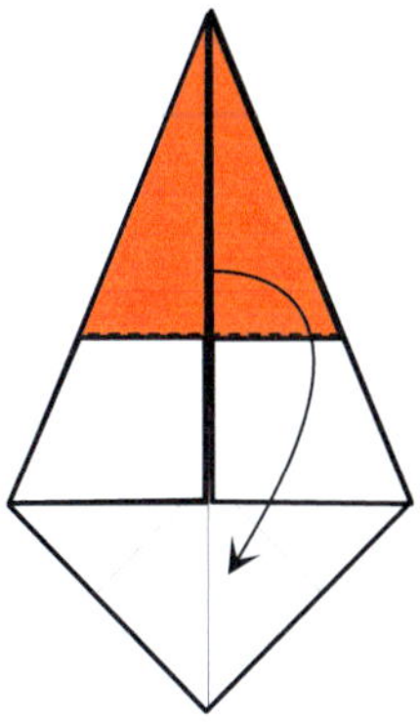

3. Valley fold the top down along the color change.

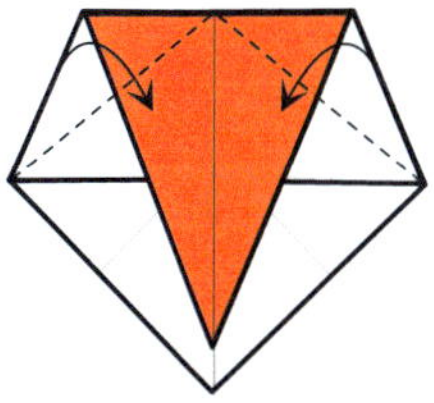

4. Valley fold from the top center to the outside corner.

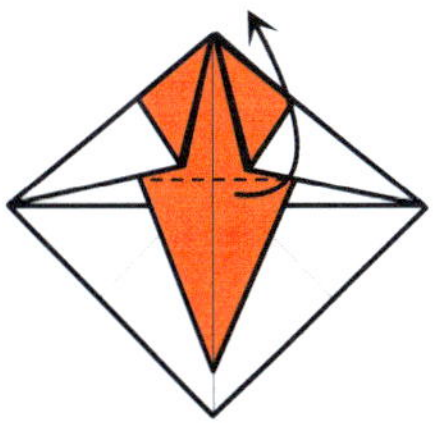

5. Valley fold the pointed flap up as far as it will go, trapping the other layers underneath.

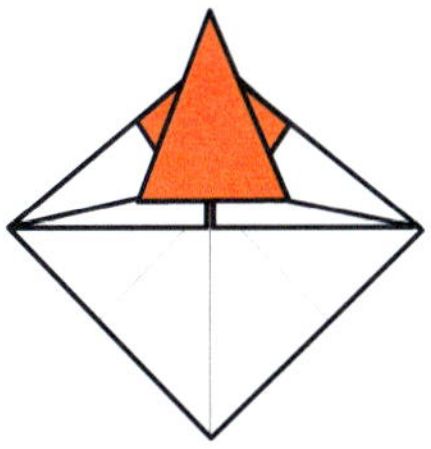

Front Piece Alpha is done.

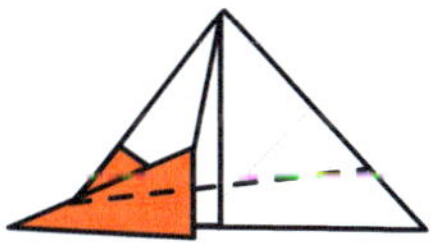

Unfortunately, there aren't a lot of landmarks for folding the wings for front piece Alpha. Don't fold from the very front, or the paper will tear. Angle the fold up a bit to get nice swoopy wings.

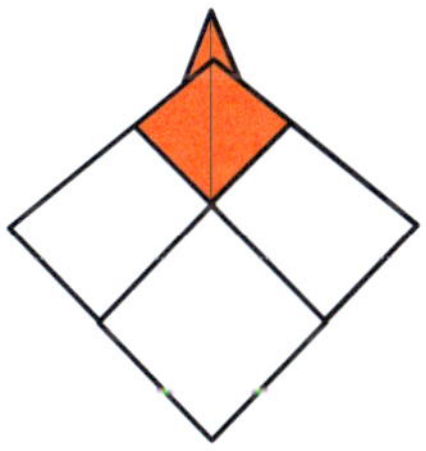

Front Piece Beta

Begin with Font Piece Alpha, step 3

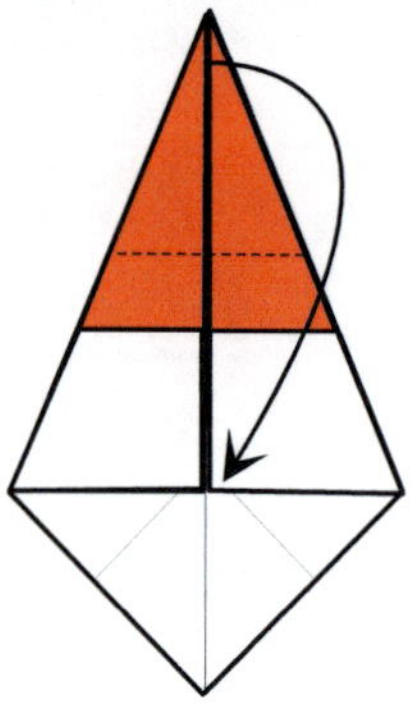

1. Valley fold the top down so that it meets the folded edges.

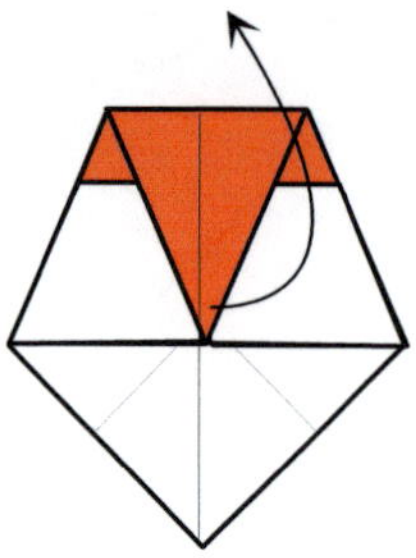

2. Unfold.

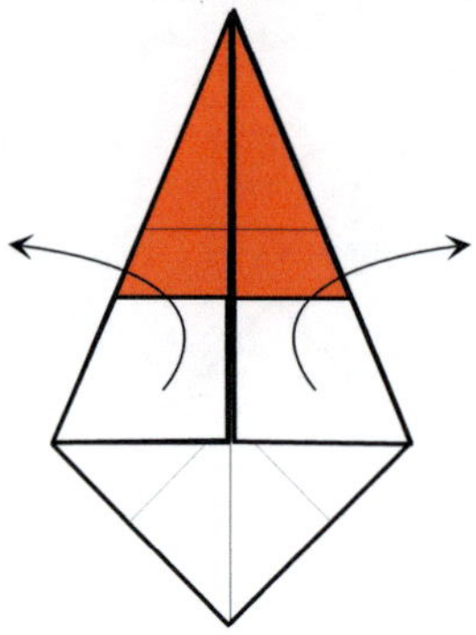

3. Unfold the sides.

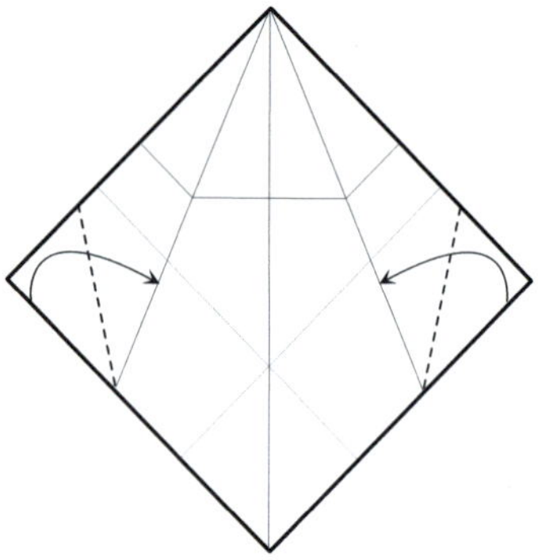

4. Fold the corners in so the lower edge meets the adjacent crease.

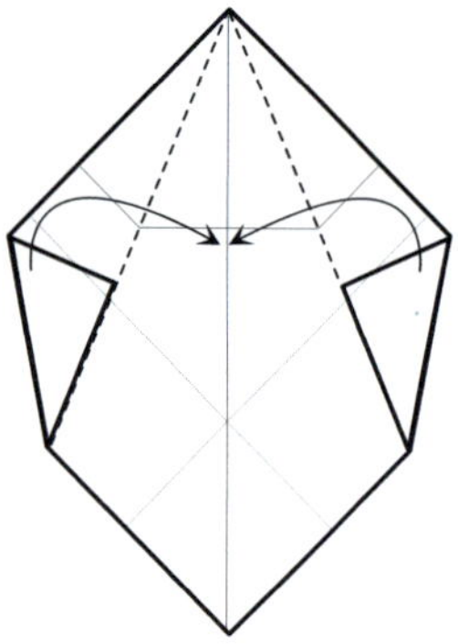

5. Fold the sides back in.

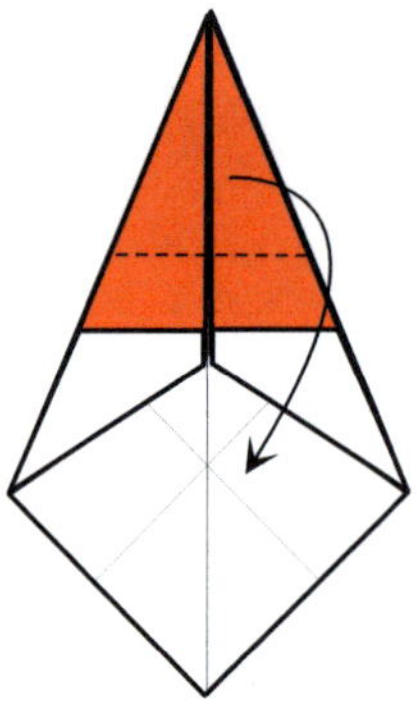

6. Valley fold the top down along a preexisting crease.

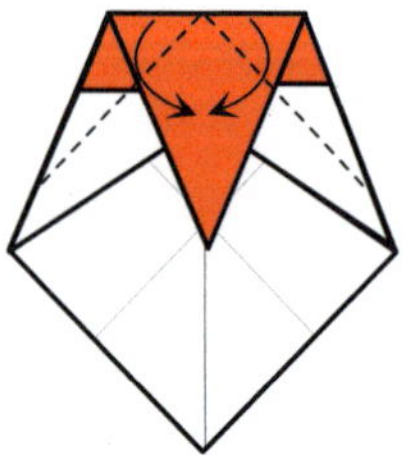

7. Valley fold so that the folded edges at the front lie on the midline.

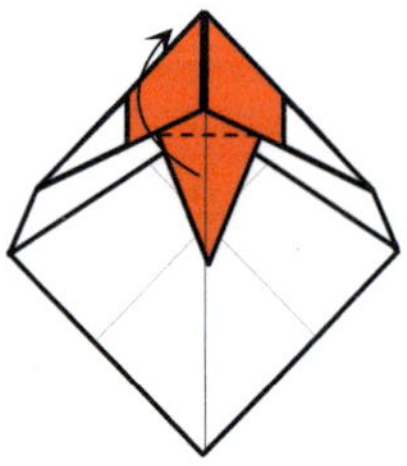

8. Valley fold the point up as far as it will go, trapping layers underneath.

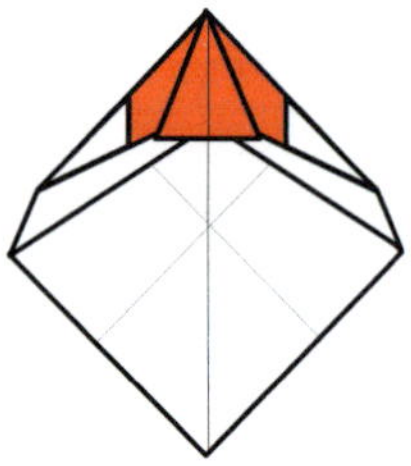

Front Piece Beta is finished.

Front Piece Gamma

Begin with step 3 of Font Piece Alpha.

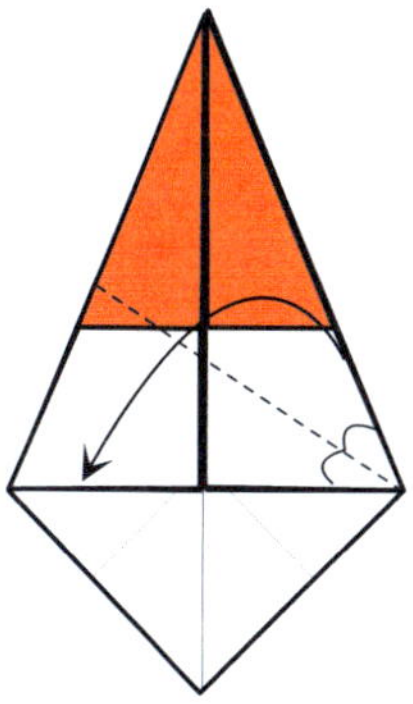

1. Valley fold the top down so that the long edge lies on the folded edges below.

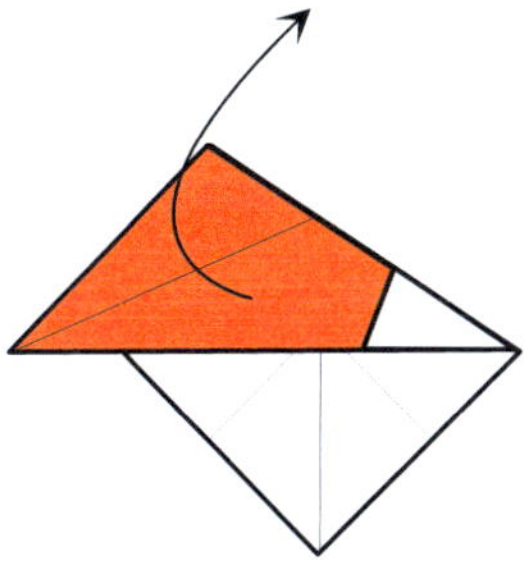

2. Unfold.

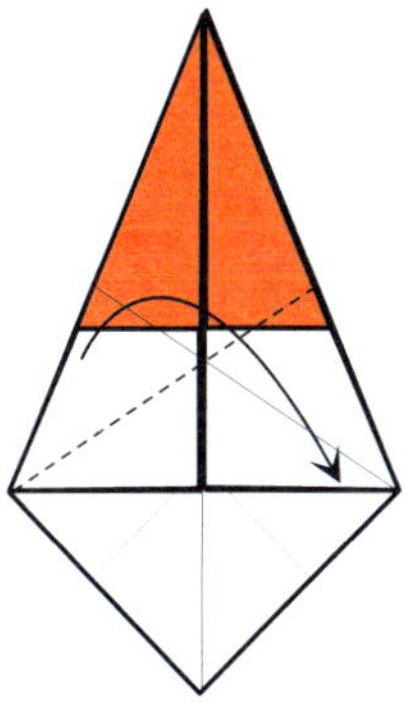

3. Repeat steps 1 and 2 on the other side.

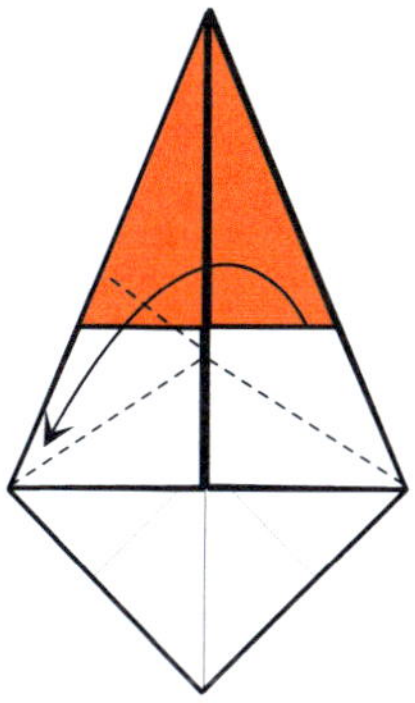

4. Rabbit ear the flap on exiting creases.

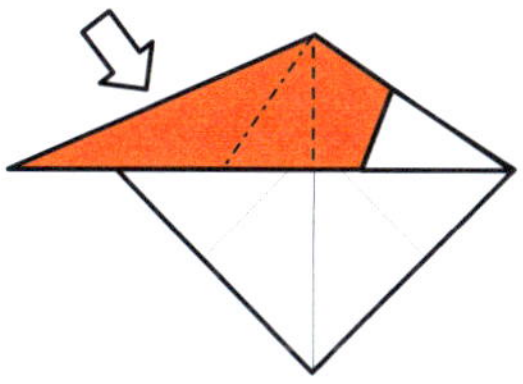

5. Squash fold the flap downwards.

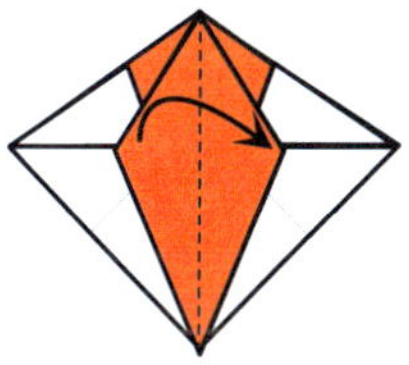

6. Valley fold the diamond flap in half.

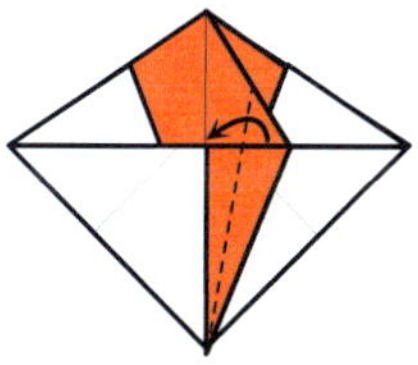

7. Valley fold the one layer in half.

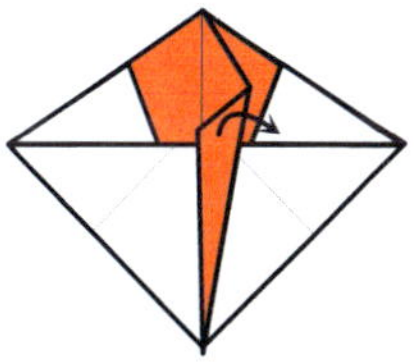

8. Unfold.

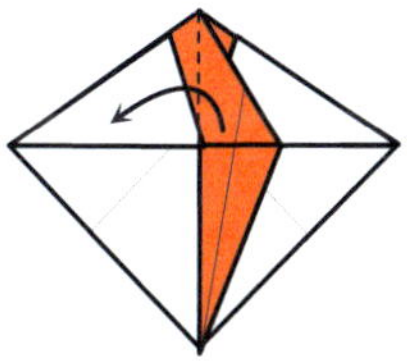

9. Valley fold 2 layers over.

Front Piece Gamma Page 2

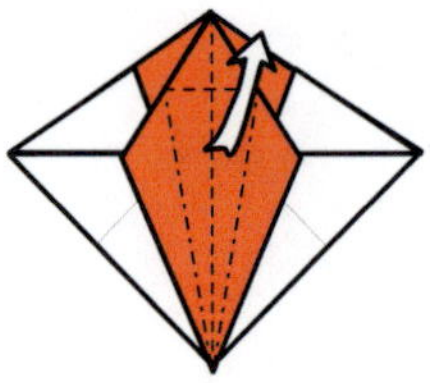

10. Petal fold upwards.

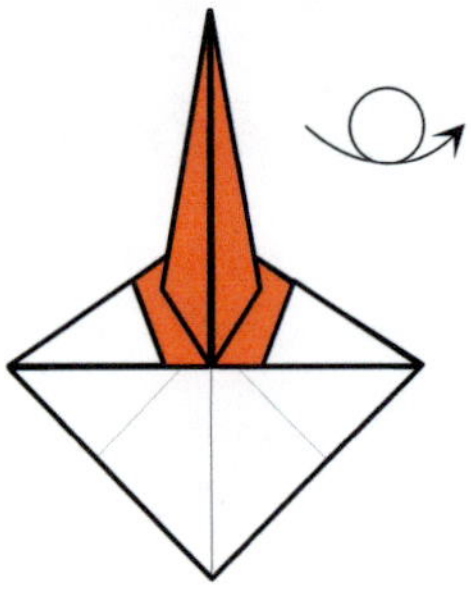

11. Turn over.

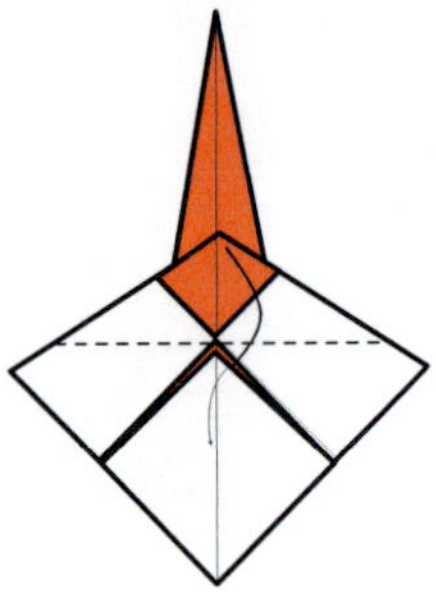

12. Valley fold the leading edge of the wing.

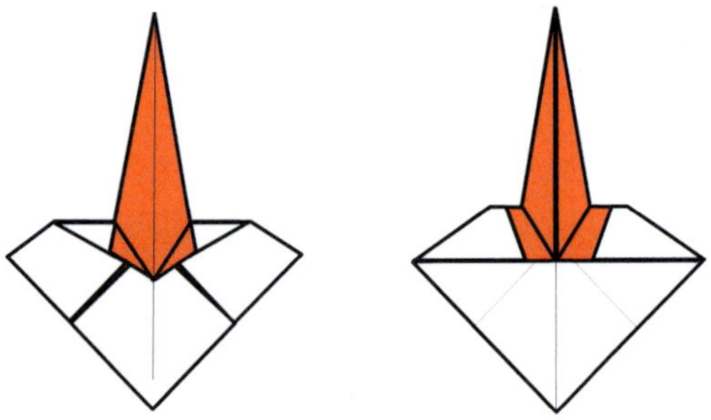

Front Piece Gamma is complete.

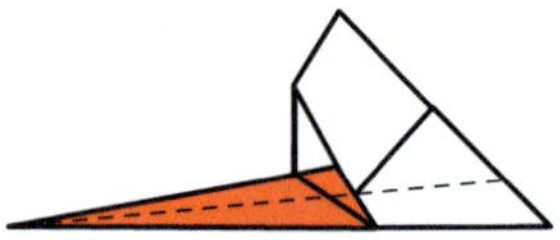

To fold wings for Front Piece Gamma, bisect the angle of the nose and keep the fold moving toward the back.

Front Piece Delta

Begin with step 6 of Font Piece Beta.

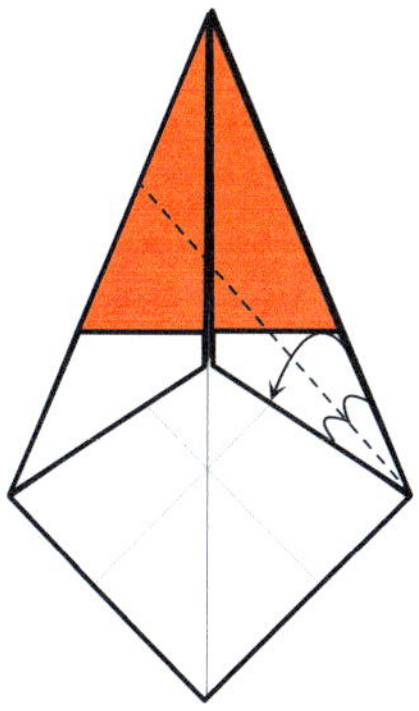

1. Valley fold so that the long folded edge at the top sits on the adjacent folded edge.

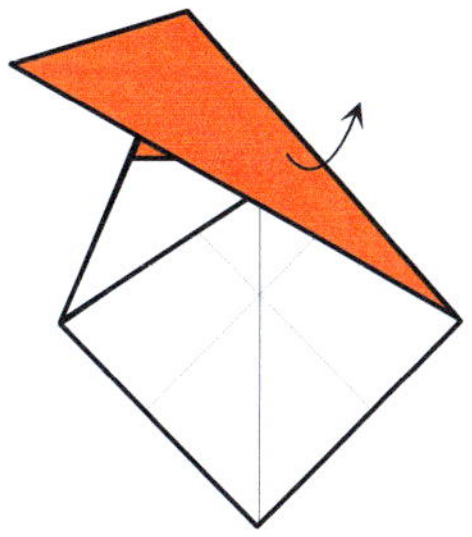

2. Unfold.

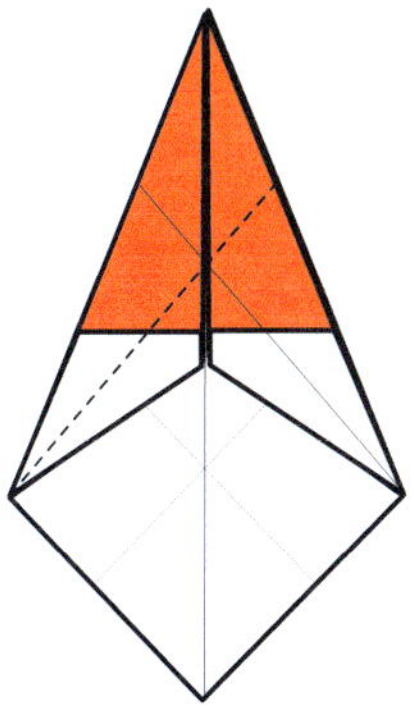

3. Repeat step 1 and 2 on the other side.

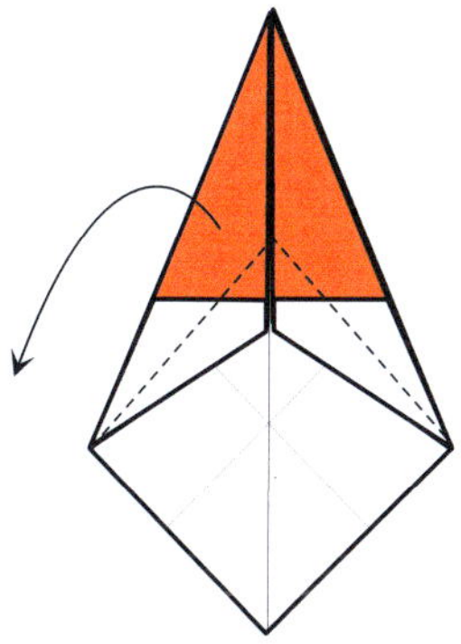

4. Rabbit ear fold the flap to the side.

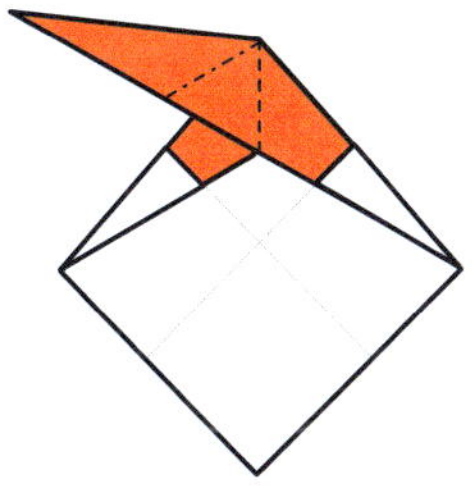

5. Squash fold the flap down.

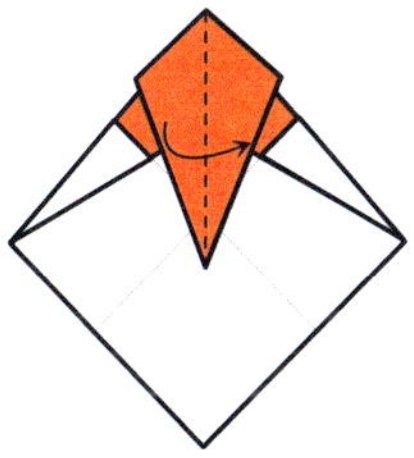

6. Valley fold the flap in half.

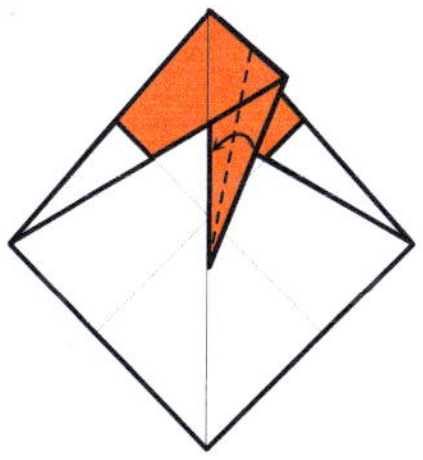

7. Valley fold the flap in half.

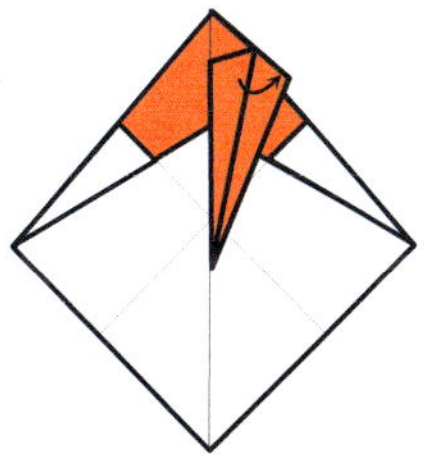

8. Unfold.

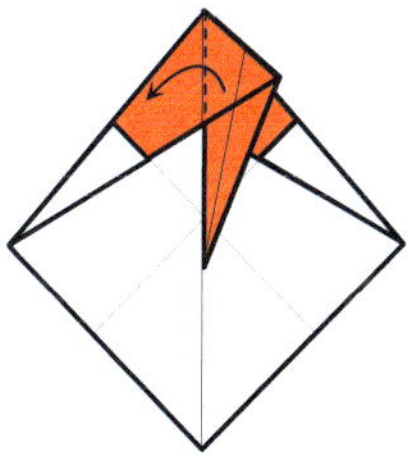

9. Fold two layers to the left and repeat steps 7 and 8.

Front Piece Delta Page 2

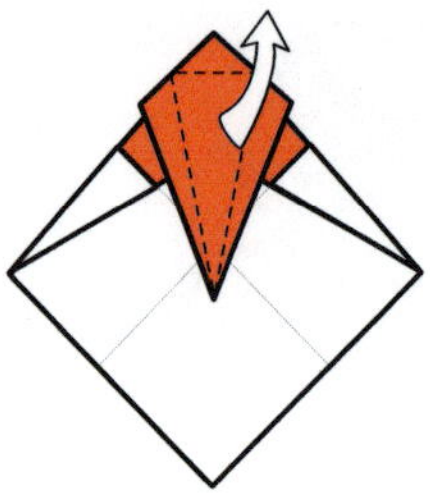

10. Petal fold the flap upwards.

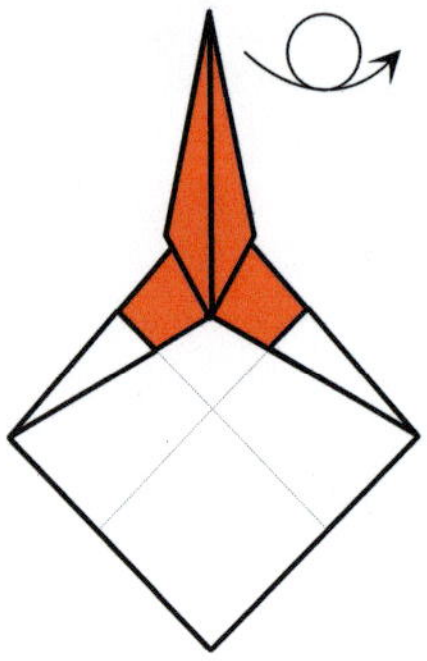

11. Turn over.

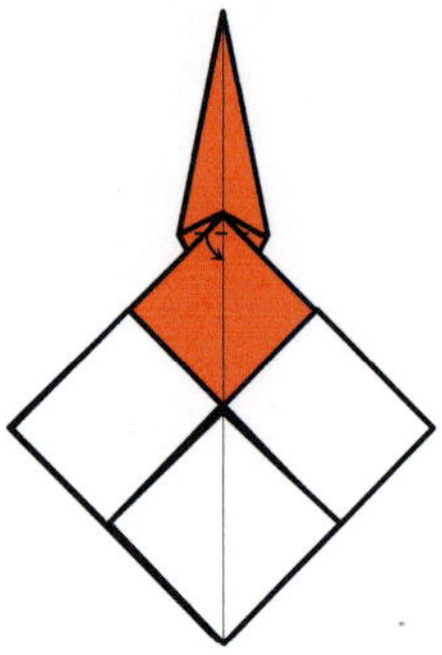

12. Valley fold small triangular flap at the top.

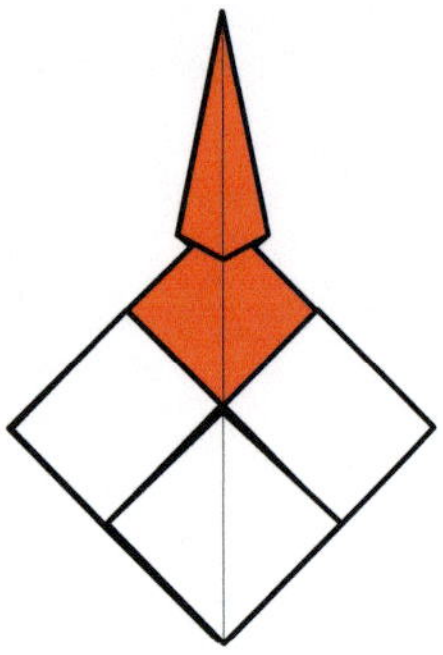

Front Piece Delta is complete.

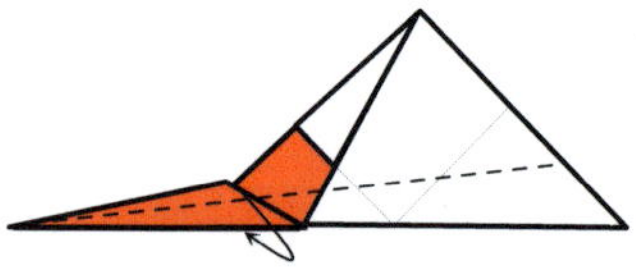

To fold the wings on Delta, fold the intersection shown to the middle. That will set the angle of the wings. Fold back, and of course fold the back wings at the same angle.

Front Piece Epsilon

Begin with step 4 of Font Piece Delta.

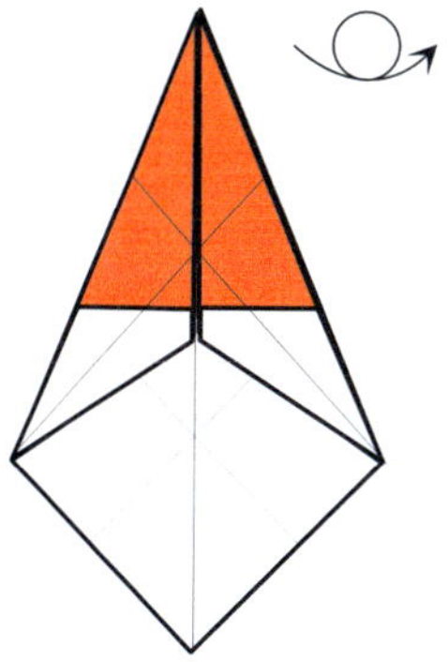

1. Turn over.

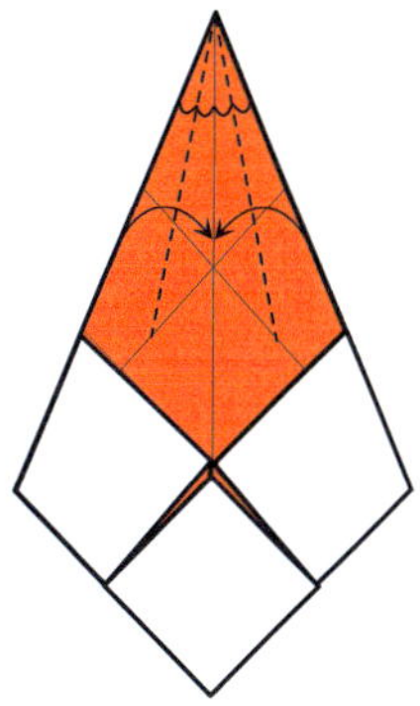

2. Partially valley fold each edge to the center. The fold should end at a diagonal

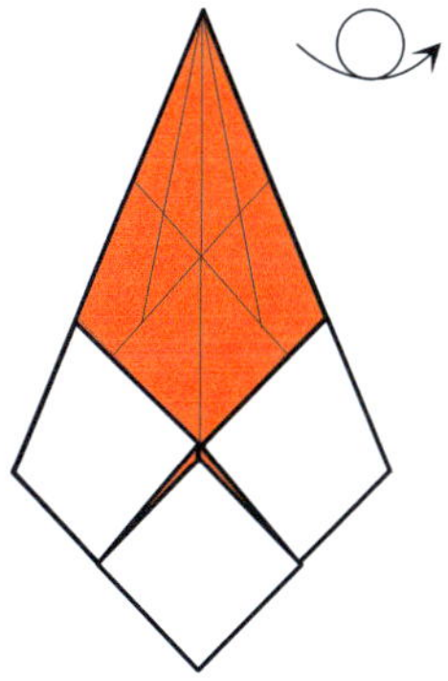

3. Turn over.

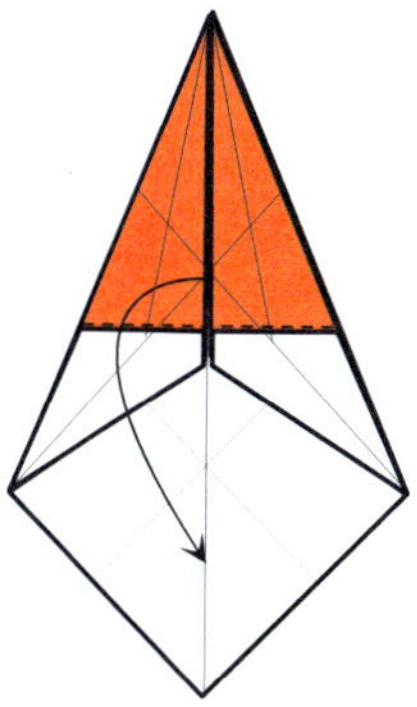

4. Valley fold at the color change.

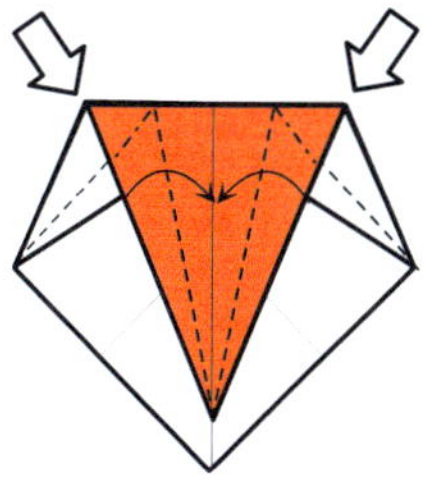

5. Squash fold each side on existing creases.

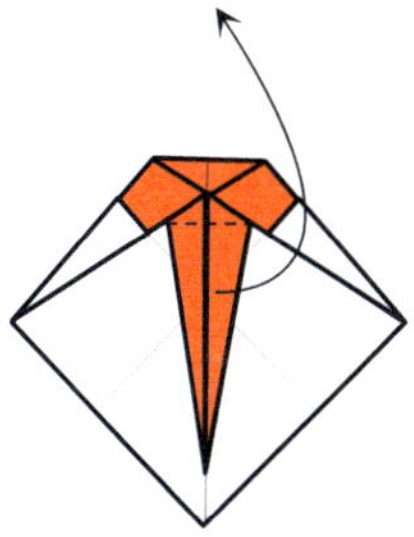

6. Valley fold the long flap upwards as far as it will go.

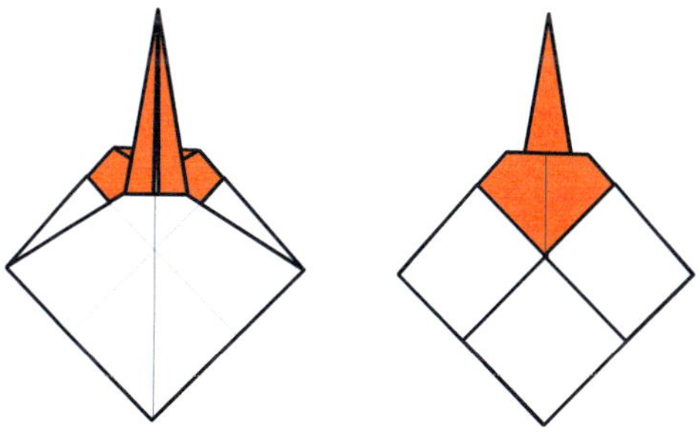

Front Piece Epsilon is done.

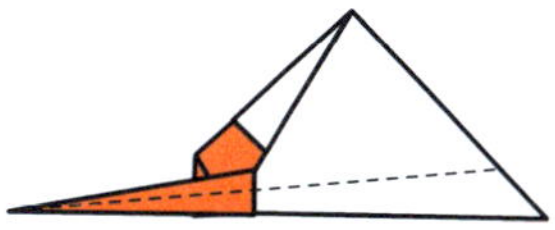

To fold the wings for Epsilon, valley fold bisecting the front angle in half. Continue the fold onto the rest of Epsilon, and then copy it onto the back piece.

Front Piece Zeta

Begin with the Front Piece Base.

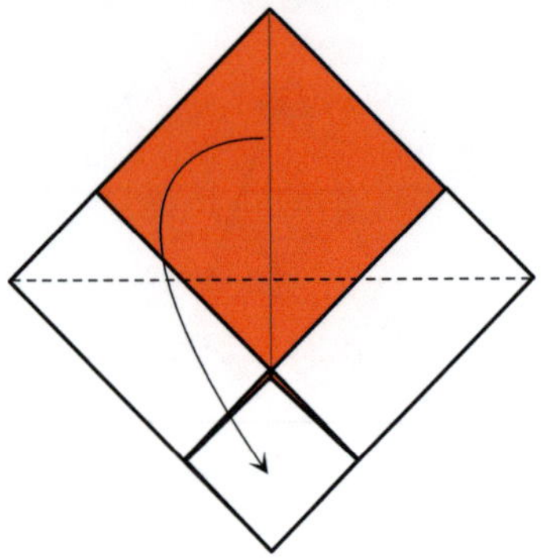

1. Valley fold in half.

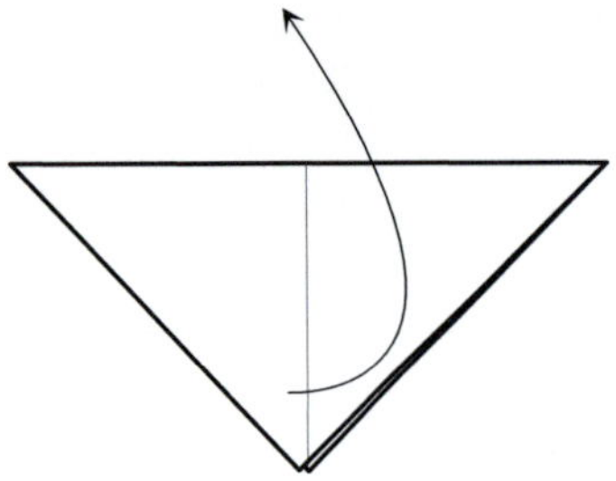

2. Unfold.

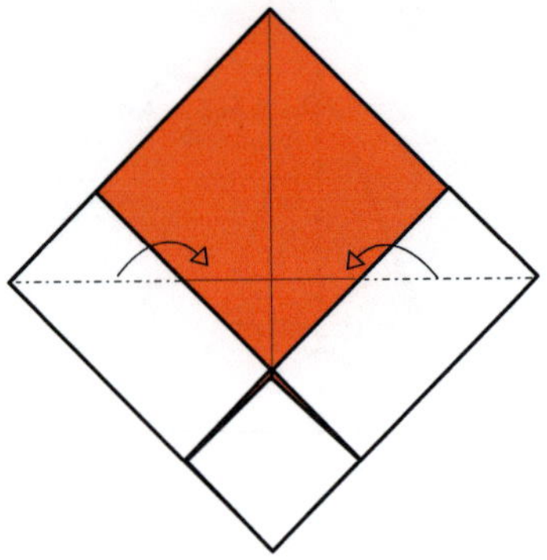

4. Mountain fold the corners behind.

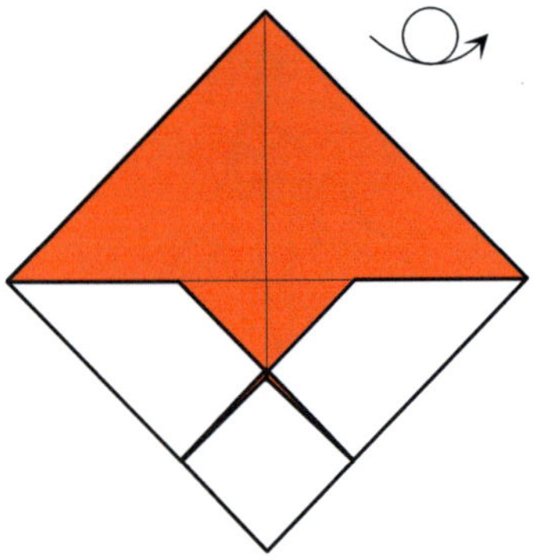

4. Turn over.

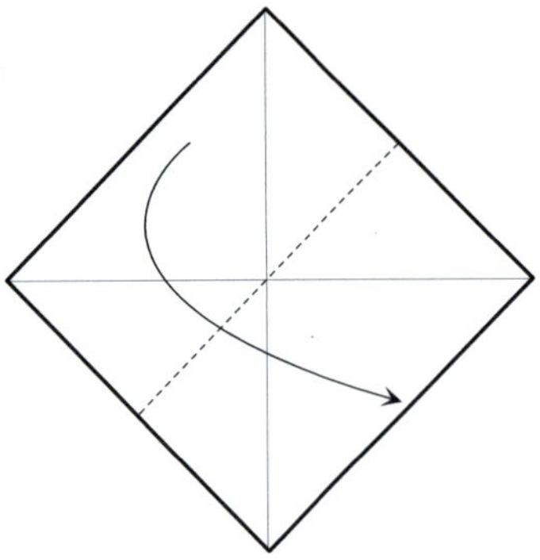

5. Valley fold in half rectangularly.

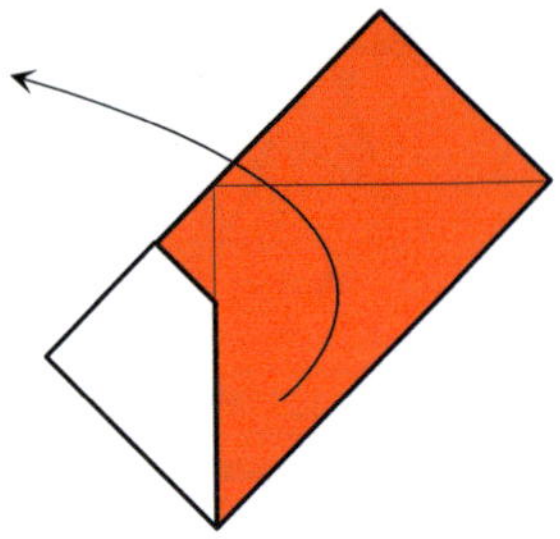

6. Unfold.

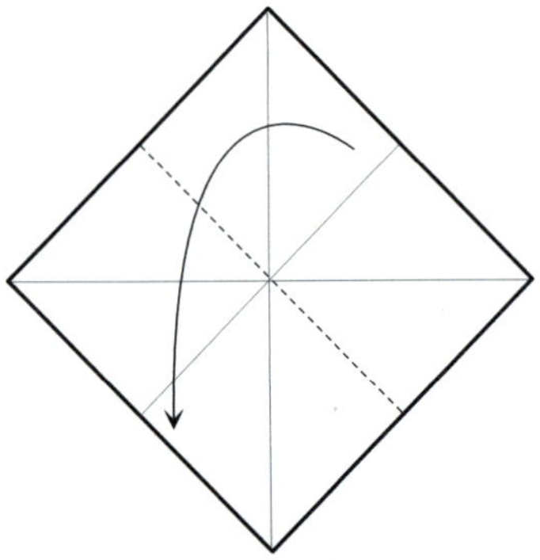

7. Repeat steps 5 and 6 on the other

Front Piece Zeta
Page 2

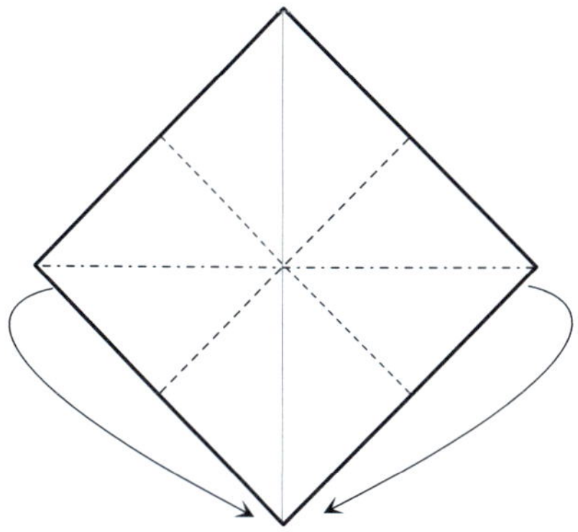

8. Bring the four corners together at the bottom.

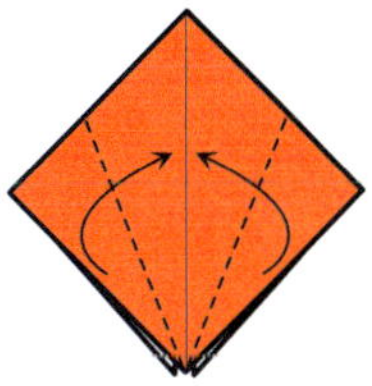

9. Valley fold the two-ply edges into the middle.

10. Unfold.

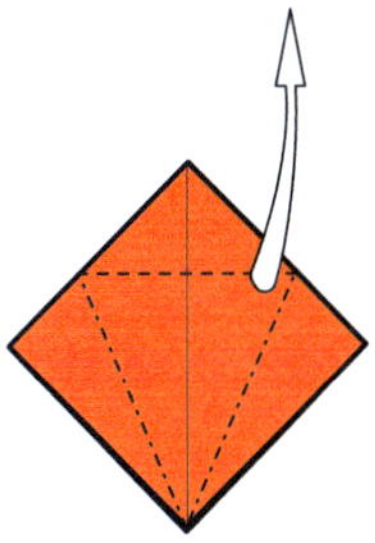

11. Petal fold the flap upwards.

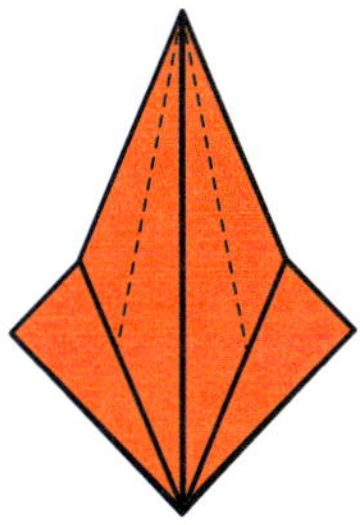

12. Narrow the front with valley folds.

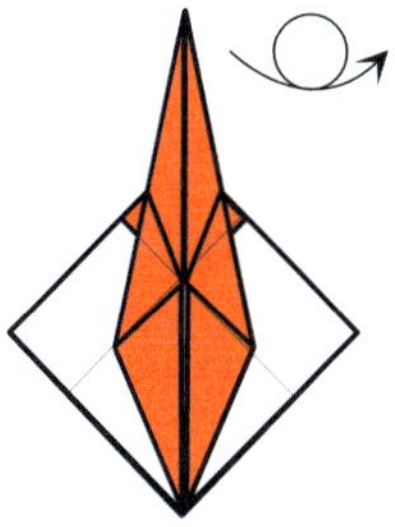

13. Turn over.

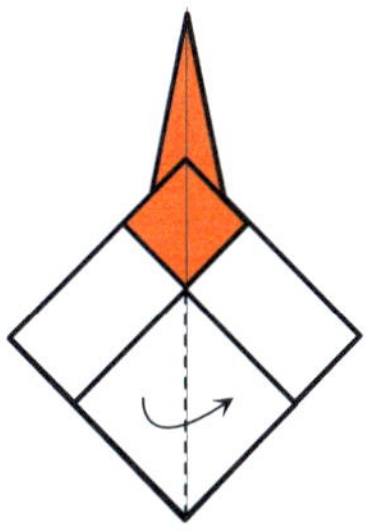

14. Valley fold one side of the square flap to the other side.

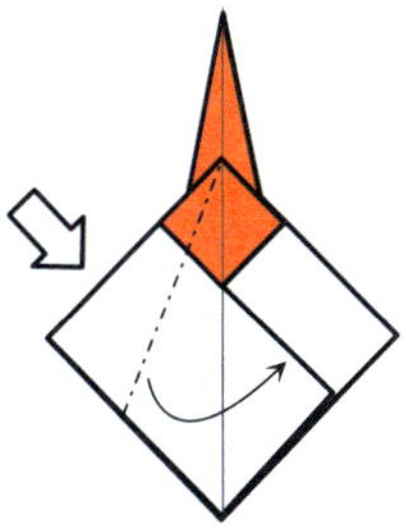

15. Squash fold the flap.

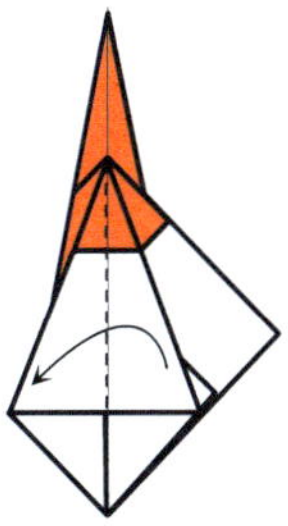

16. Fold two flaps back to the other side.

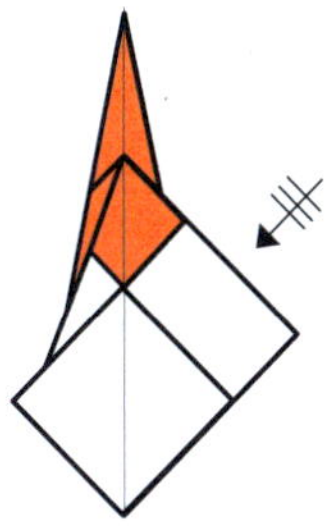

17. Repeat steps 14-16 on the other side.

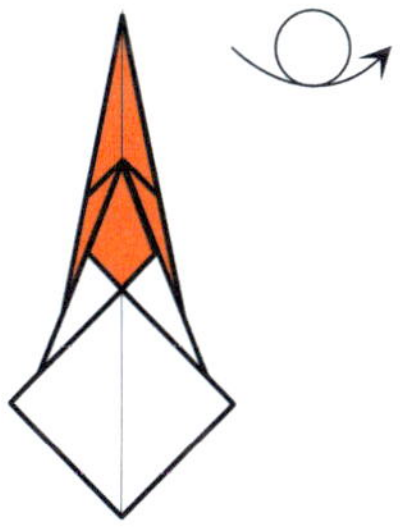

18. Turn over.

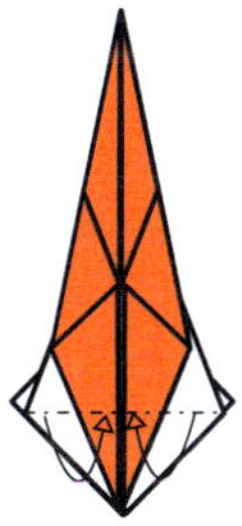

19. Mountain fold the lower flaps, trapping two layers of paper and locking the bottom together.

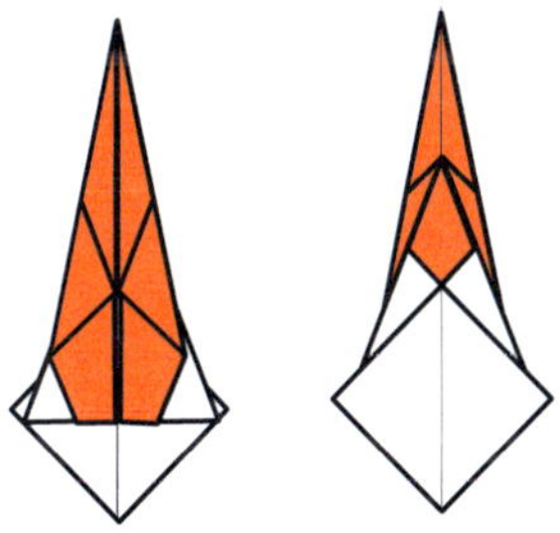

Front Piece Zeta is at last finished. It hasn't a lot of aerodynamic heft, but makes a nice weight at the front of an airplane.

Front Piece Kappa

Begin with the Front Piece Base.

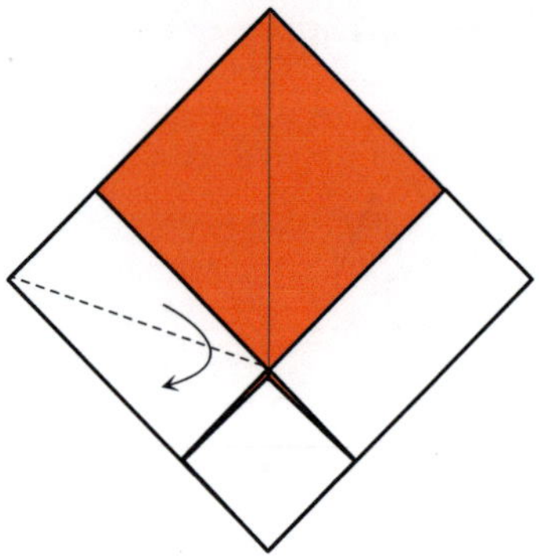

1. Valley fold along the diagonal.

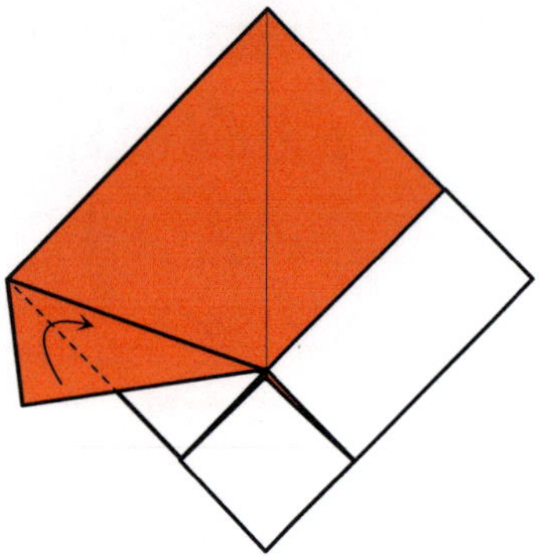

2. Valley fold the overhang back.

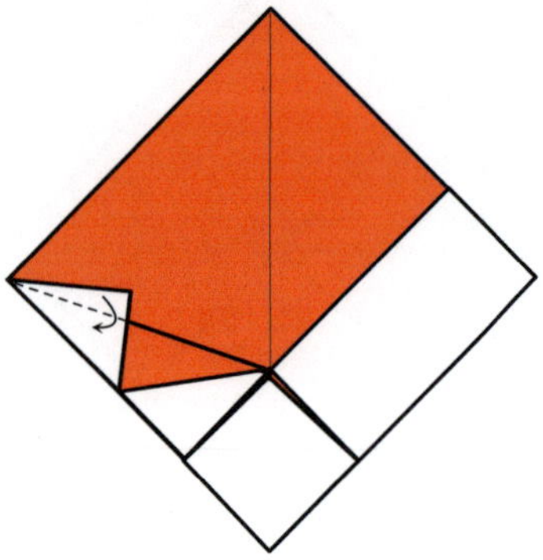

3. Valley fold the overhang back again.

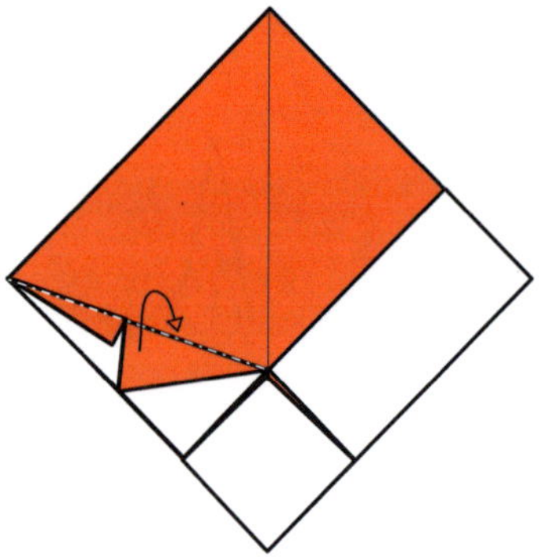

4. Mountain fold the flap behind.

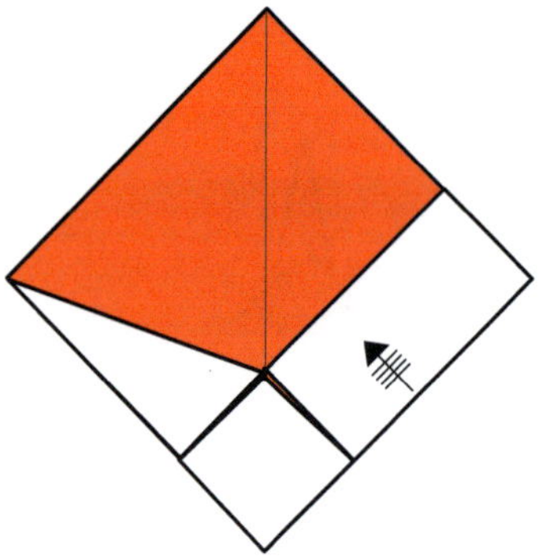

5. Repeat steps 1-4 on the other side.

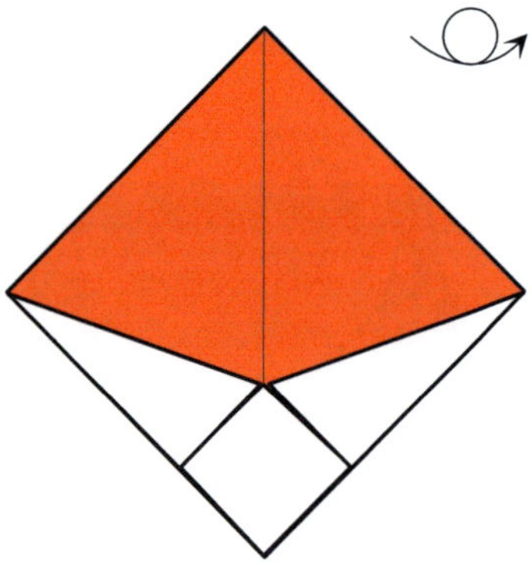

6 Turn over.

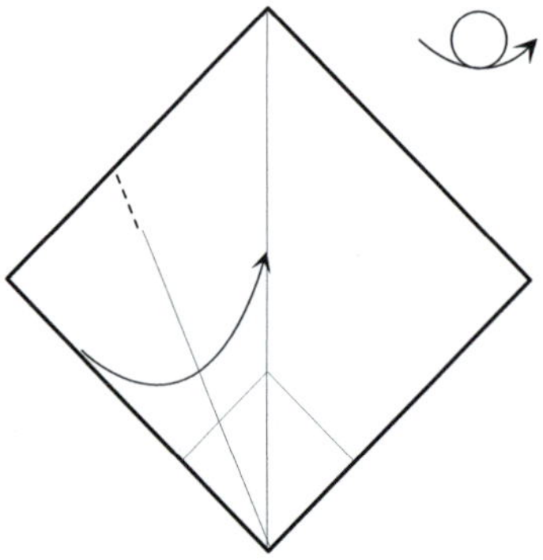

7. Valley fold so that the lower side hits the middle. Crease only at the top. Turn over.

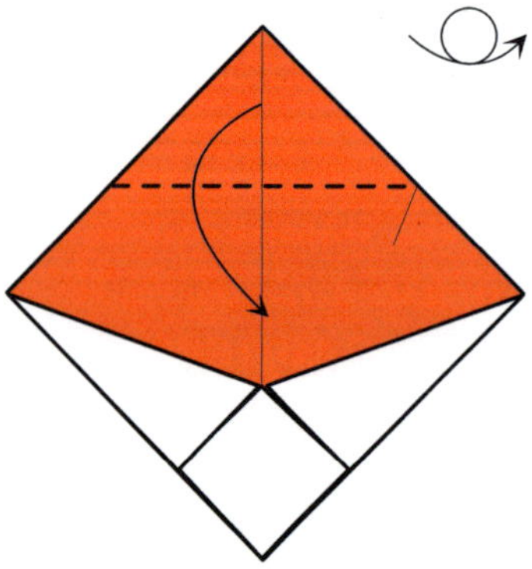

8. Valley fold the top where the crease hits the edge.

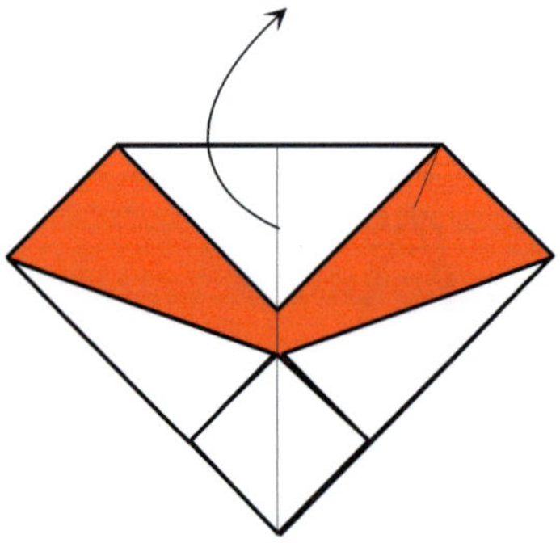

9. Unfold.

Front Piece Kappa Page 2

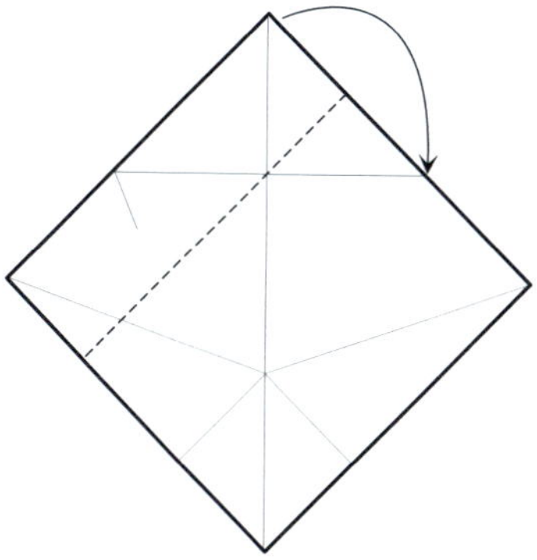

10. Valley fold the edge so the top corner hits the crease made in step 3.

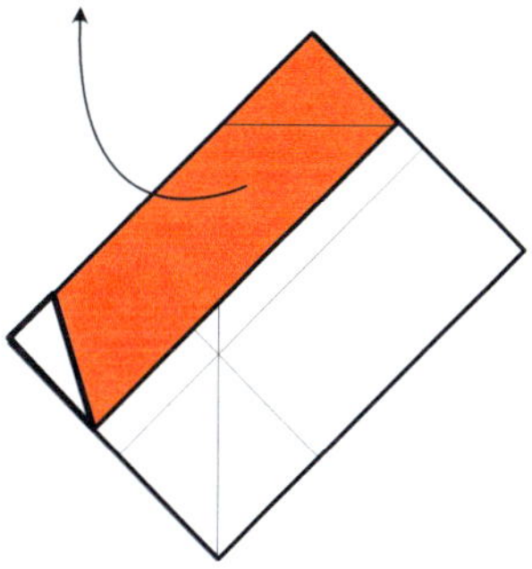

11. Unfold.

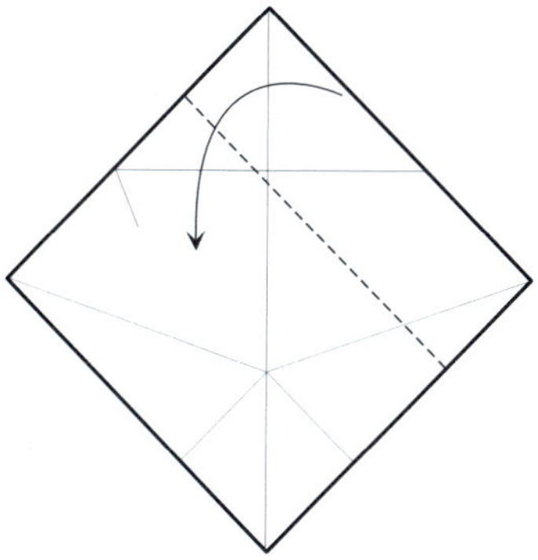

12. Valley fold the other edge so the top corner hits the crease made in step 3.

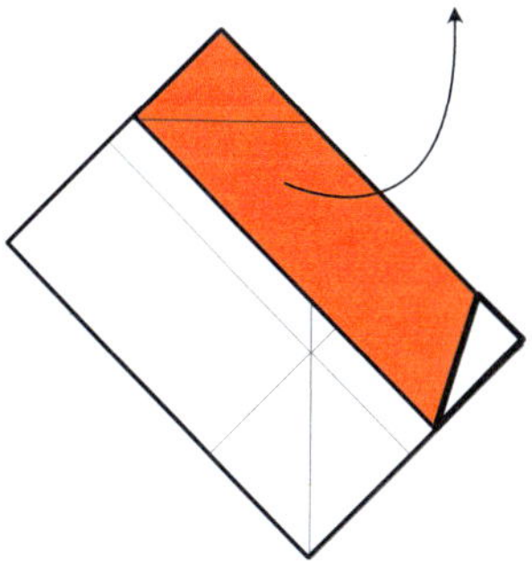

13. Unfold.

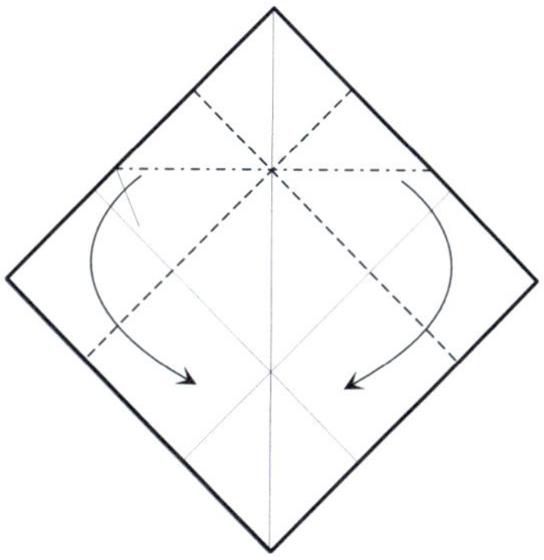

14. Fold on all three creases at once, bringing the top down.

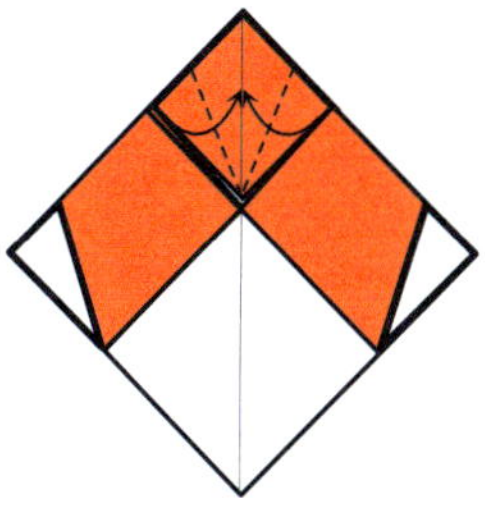

15. Valley fold so that the two-ply edges at the top lie on the center.

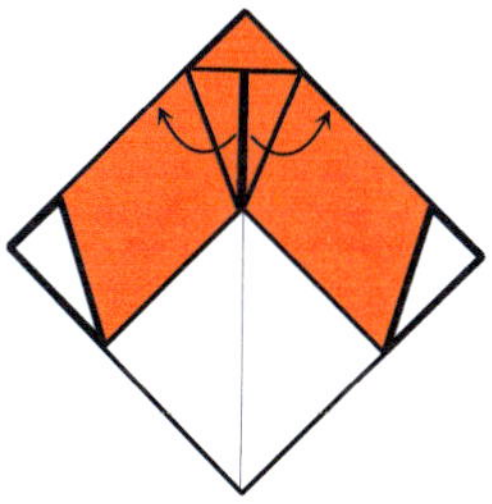

16. Unfold.

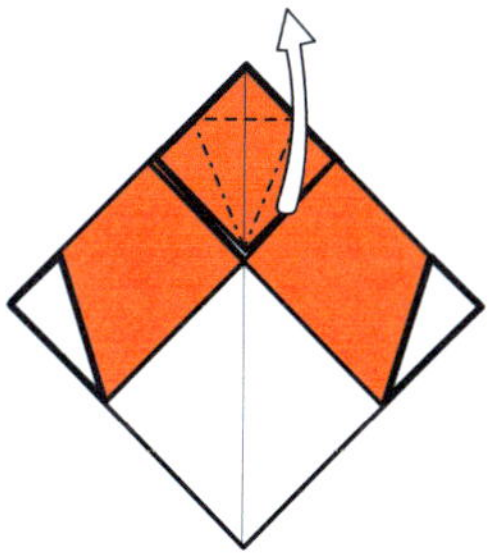

17. Petal fold the flap upwards

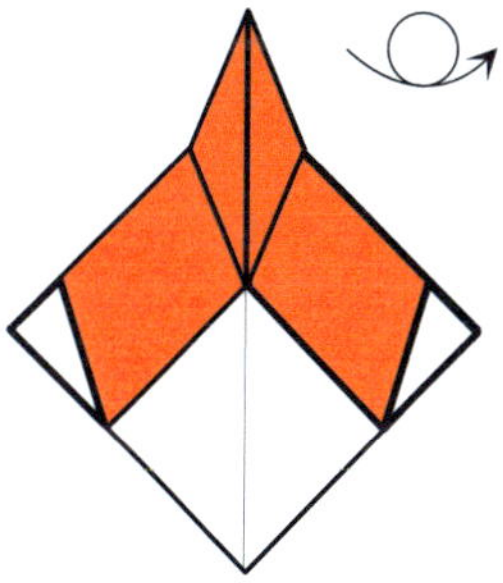

18. Turn over.

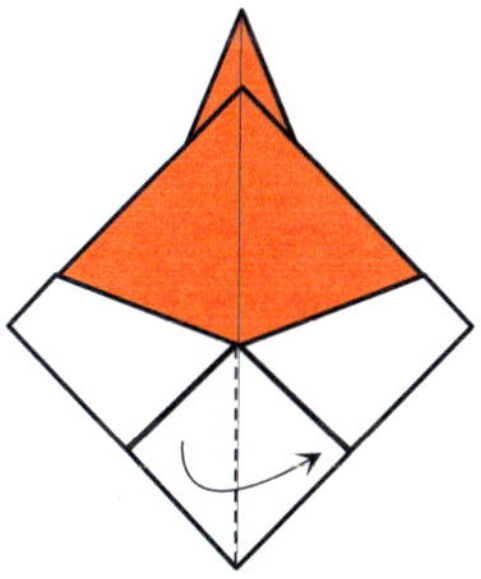

19. Valley fold one small flap to the other side.

Front Piece Kappa Page 3

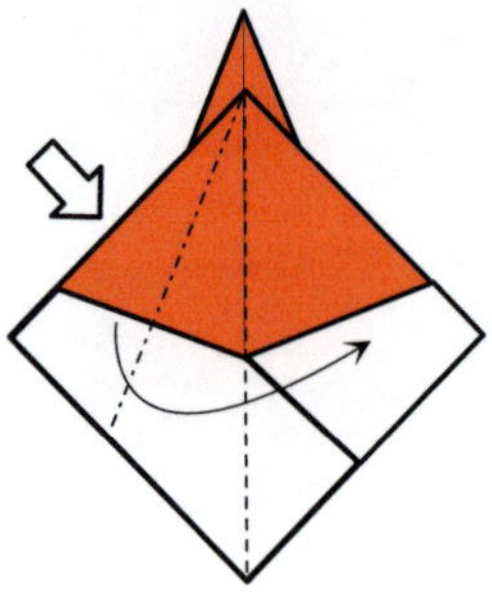

20.Squash fold.

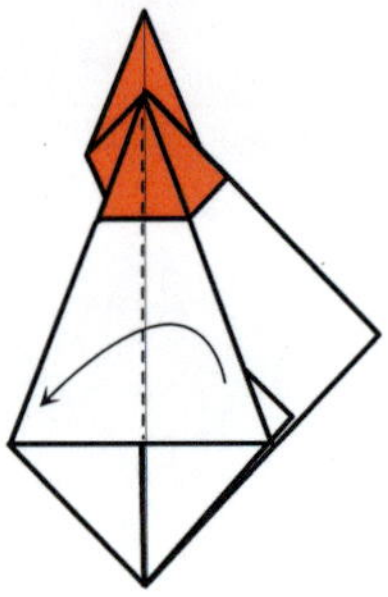

21. Fold two flaps back to the other side.

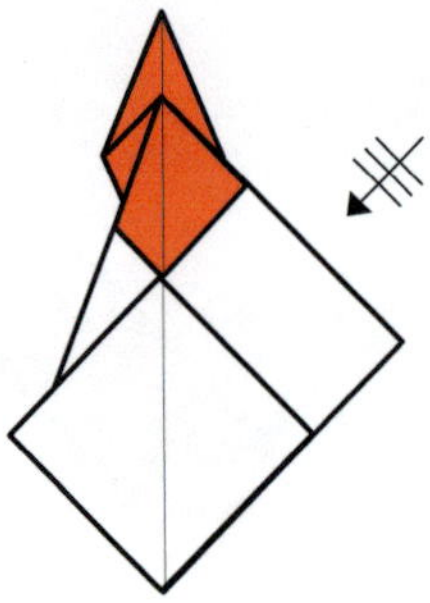

22. Repeat steps 14-16 on the other side.

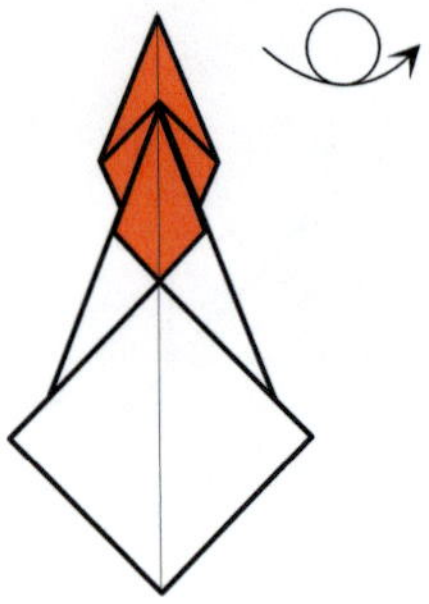

23. Turn over.

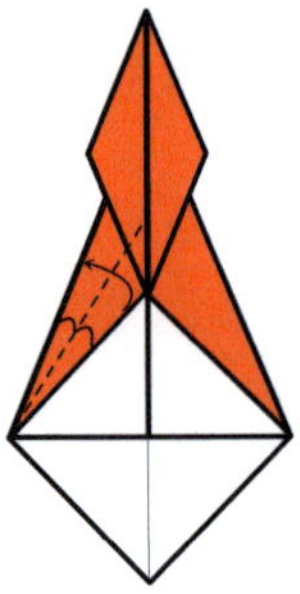

24. Valley fold the top layer in half.

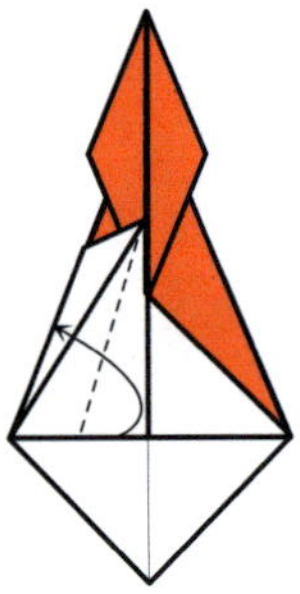

25. Valley fold the middle flap so its corner touches the outside edge.

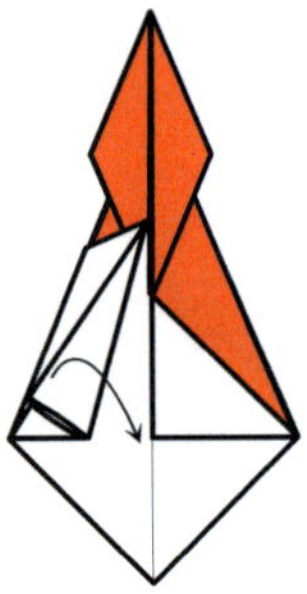

26. Unfold.

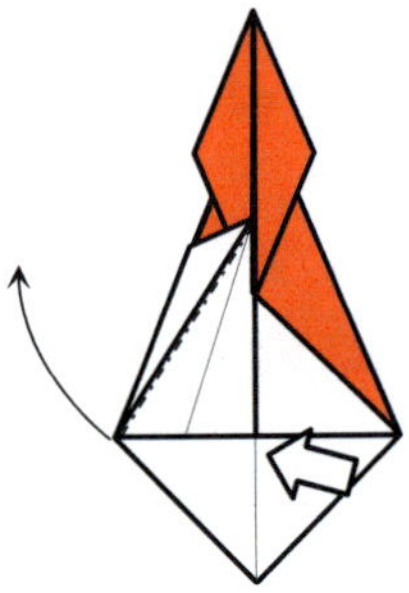

27. Reverse fold the wing out. One fold will be adjacent to the top folded layer, the other a hidden crease.

28. Narrow the front with a valley fold.

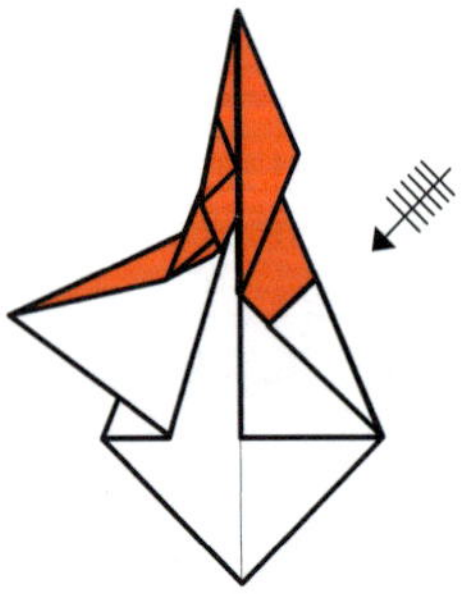

29. Undo a pocket and flatten. Narrow the front.

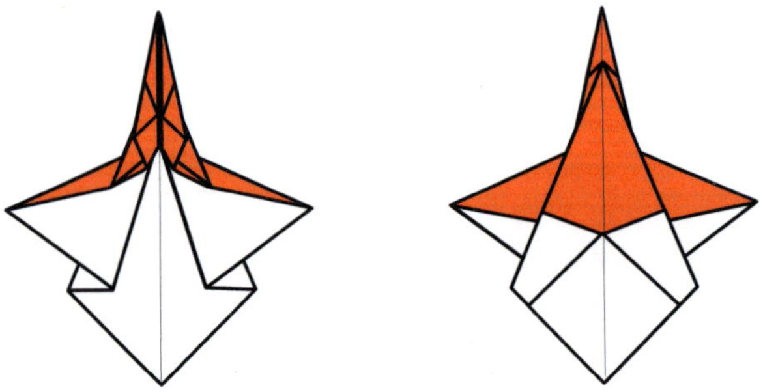

Front Piece Kappa is complete.

Front Piece Lambda

Begin with step 6 of the Front Piece Beta. You can omit the valley fold in step 1.

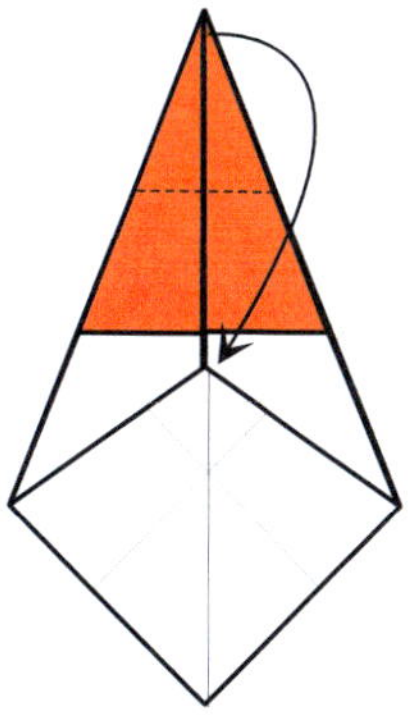

1. Valley fold the top point down to the intersection shown.

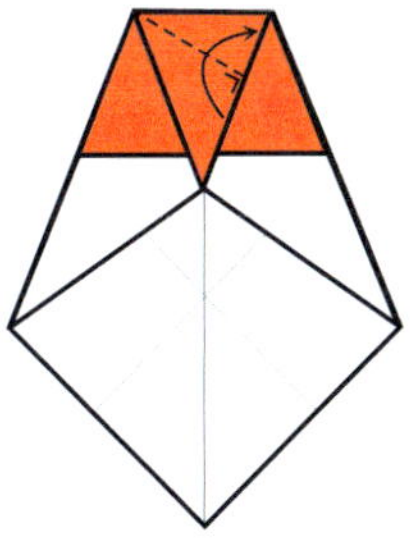

2. Valley fold the flap upward at an angle perpendicular to its folded edge.

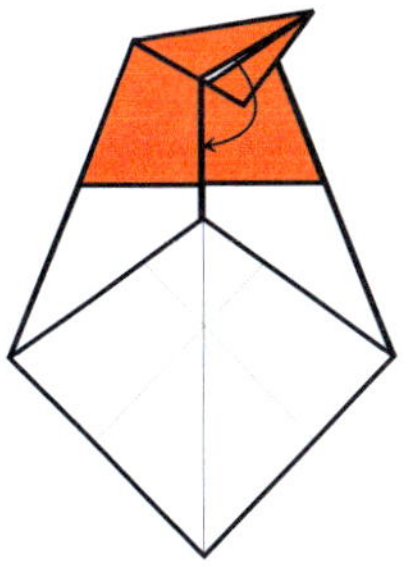

3. Unfold.

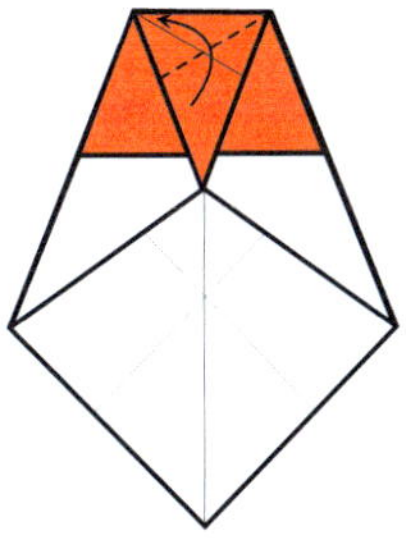

4. Repeat steps 2 and 3 on the other side.

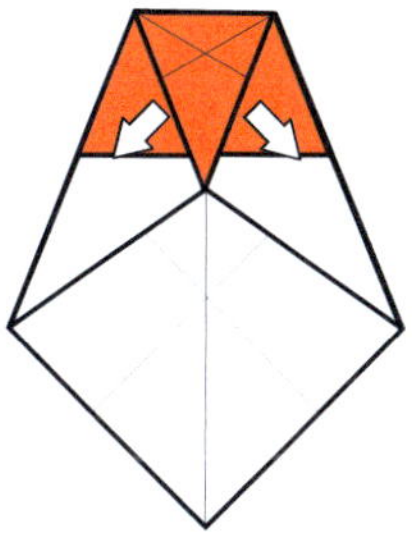

5. Pull out and squash fold the hidden paper. There will be creases to guide you.

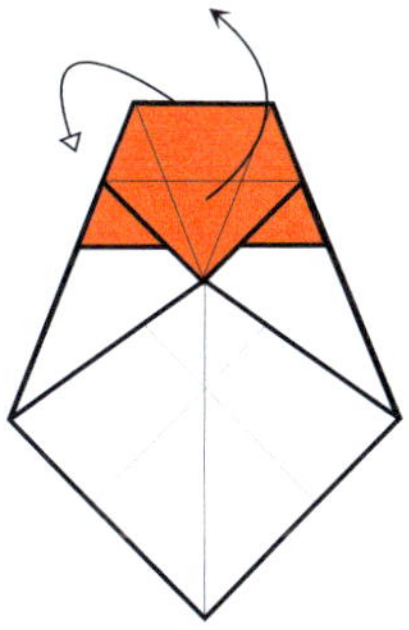

6. Note the edges of the paper you pulled out lie right on the outer edge. Flip the top, you'll make a hidden mountain fold.

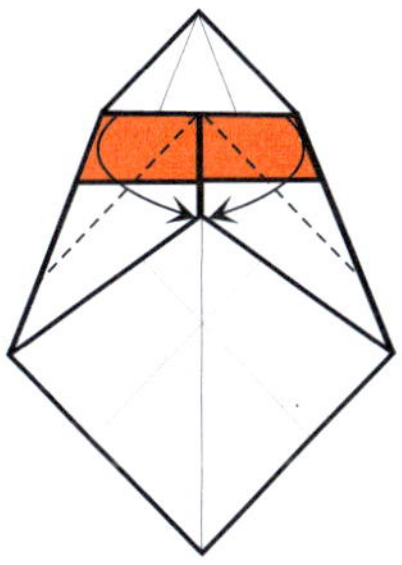

7. Valley fold so the edges in front lie on the center.

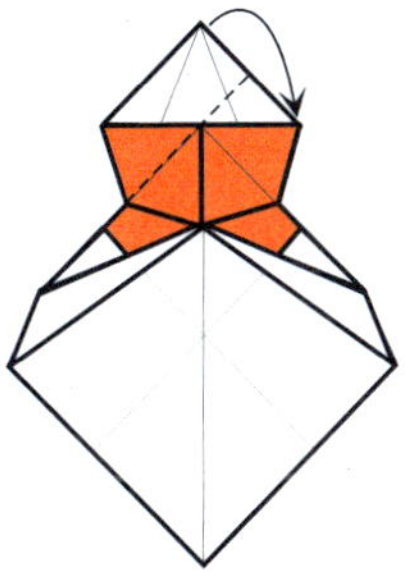

8. Valley fold the top diagonally.

Front Piece Lambda Page 2

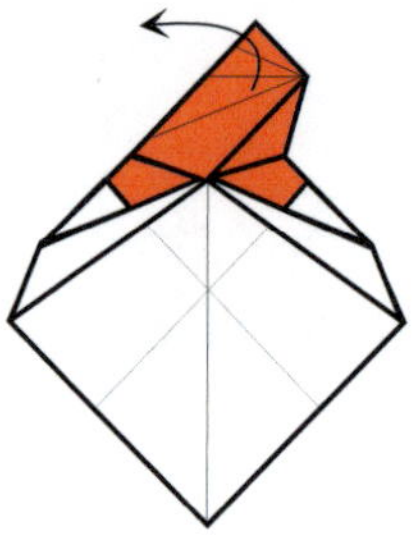

9. Unfold.

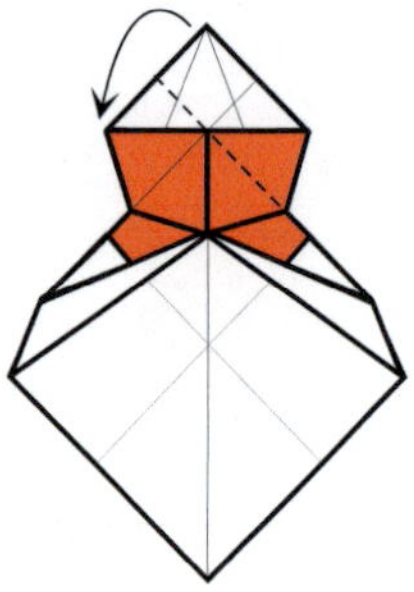

10. Repeat on the other side.

11. Mountain fold the top behind and unfold.

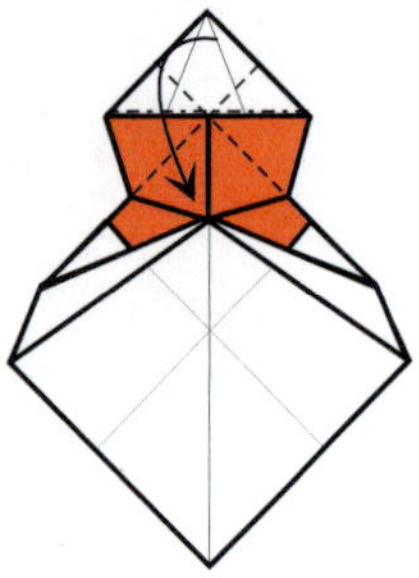

12. Collapse the front.

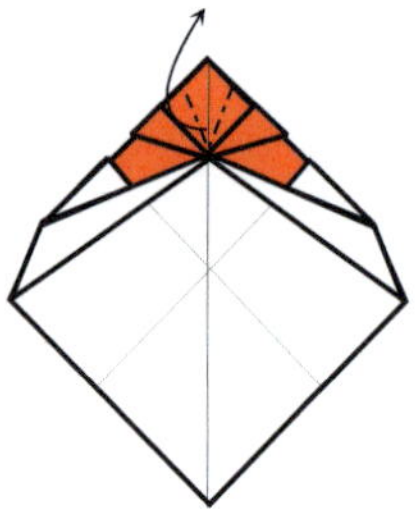

13. Petal fold upwards.

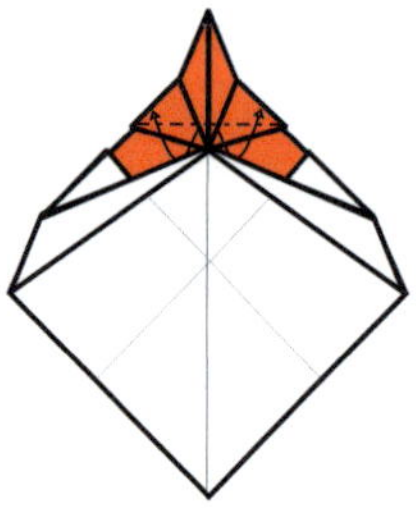

14. Mountain fold the bottom of the flaps behind to lock up the front.

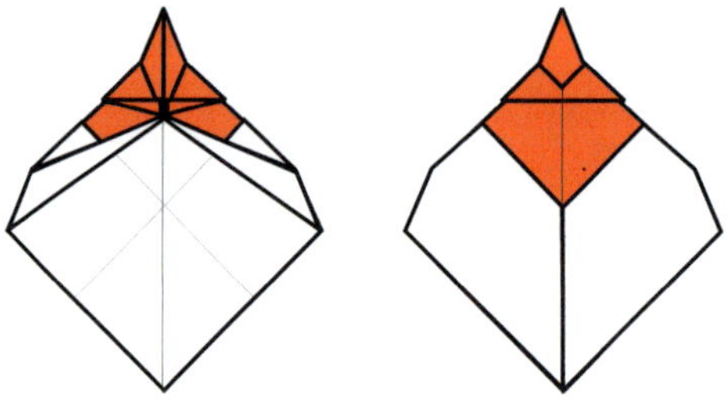

Front Piece Lambda is done.

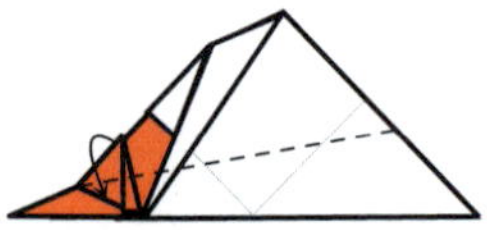

The landmark for folding the wings for Lambda is to line up the front outside edge with the inner folded edge of the nose. Use that to set the angle and fold back.

Front Piece Omicron

Begin with the Front Piece base.

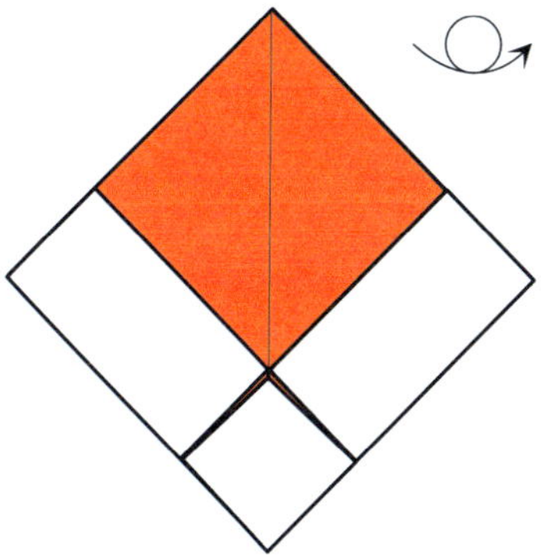

1. Turn over.

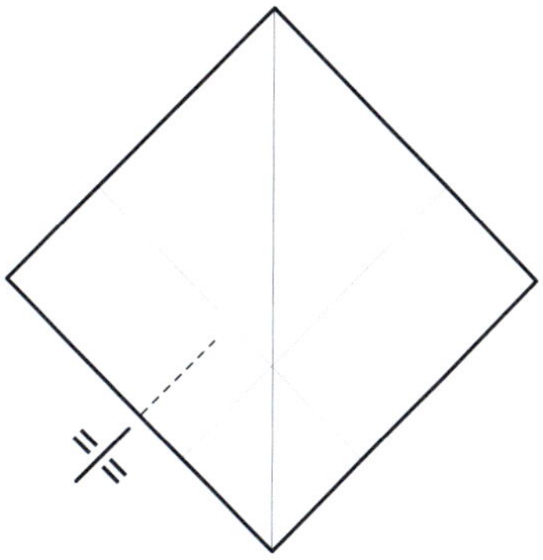

2. Make a partial crease on the middle of one edge.

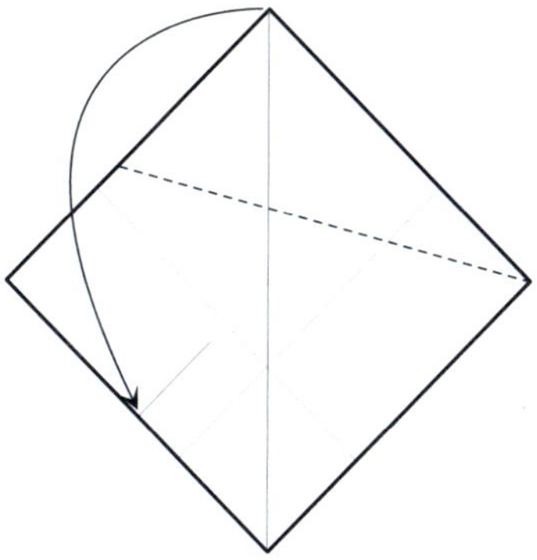

3. Valley fold so that the top corner rests on the midline crease made in step 2.

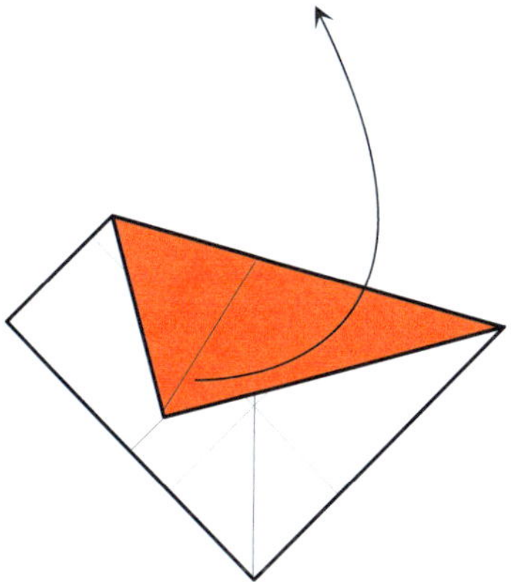

4. Unfold.

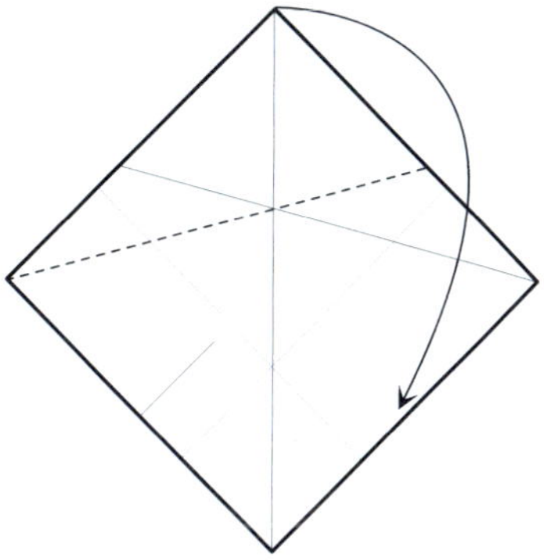

5. Make the same fold going the other way.

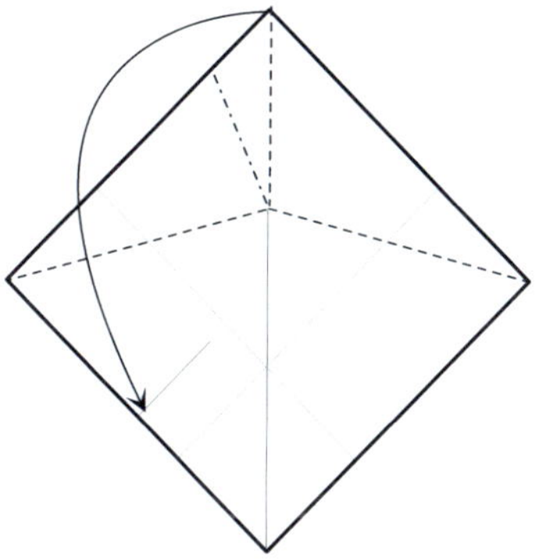

6. Rabbit ear fold along the creases.

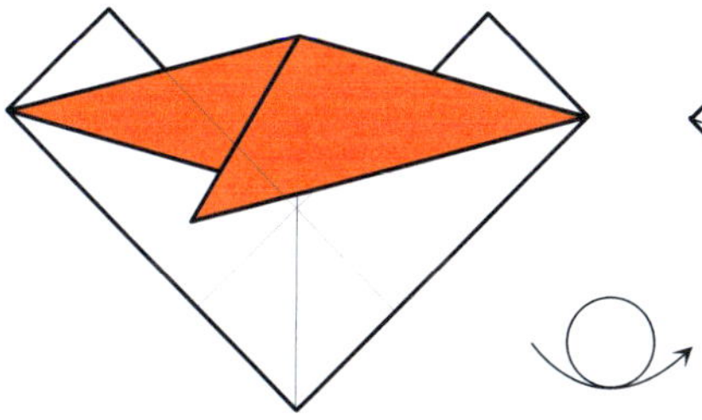

7. Turn over.

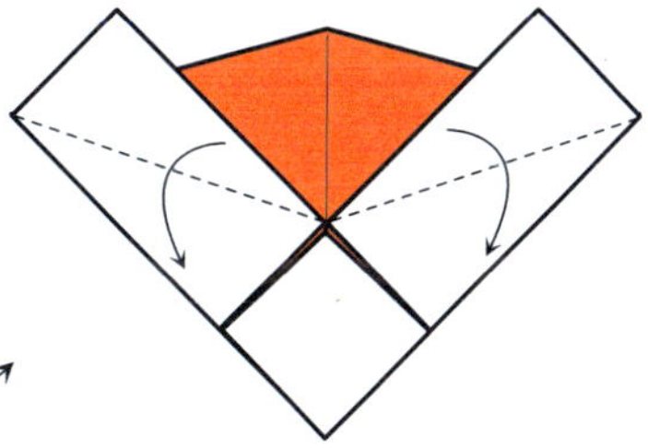

8. Valley fold the flaps down.

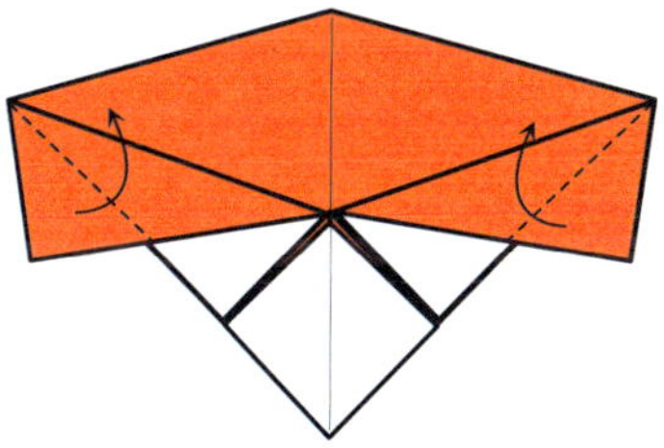

9. Valley fold up along the back edge.

Front Piece Omicron Page 2

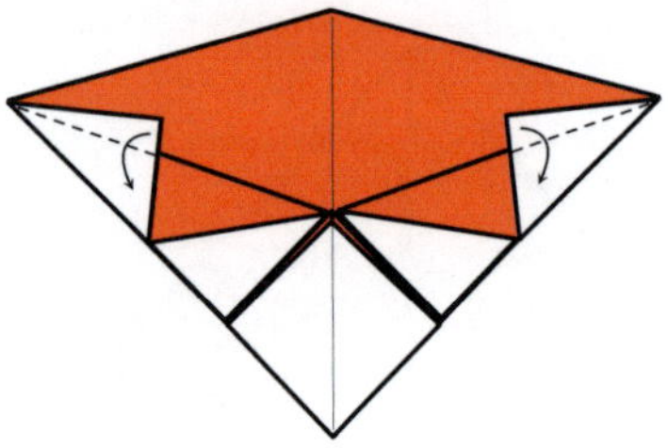

10. Valley fold back down along the edge.

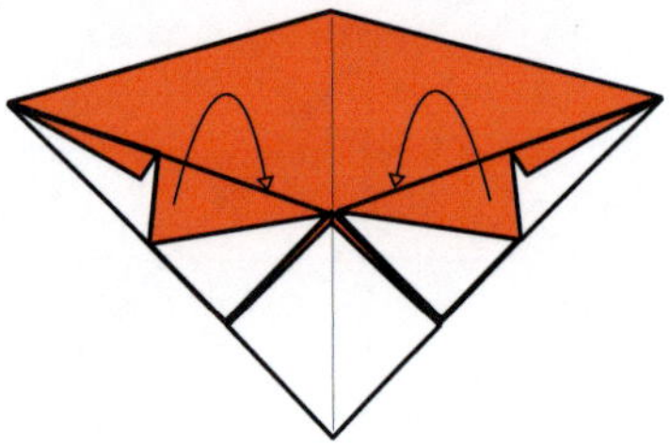

11. Mountain fold the whole flap behind to get it out of the way.

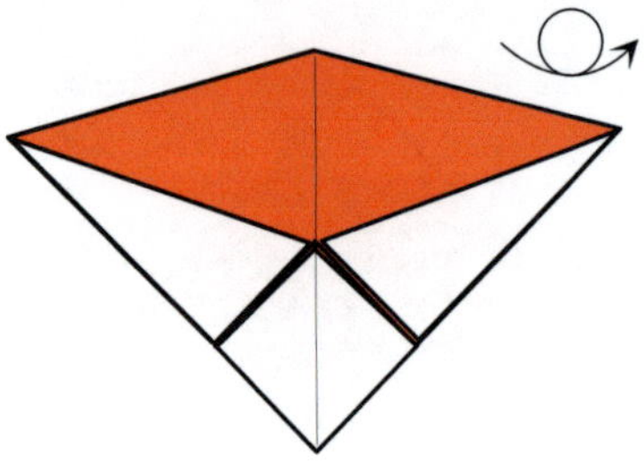

12. Turn over.

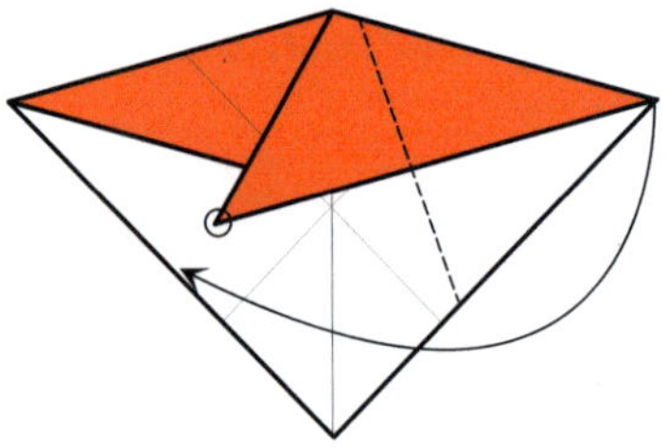

13. Valley fold so that the outside corner meets the opposite side. The front edge should touch the circled point.

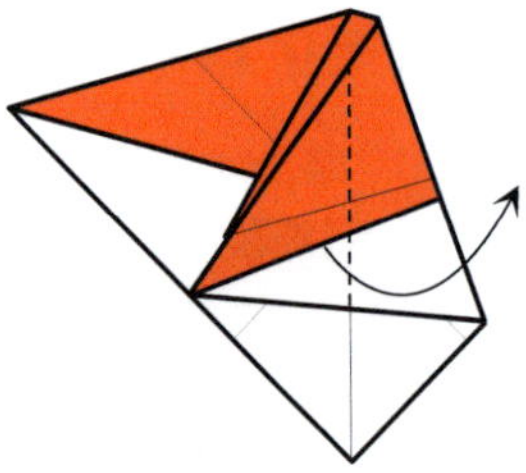

14. Valley fold back along the midline.

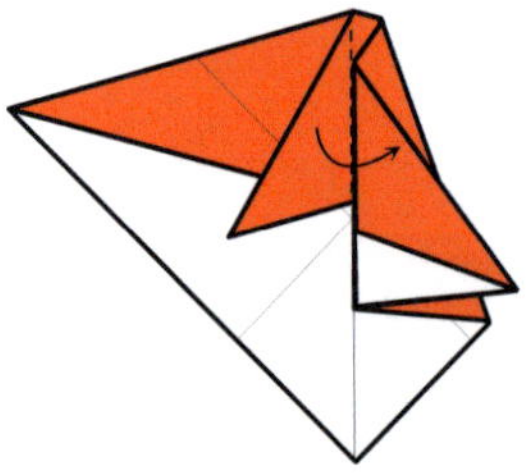

15. Fold the flap to the other side.

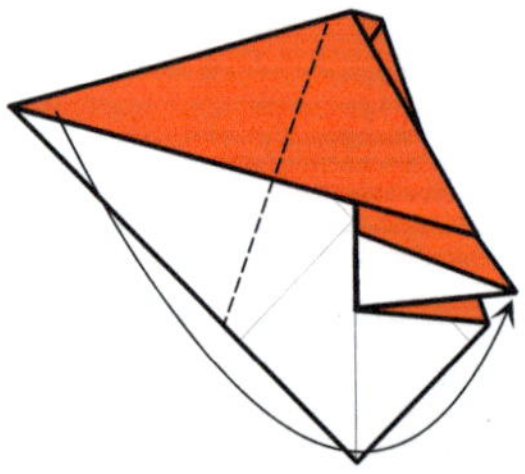

16. Fold the other side to match.

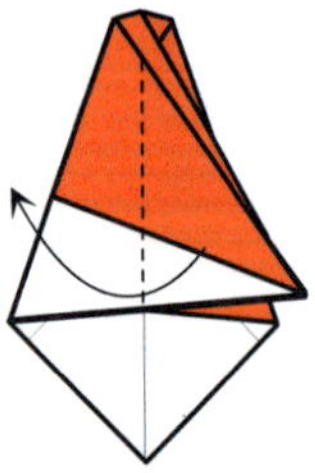

17. Fold back along the midline.

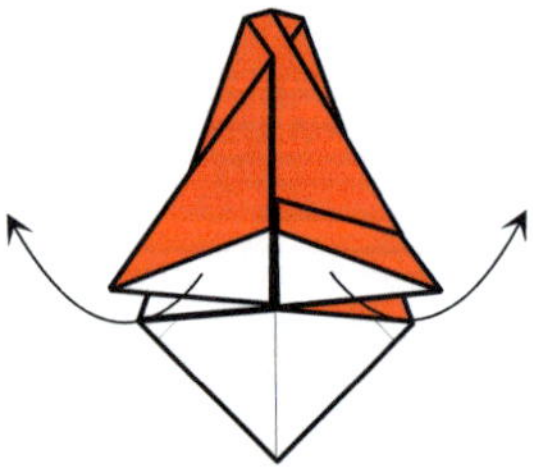

18. Unfold to step 12.

Front Piece Omicron Page 3

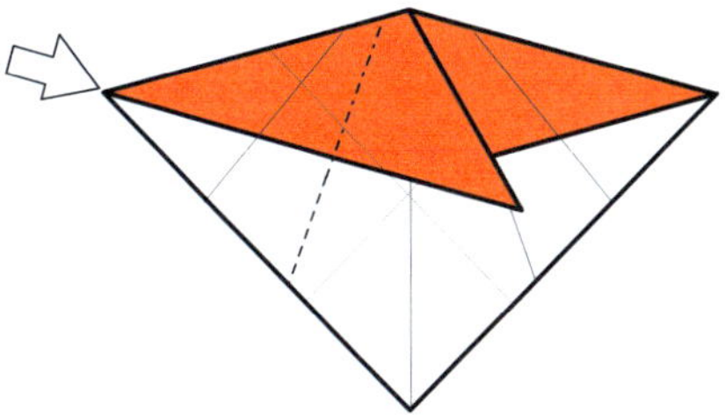

19. Reverse fold in along a preexisting crease.

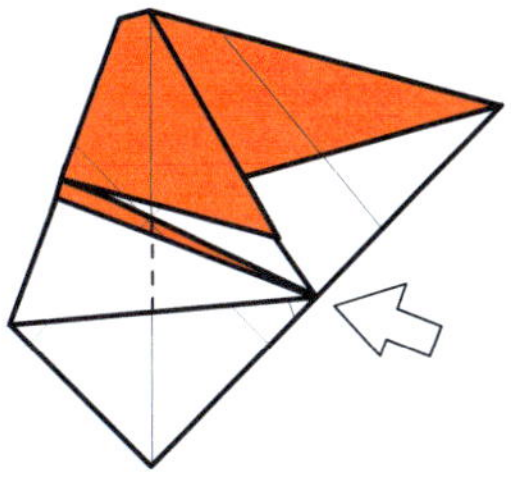

20. Reverse fold back out along the next crease.

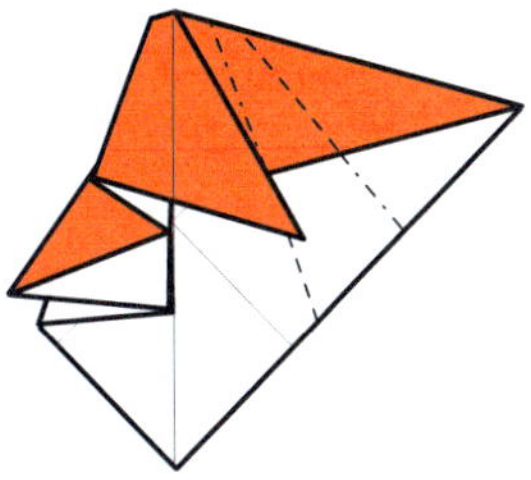

21. Repeat steps 19 and 20 on the other side.

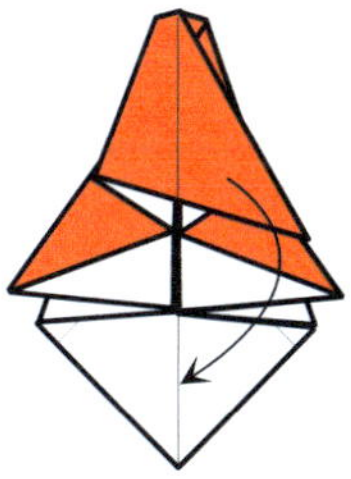

22. Pull the front flap down and squash fold the top.

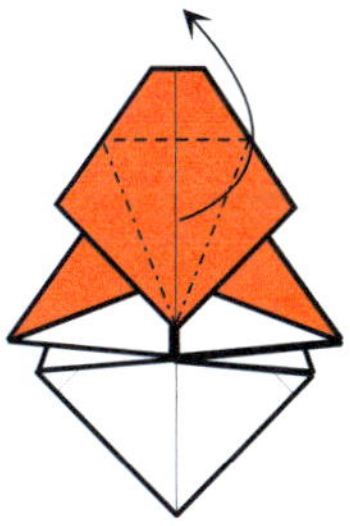

23. Petal fold the flap upwards.

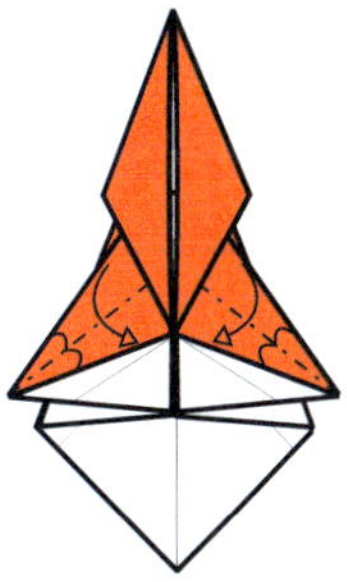

24. Mountain fold the leading edges of the wings behind.

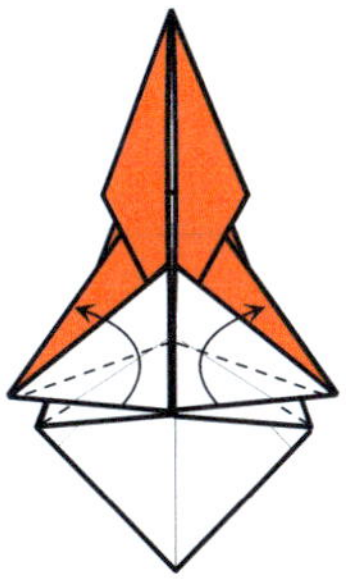

25. Squash fold the flap so that the edge hits the leading edge of the wing.

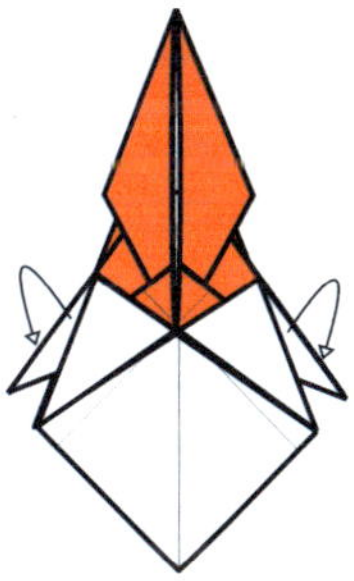

26. Tuck the flaps inside the flaps beneath them.

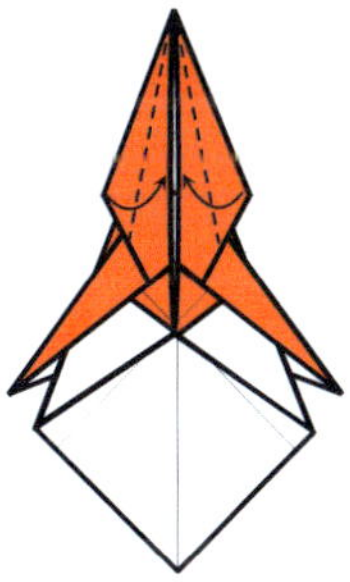

27. Narrow the front with valley folds.

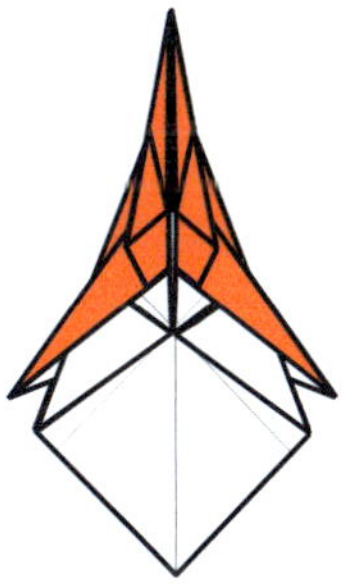

Front Piece Omicron is done

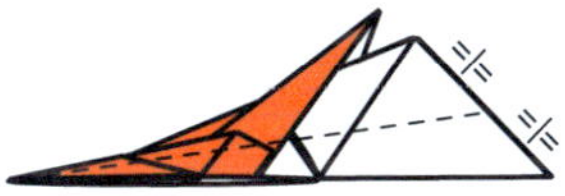

To fold wings for Omicron, bisect the rear. The fold should go through the very front. Continue the fold angle on the back piece.

Back Piece One Airplanes

Alpha One

Beta One

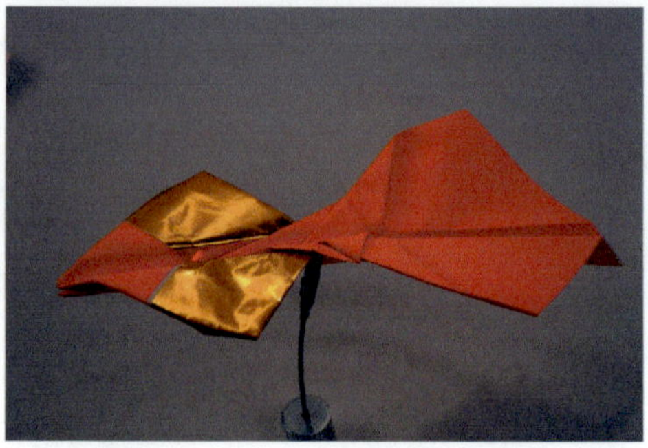

Gamma One

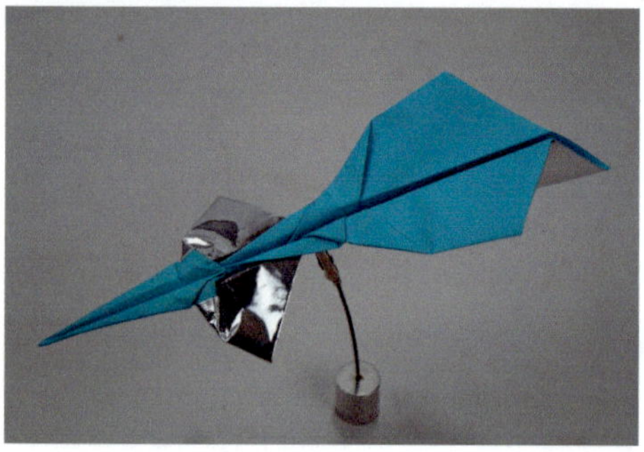

Delta One

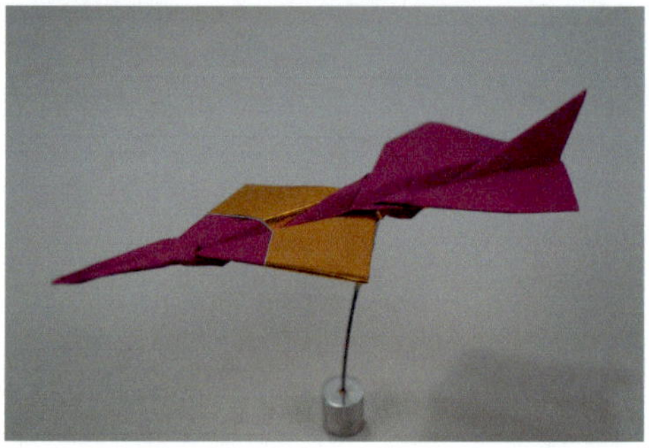

Epsilon One

Zeta One

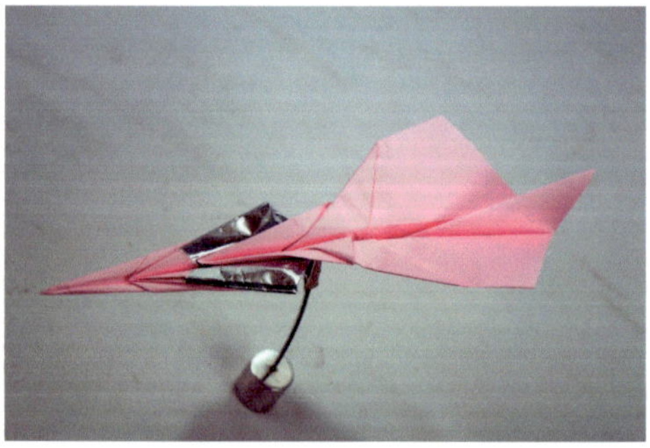

Lambda One

Back Piece One

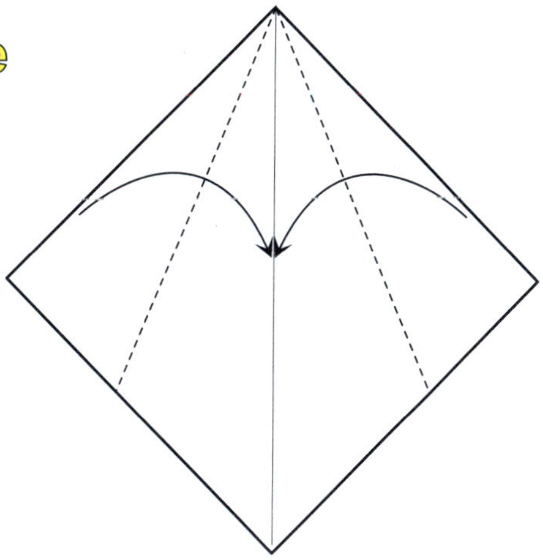

1. Valley fold both sides into the middle.

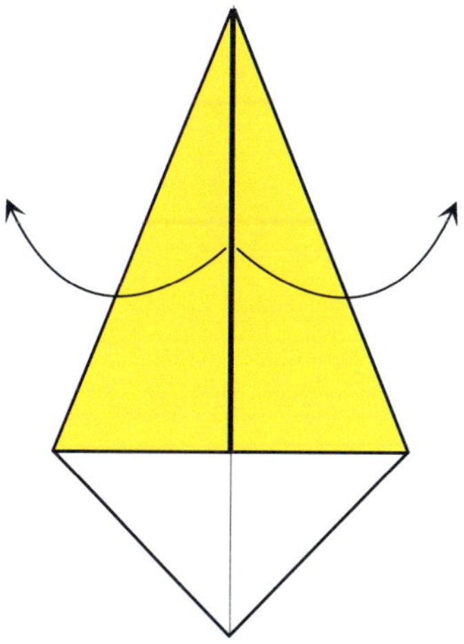

2. Unfold.

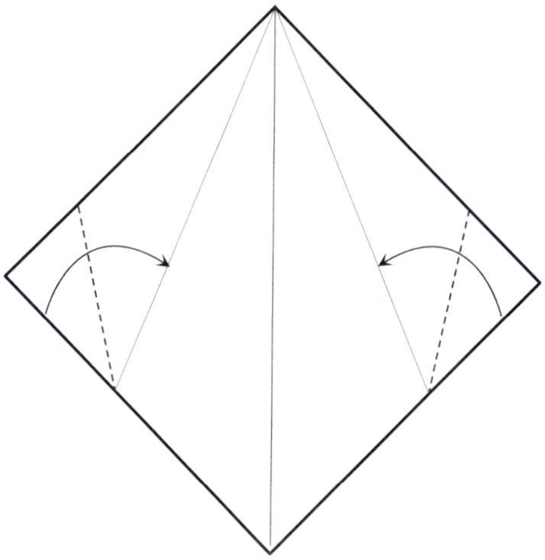

3. Valley fold both sides so that the outside edges lie on the creases.

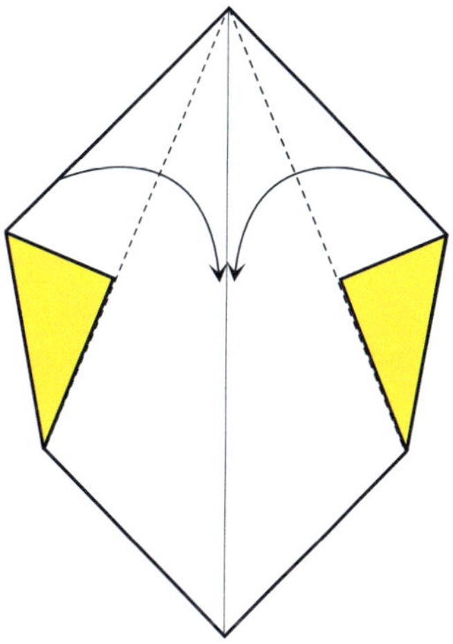

4. Refold on the creases made earlier.

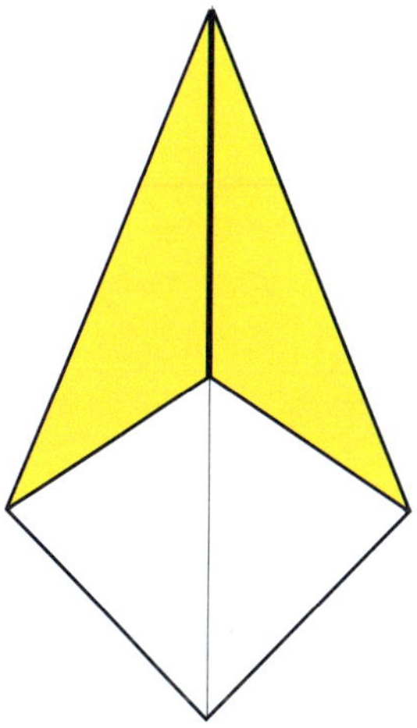

Back Piece One is finished. The assembly is a bit involved, though.

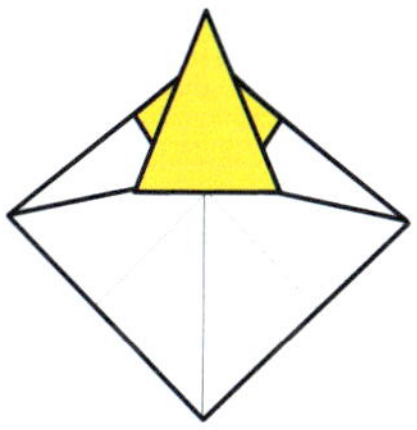

1'. We'll begin with front piece Alpha. But we have to modify it just a bit for this base. Turn it over.

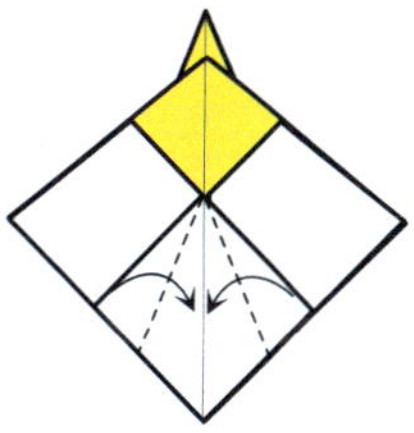

2'. Valley fold the two-ply edges of the inner square to the middle.

Back Piece One Page 2

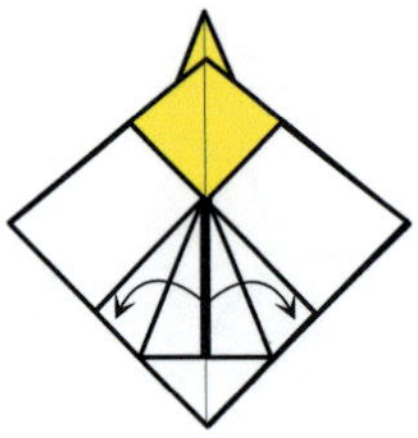

3'. Unfold.

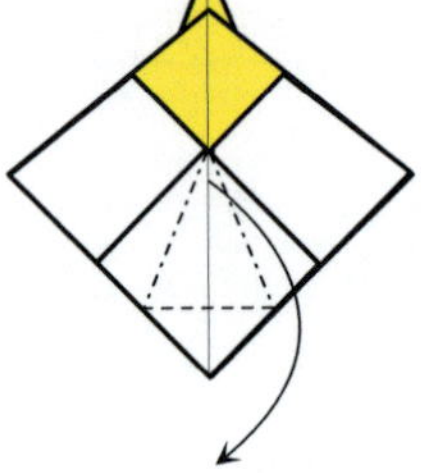

4'. Petal fold along the creases.

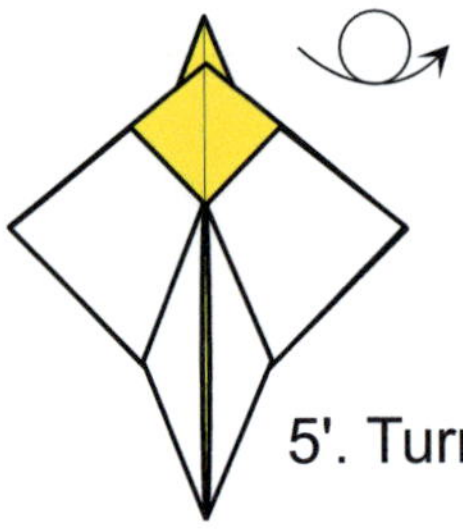

5'. Turn over.

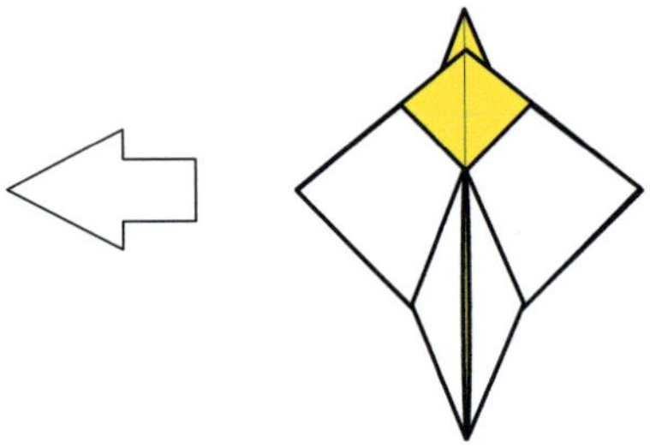

Front Piece Alpha is ready for assembly. This won't be required for all the back pieces, but is for this one.

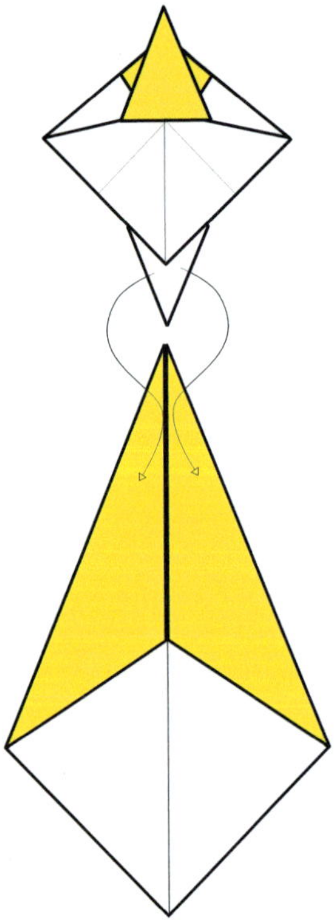

6'. The diamond shaped flap in back of the Alpha Front Piece gets inserted beneath the flaps of the Back Piece One.

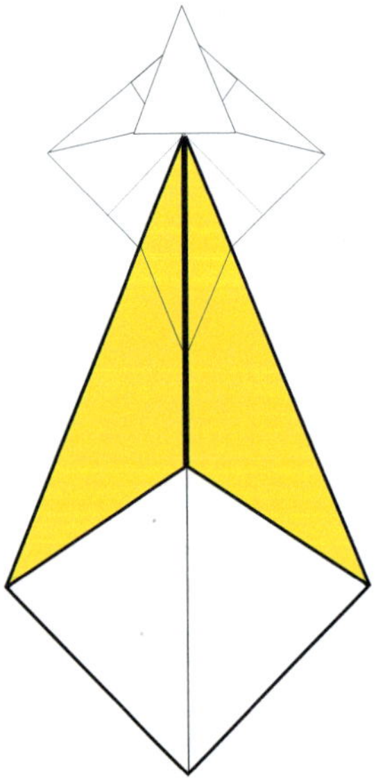

7'. Slide the diamond shaped flap as far forward as it will go, to the front of the Back Piece.

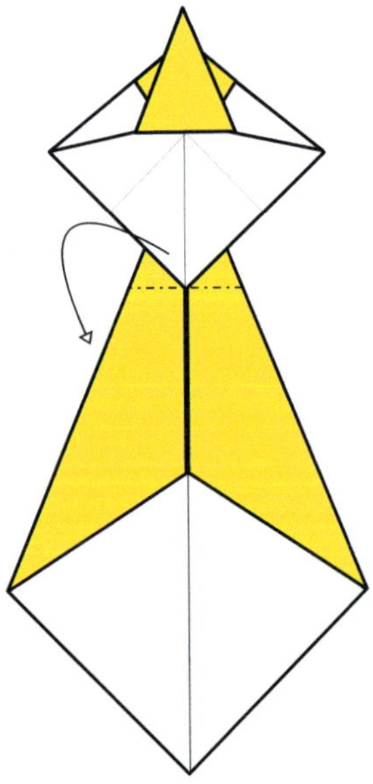

8'. Mountain fold at the base of the Front Piece Alpha.

Back Piece One Page 3

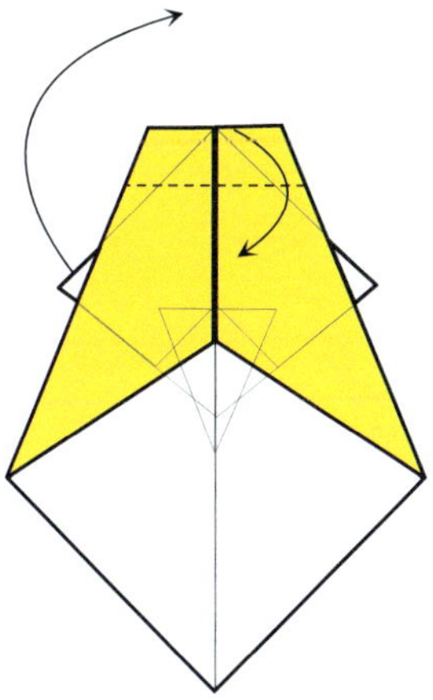

8'. Valley fold a little below the edge and flip the front piece back up. No good landmark for this, sorry.

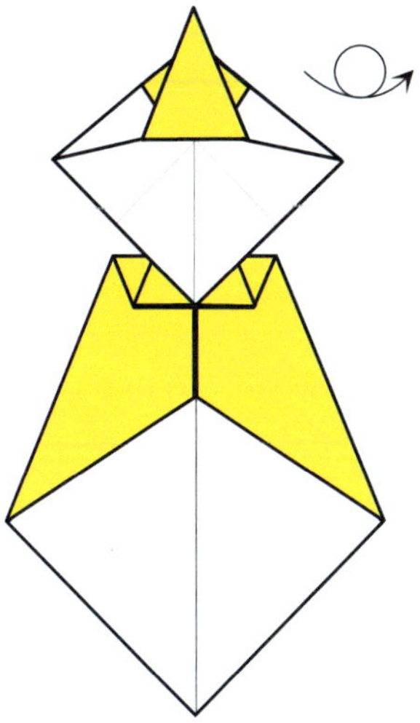

9'. Turn over.

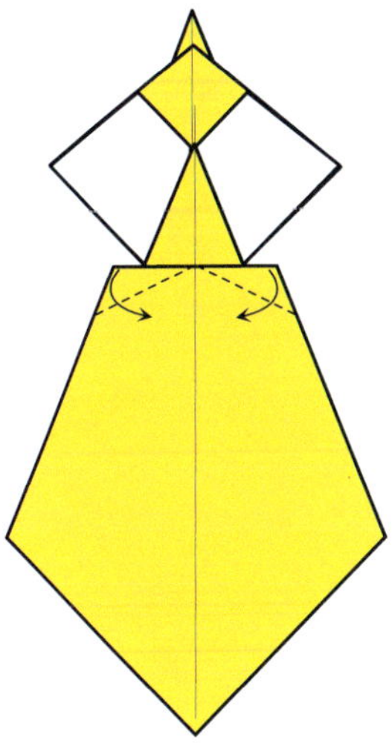

10'. Valley fold the top edges down as far as you can.

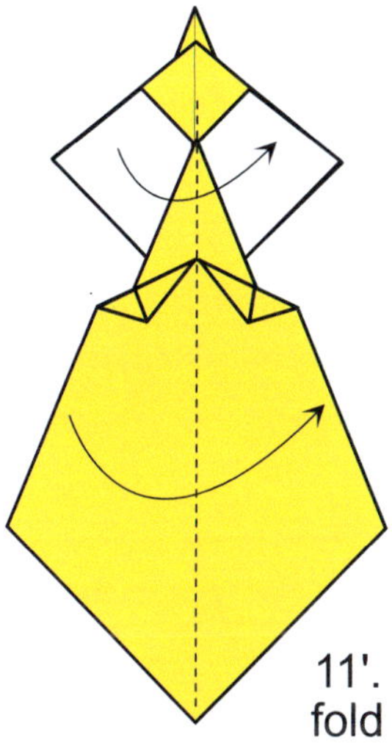

11'. Valley fold in half.

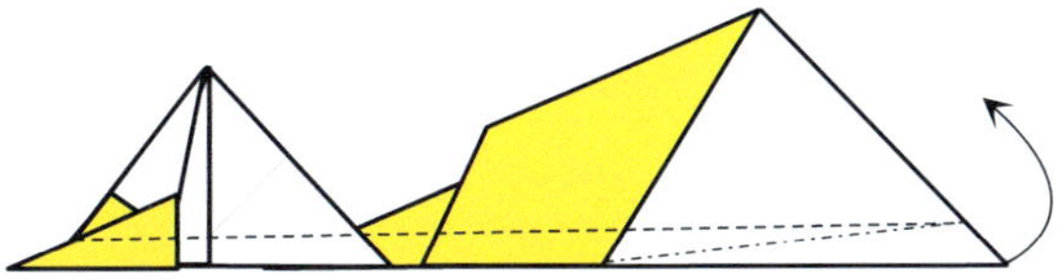

To fold the wings, fold at the intersection in front. You don't want fold from the very front with Alpha, since the pressure could tear the paper. The fold for the wings goes just below the intersection of the two pieces. Be careful to keep the pitch angle the same for front and back. Aircraft made with this back piece tend to fly better if given a tail.

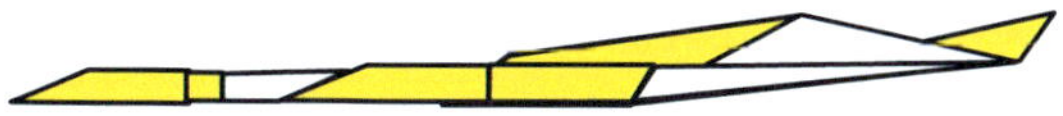

Side View

Front View

I will provide Front, Side, and Top views for all the airplanes. I won't supply attachment instructions for each, since that would be very repetitious. But you should have enough information to determine how the aircraft should be configured for flight.

For every aircraft you will get a front view. In this case the front (canard) wings should be level, and the main wings will be at a slightly elevated angle. Front views will show you the angle of the wings and dihedral.

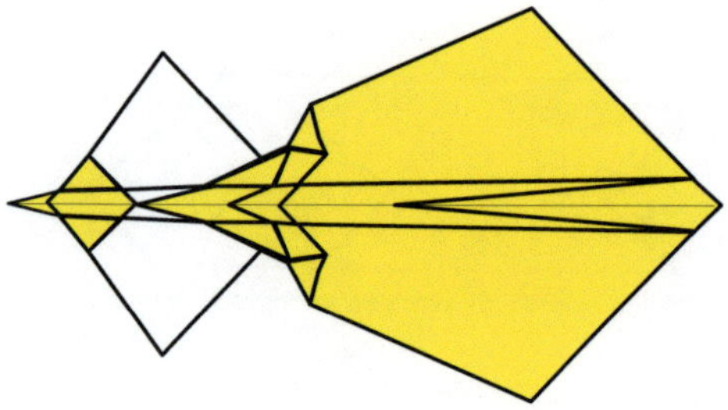

Top View.

Beta One

Front View

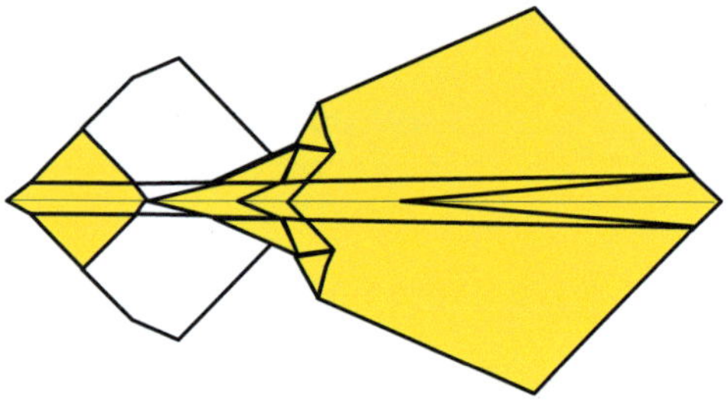

Top View.

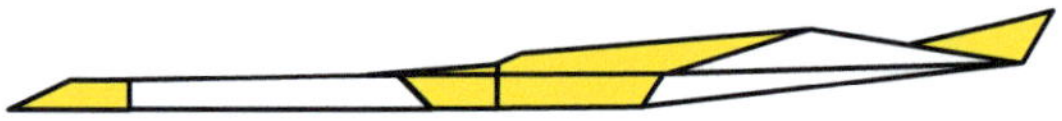

Side View.

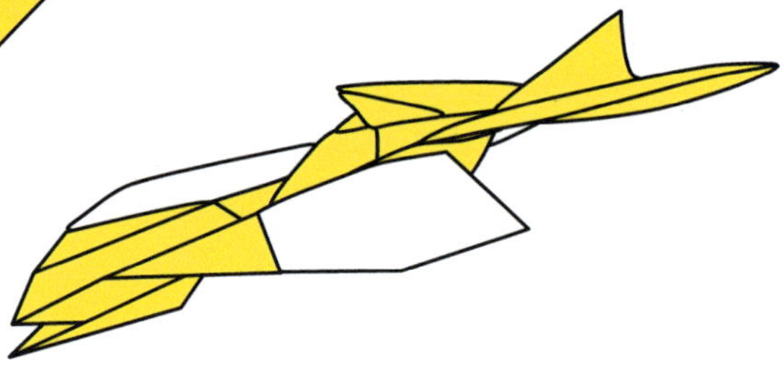

Gamma One

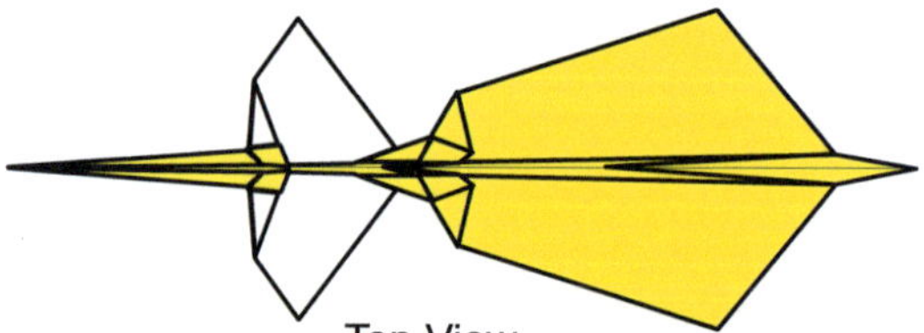

Top View.

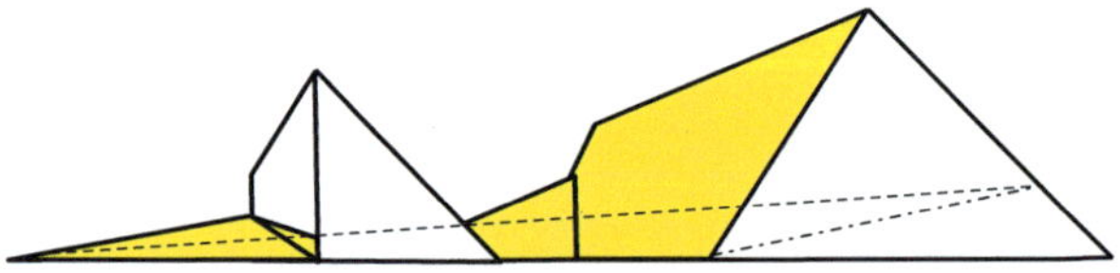

Because of its extended nose section the wings of Gamma One are folded differently. The front section should be roughly bisected by the wing folds.

Side View.

Front View

Delta One

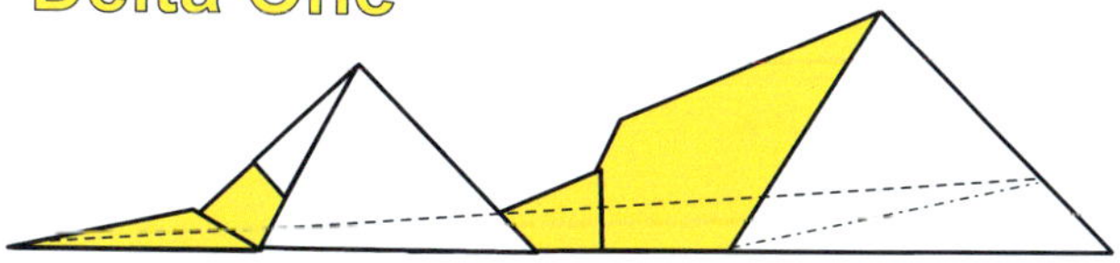

The wing fold roughly bisects the angle in front. Remember to keep the wing angle the same between the front wings and rear.

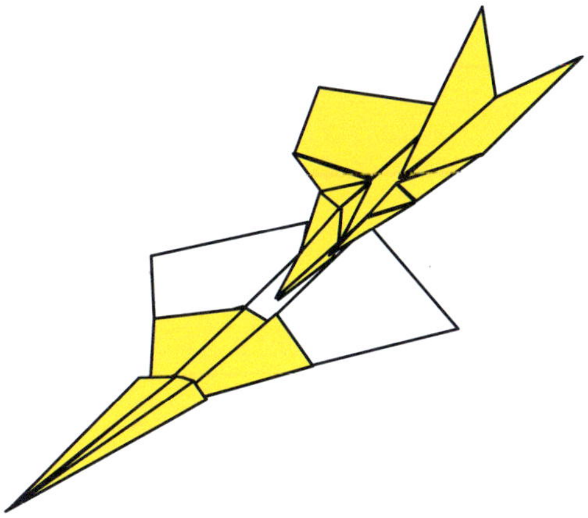

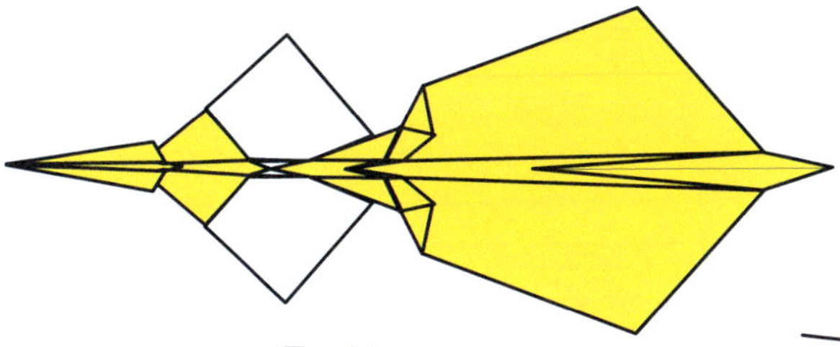

Top View.

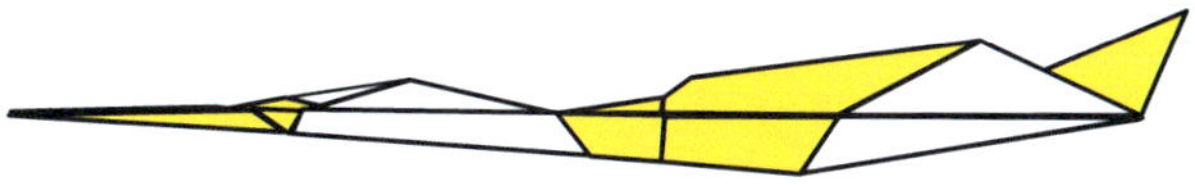

Side View.

Front View

Epsilon One

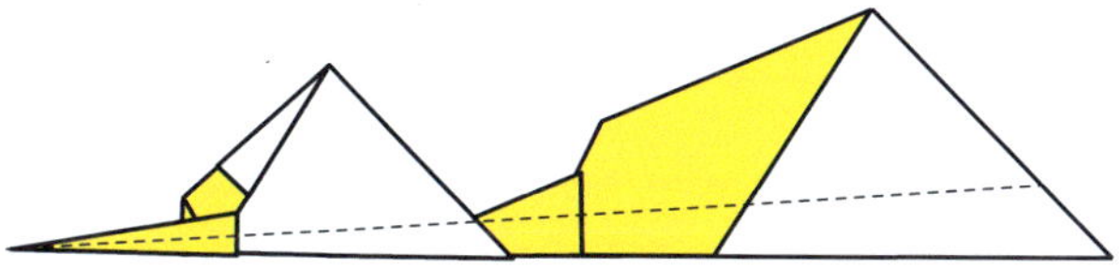

Front View

The wing fold roughly bisects the angle in front. This one tends to work a bit better without the tail.

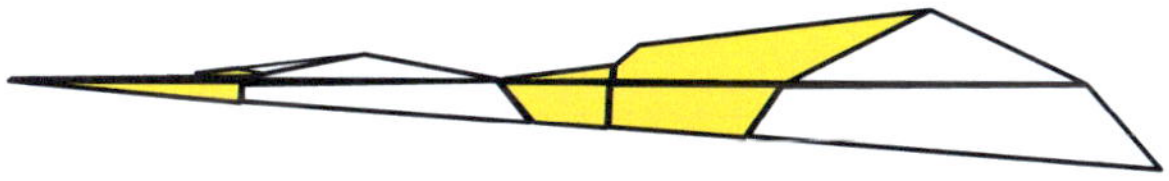

Side View.

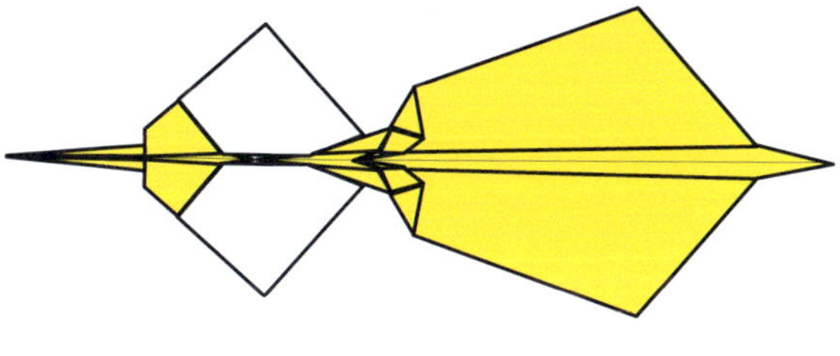

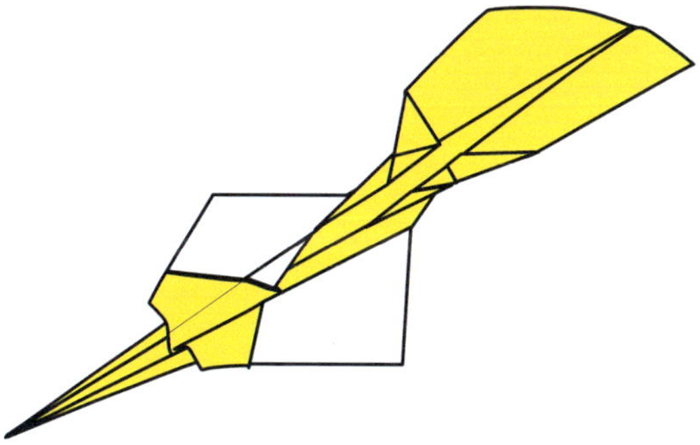

Zeta One

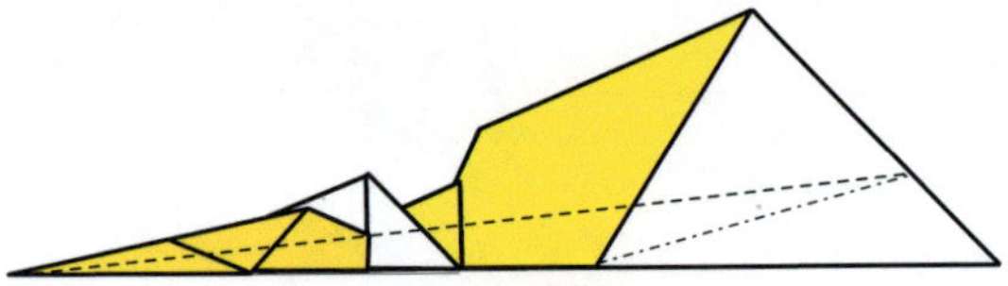

Front View

Side View.

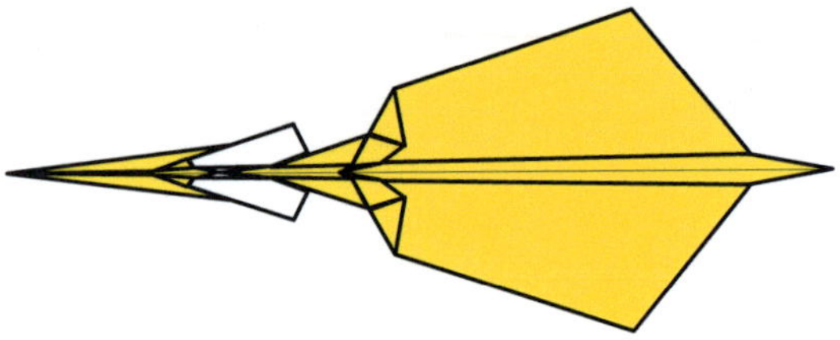

Top View.

Lambda One

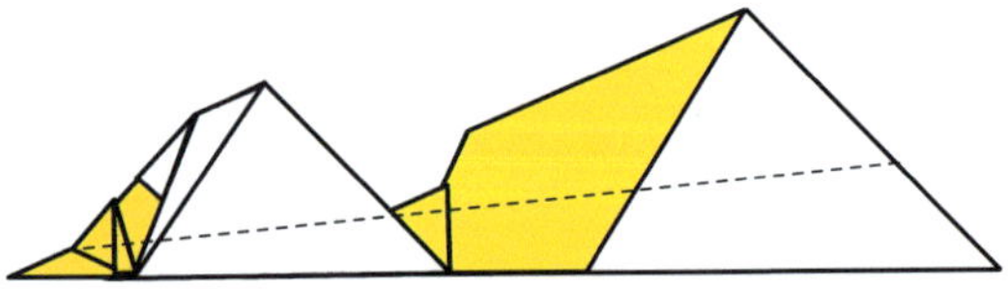

Front View

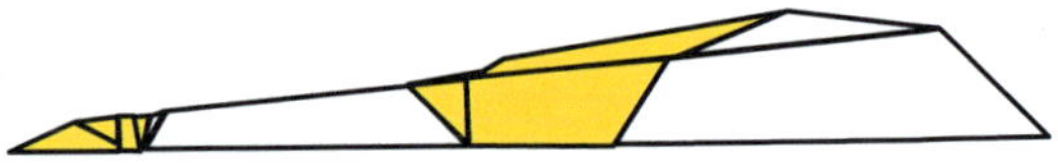

Side View.

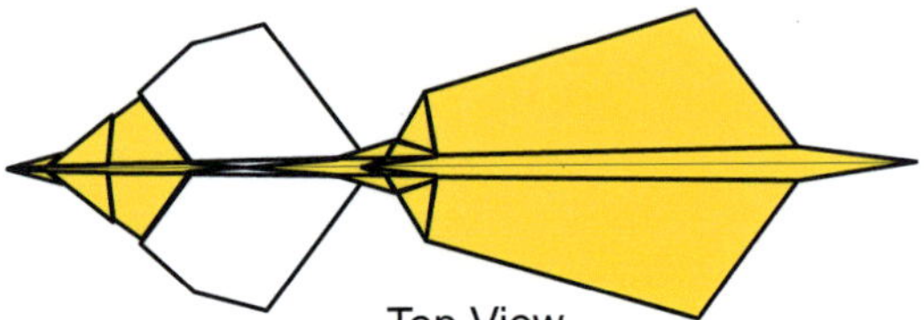

Top View.

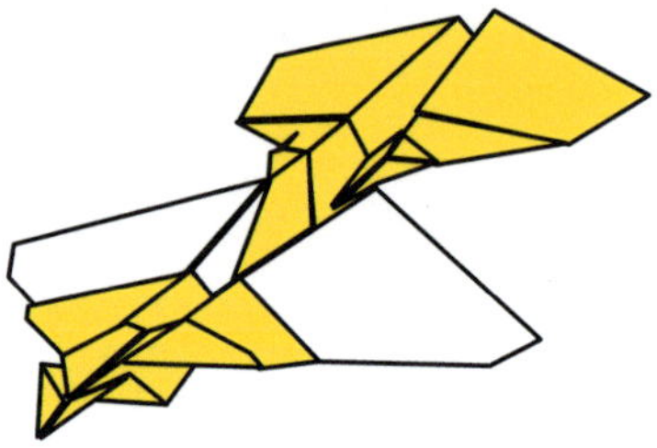

Back Piece Two Airplanes

Alpha Two

Beta Two

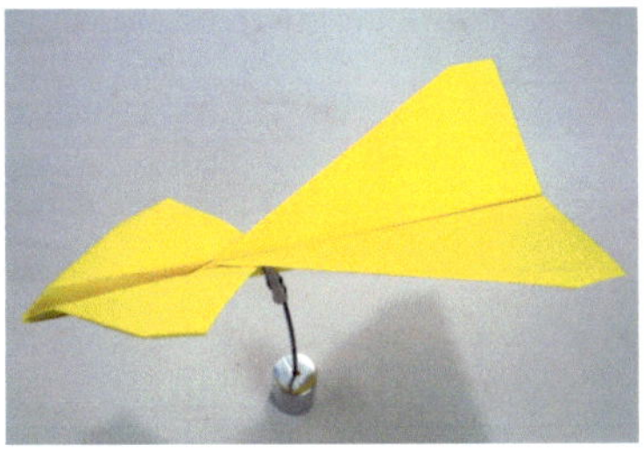

Delta Two

Epsilon Two

Iota Two

Zeta Two

Omicron Two

Kappa Two

Lambda Two

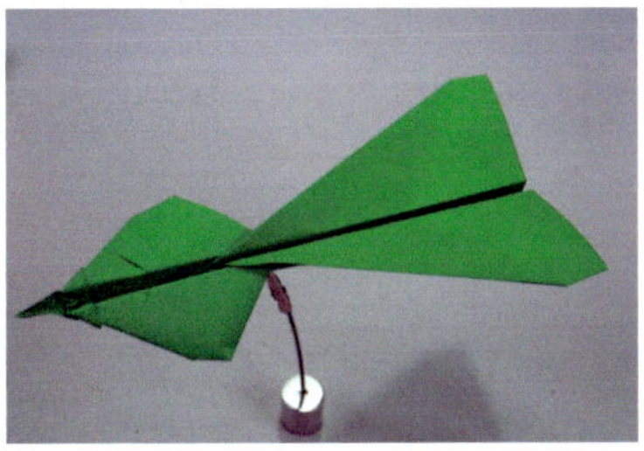

Back Piece Two

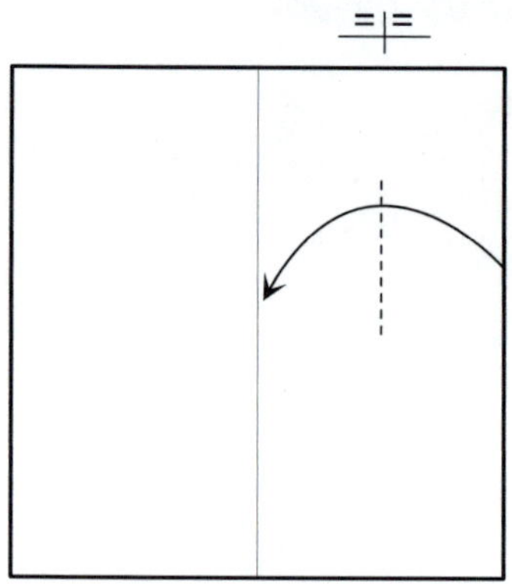

1. Valley fold so that the outer edge hits the midline. Crease only part way and unfold.

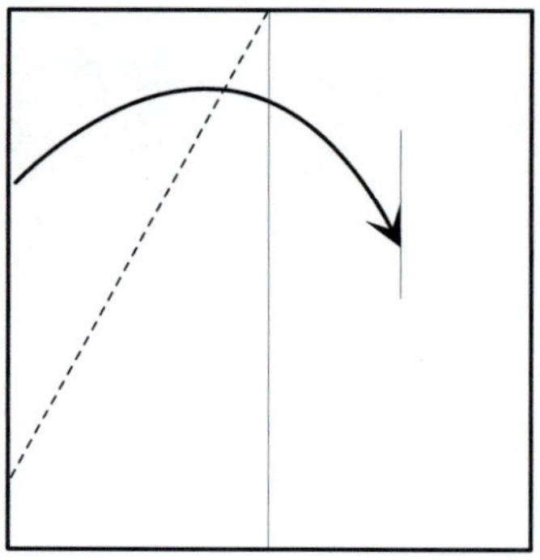

2. Valley fold so that the fold originates at the top center, and the corner hits the crease made in step 1.

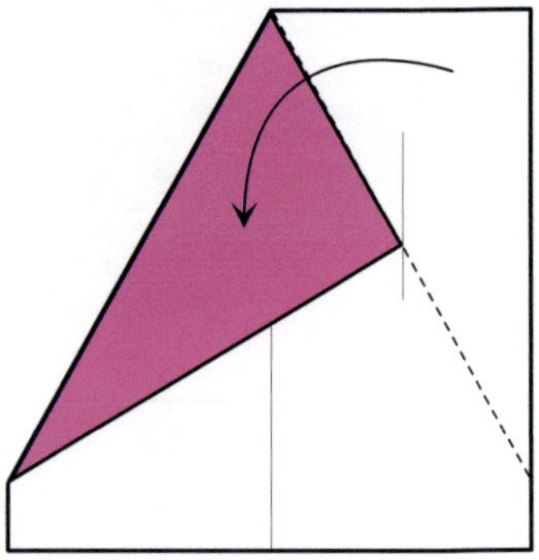

3. Valley fold along the edge of the colored flap.

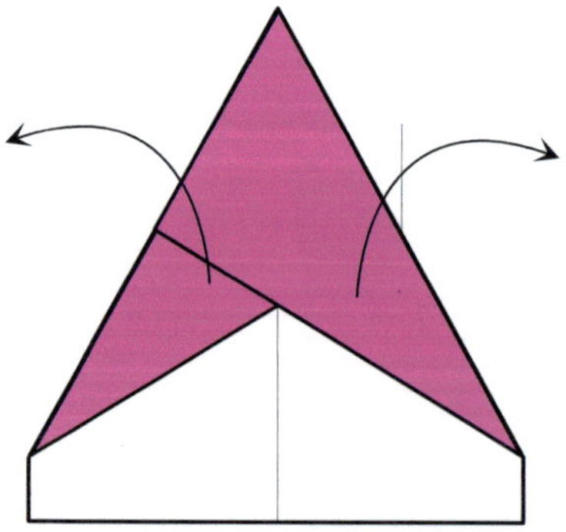

4. Unfold both flaps.

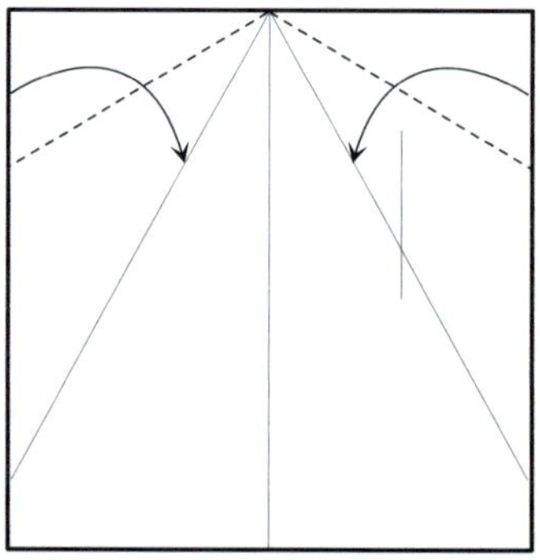

5. Valley fold so that the top edge lies along the creases made in the previous steps.

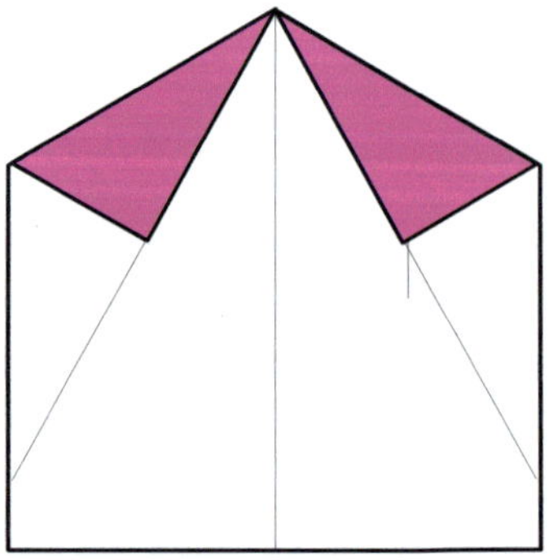

6. Done for now.

Back Piece Two Page Two

We'll attach Front Piece Beta

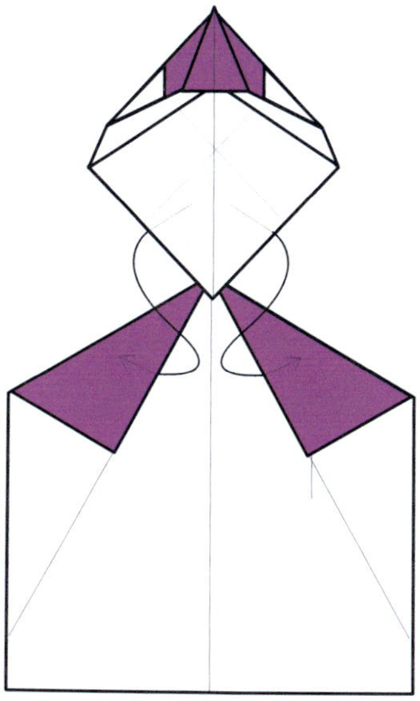

7. Place the square flaps in the back of the front piece underneath the colored flaps of the back piece.

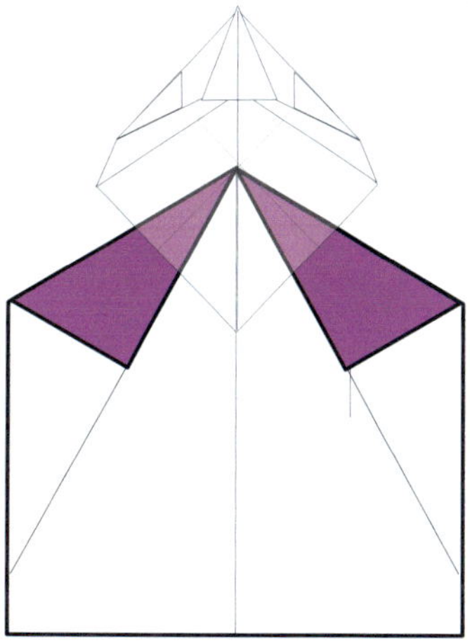

8. Pull the front piece all the way up to the front of the back piece.

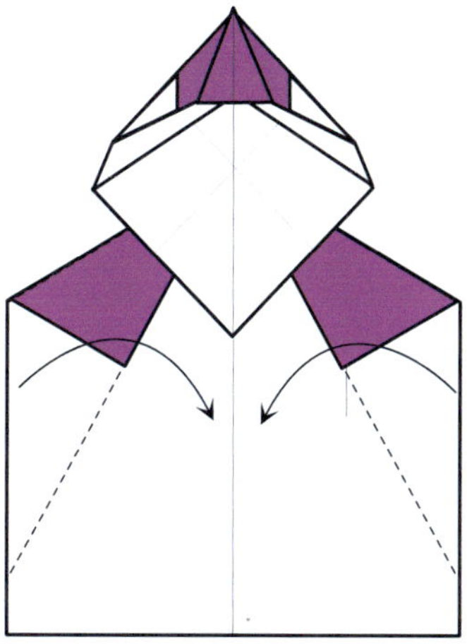

9. Valley fold the two flaps in along preexisting creases.

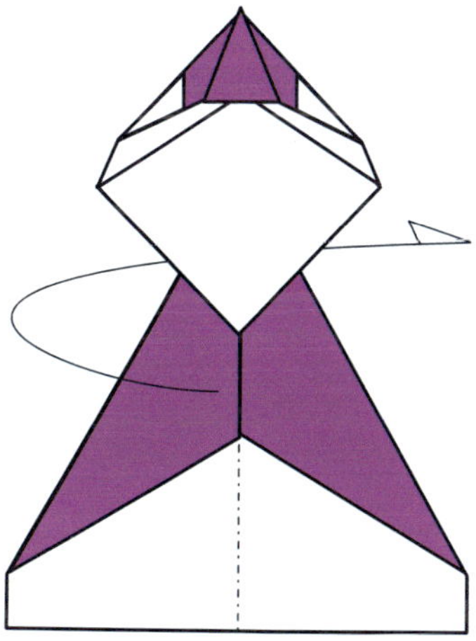

10. The front piece should be securely attached. Mountain fold in half.

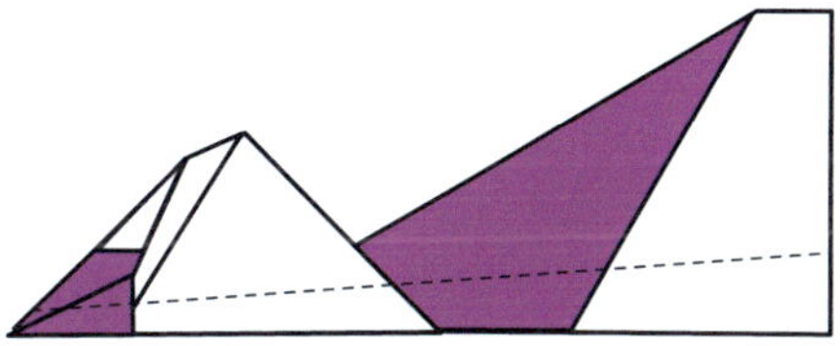

Sadly, there isn't a really good landmark for folding the wings on the Beta front piece. You don't want to fold from the very front, or you can tear Front Piece Beta. A bit of a slope to the fold works best.

Beta Two

Front view

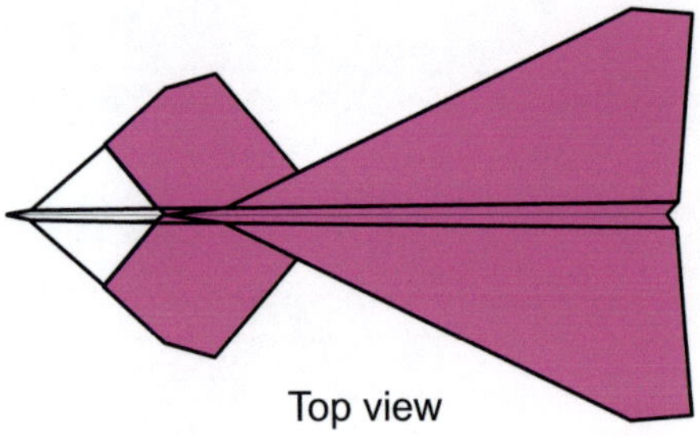

Top view

Side view

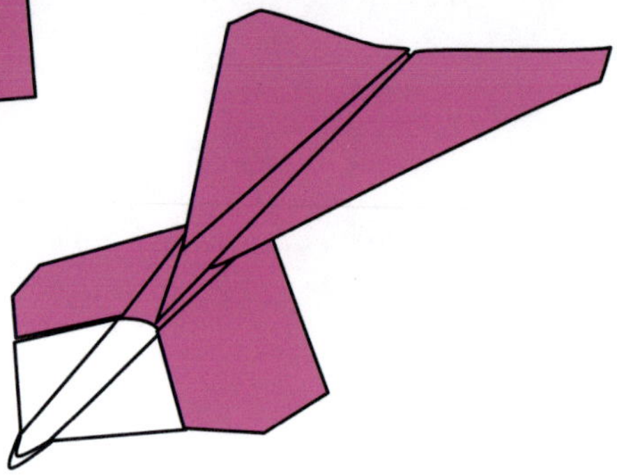

Alpha Two

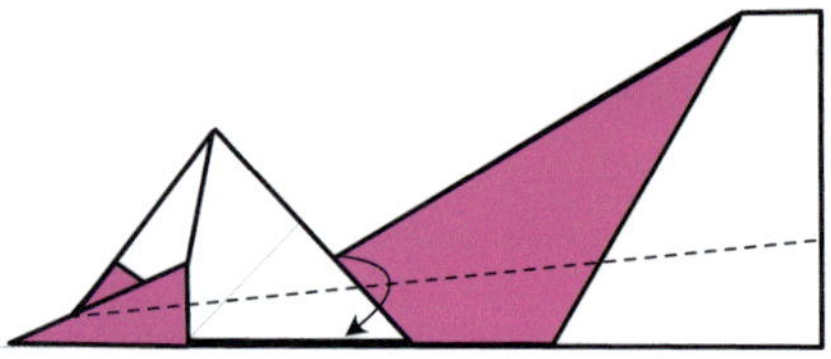

FP Alpha is folded just off the top of the front. If you fold so that the join angle hits the bottom your airplane will really fly.

Front View

Side View

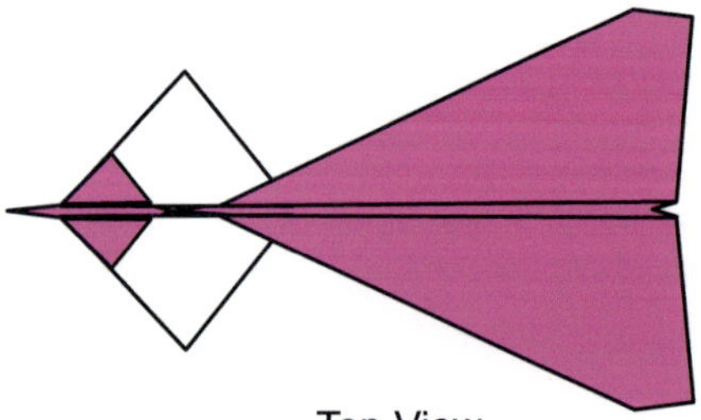

Top View

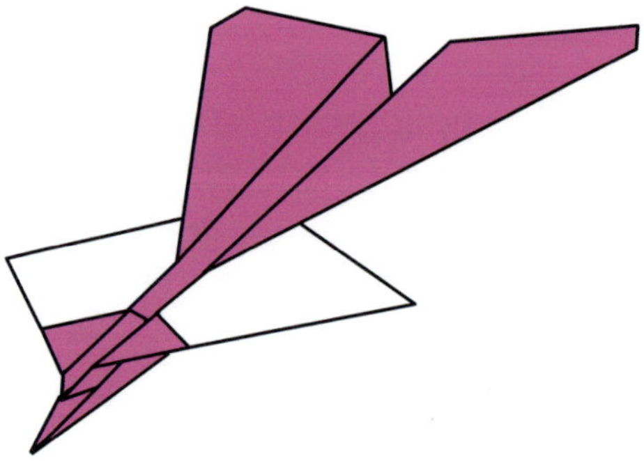

EpsilonTwo

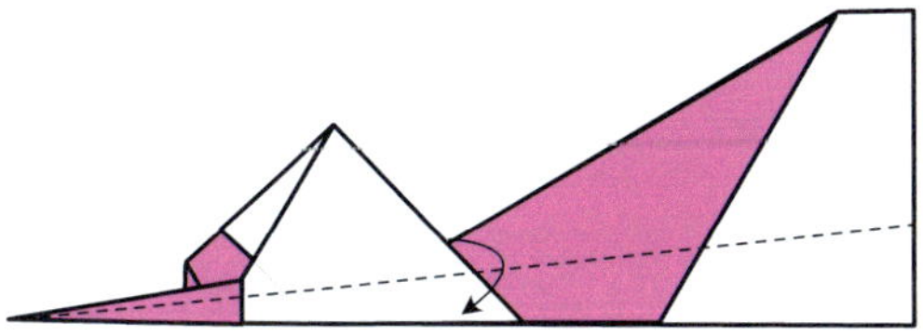

Front View

Side View

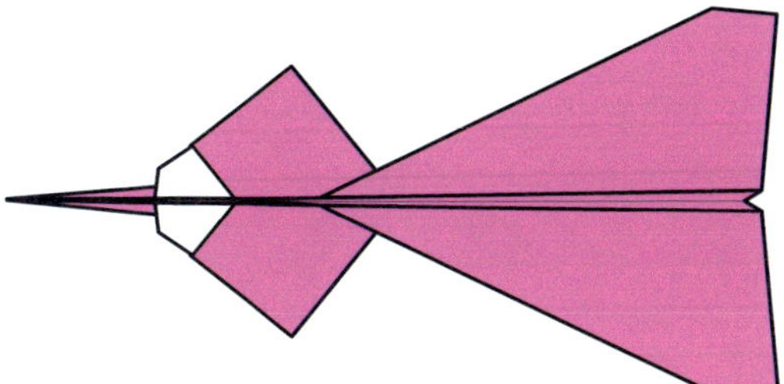

Top View

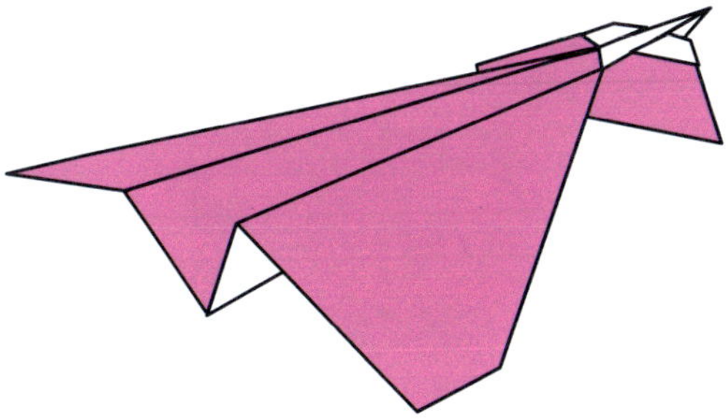

ZetaTwo

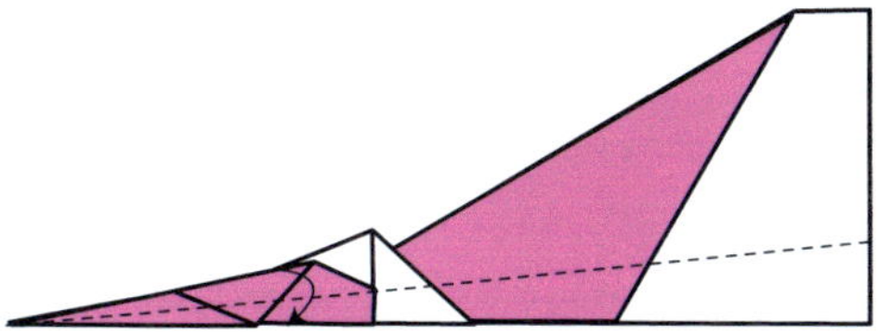

Bisect the angle in front of the Zeta front piece, and fold the main wings to match.

Front View

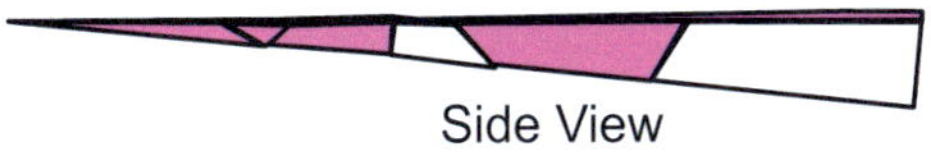

Side View

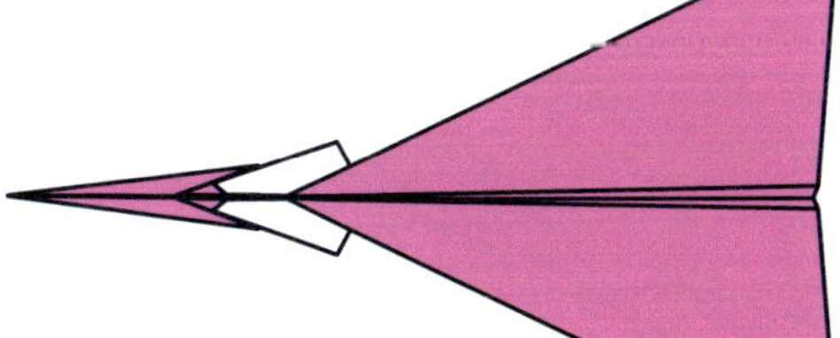

Top View

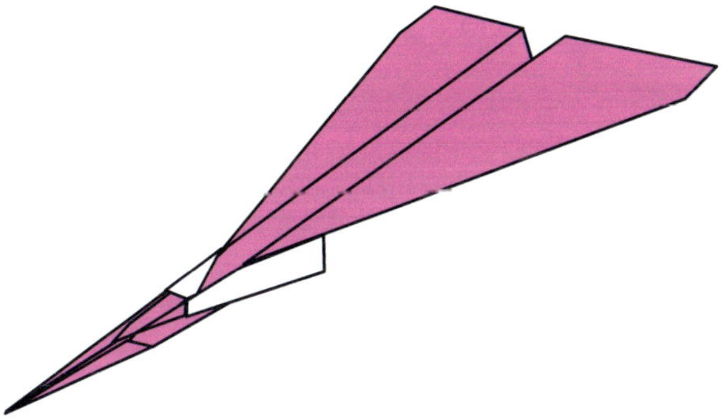

IotaTwo

Front View

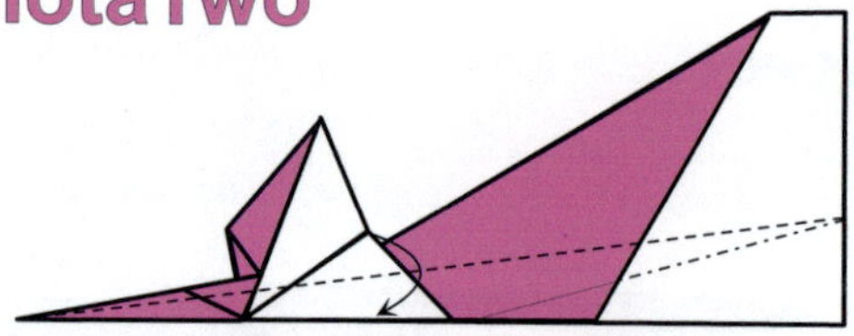

Valley fold the front so the angle shown hits the midline. Fold the back piece to match. Reverse fold a tail.

Side View

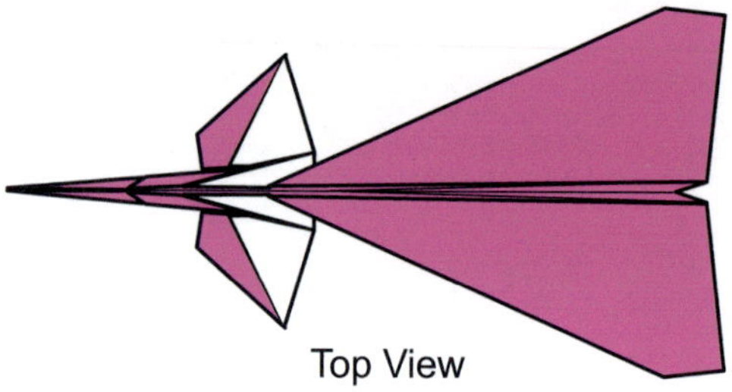

Top View

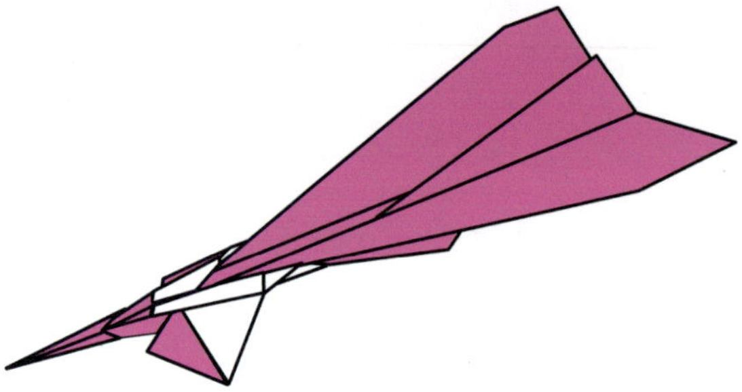

Front View

KappaTwo

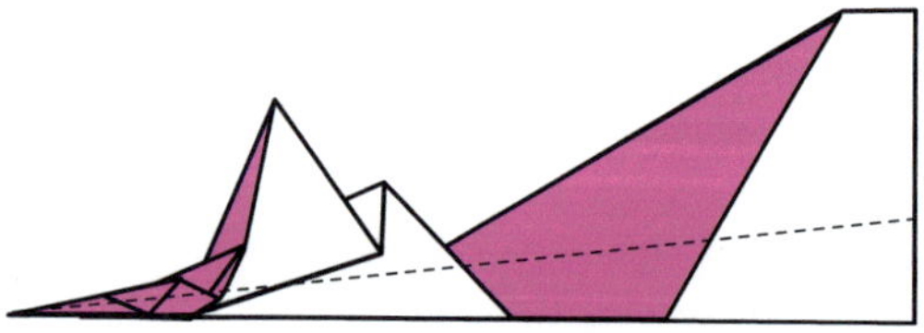

Side View

I haven't a good landmark to fold the wings, sorry. The wing fold should be just beneath the intersection of the two pieces.

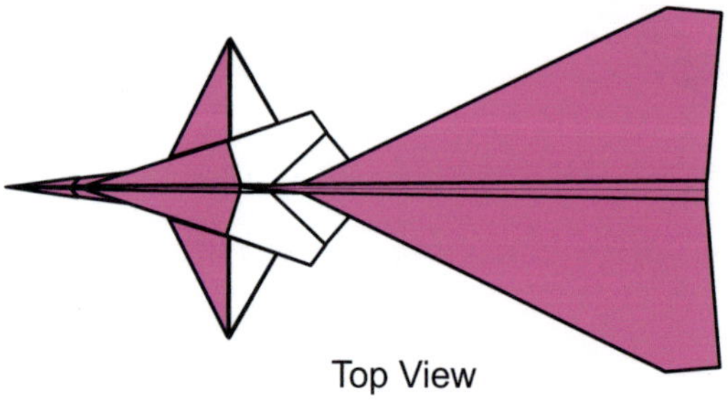

Top View

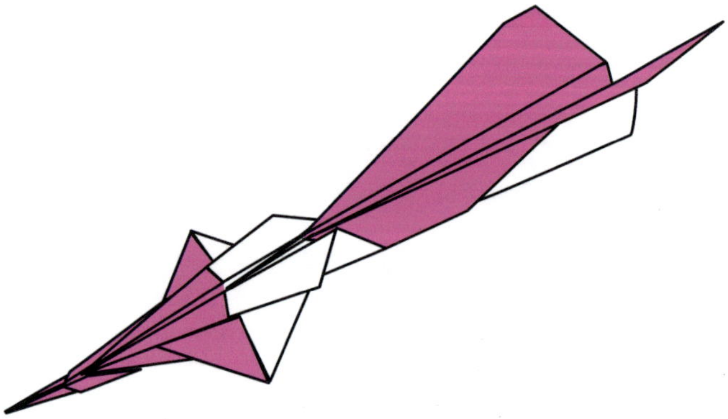

LambdaTwo

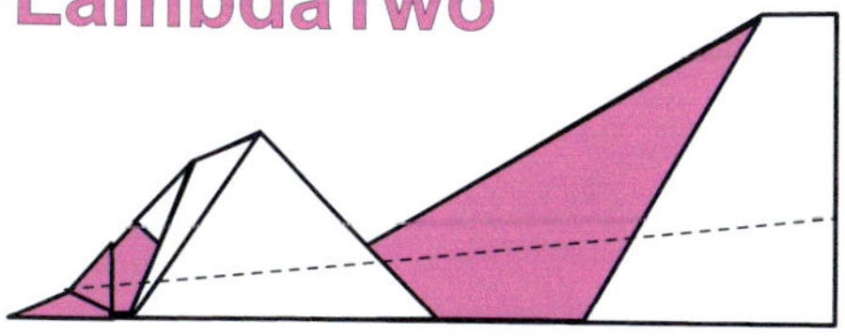

No good landmark for folding the wings. Just below the intersection of the two creases gives a nicely flying airplane.

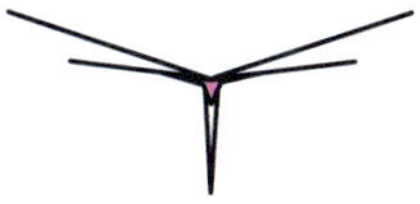

Front View

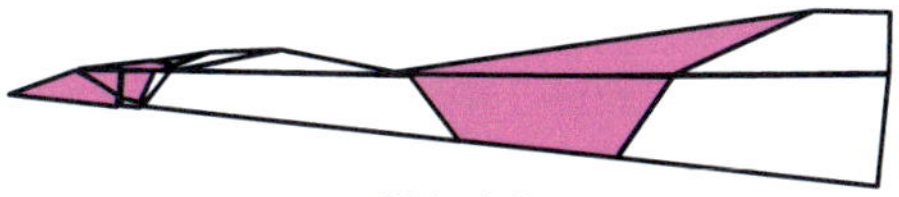

Side View

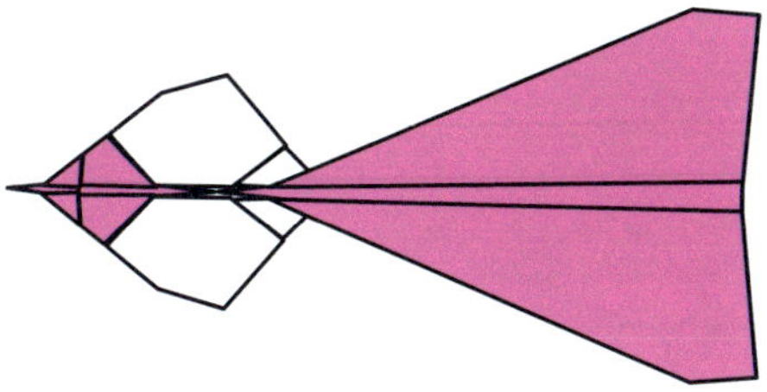

Top View

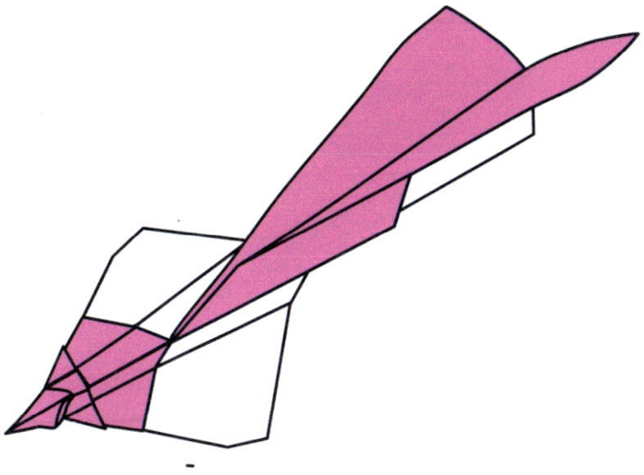

Omicron Two

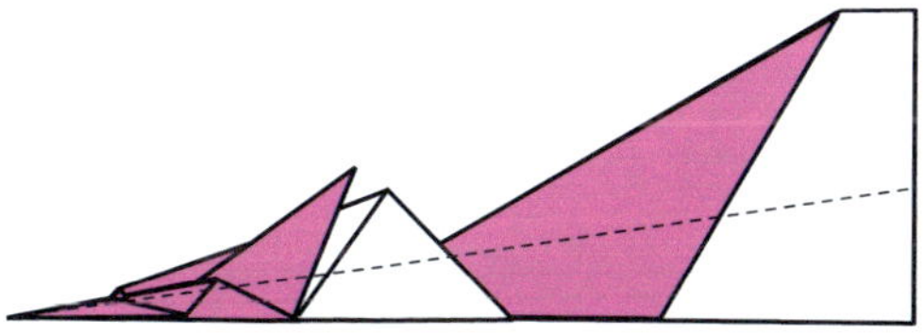

Front View

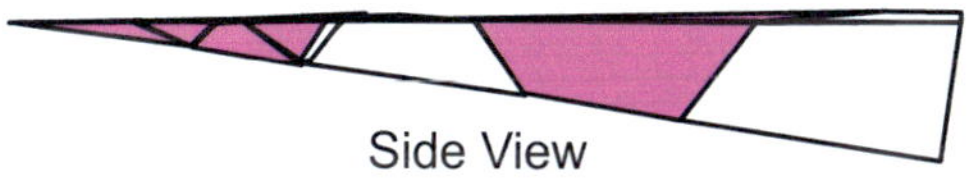

Side View

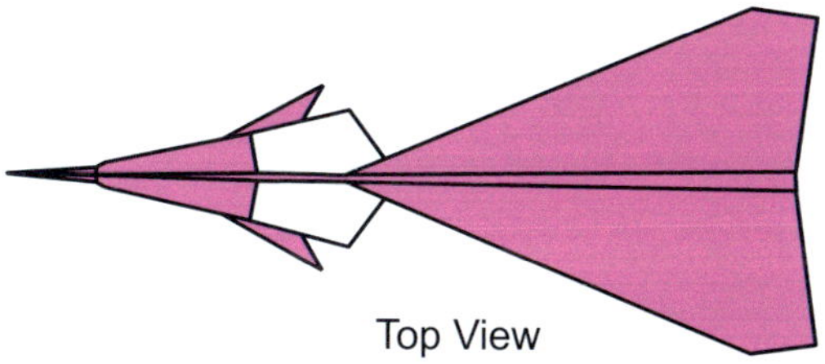

Top View

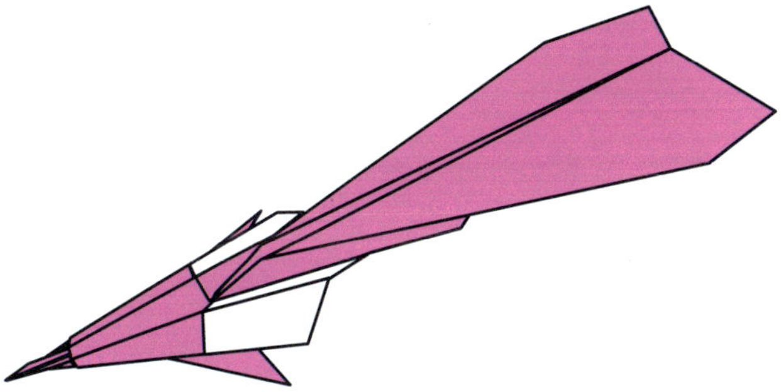

Delta Two

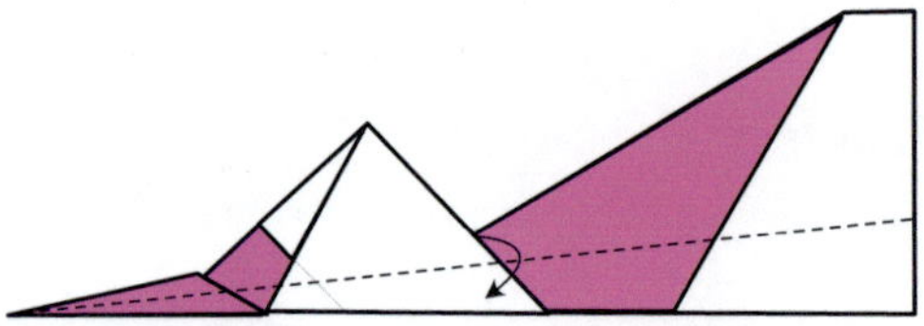

FP Delta is folded from the very front. Fold the wings so that the join angle just falls short of the bottom. Too much wing area and the aircraft won't fly to its maximum potential.

Front View

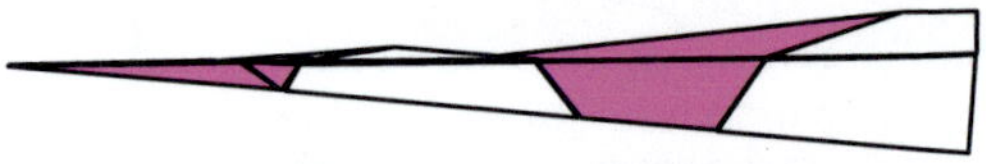

Side View

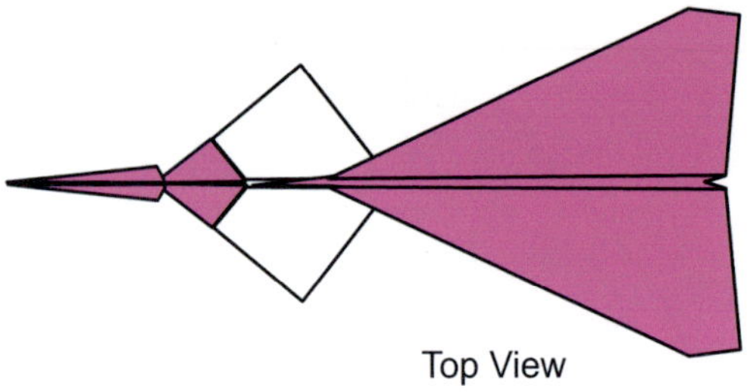

Top View

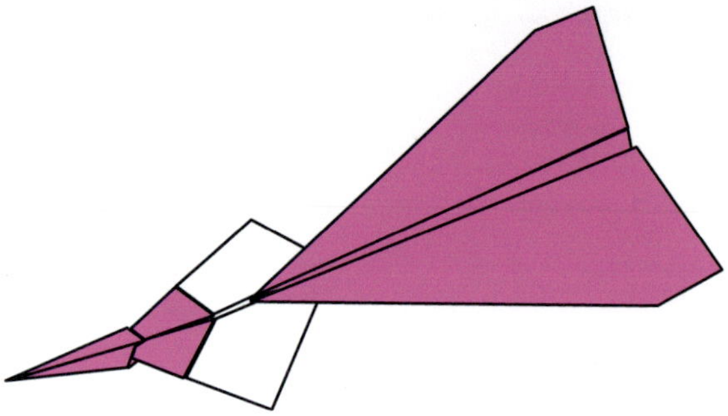

Back Piece Three Airplanes

Alpha Three

Beta Three

Gamma Three

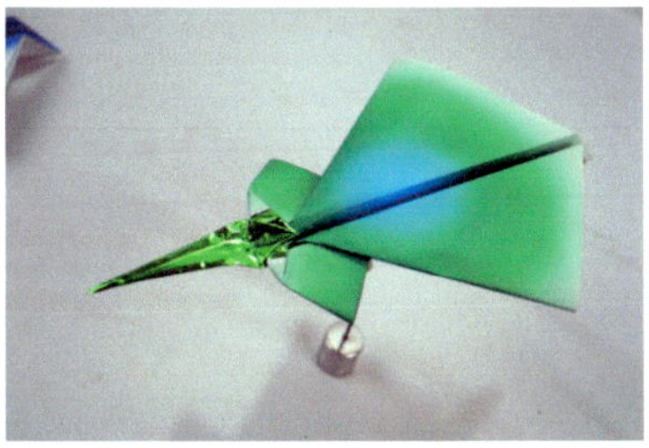

Delta Three

Epsilon Three

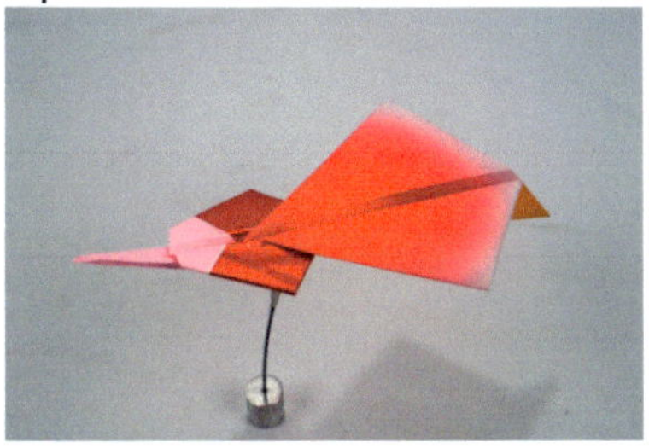

Zeta Three

Iota Three

Kappa Three

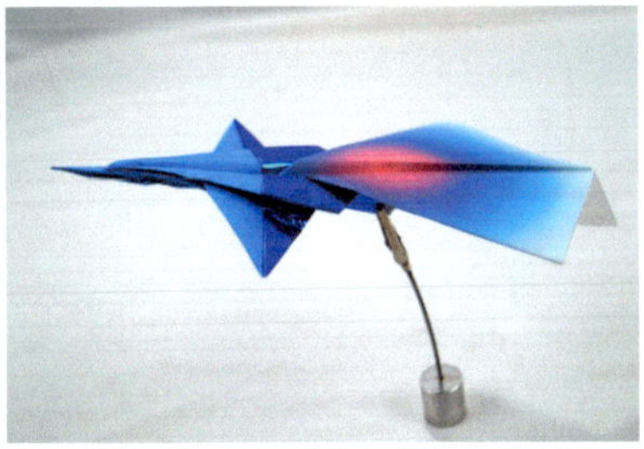

Omicron Three

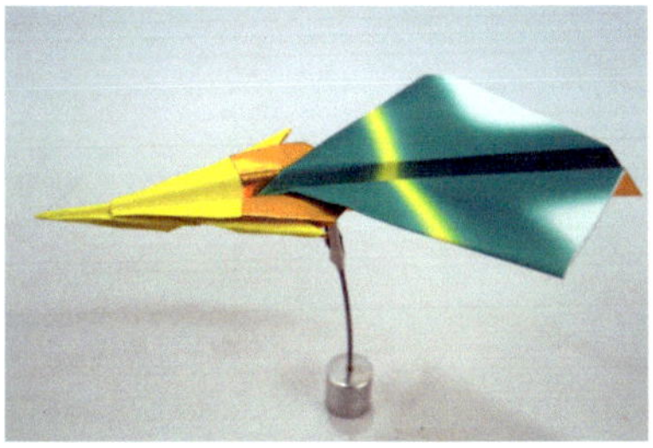

Back Piece Three

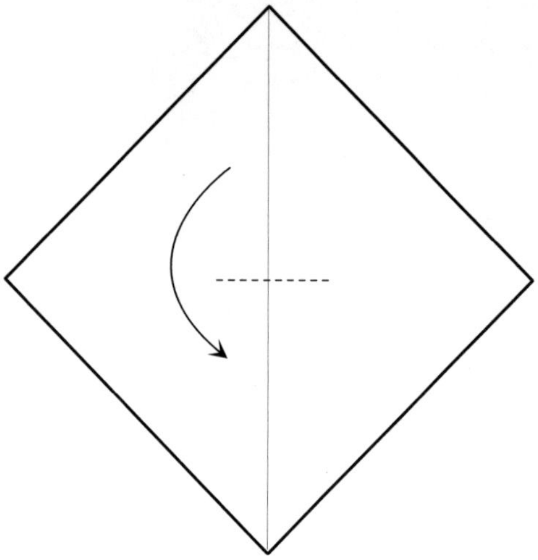

1. Valley fold the second diagonal. Crease only in the middle.

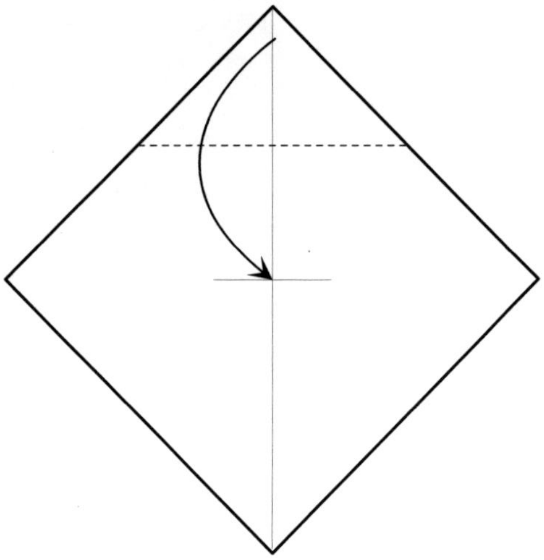

2. Valley fold the top to meet the diagonal.

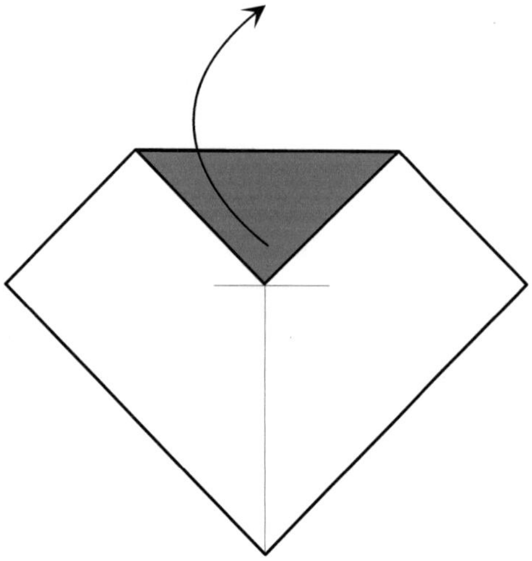

3. Unfold.

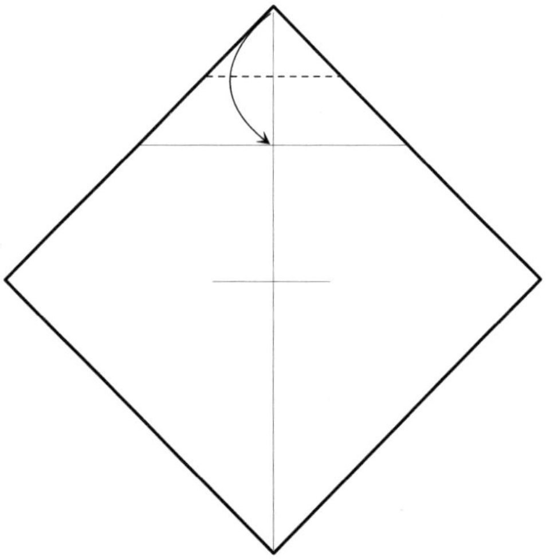

4. Valley fold the top point so it hits the crease made in step 3.

Back Piece Three Page 2

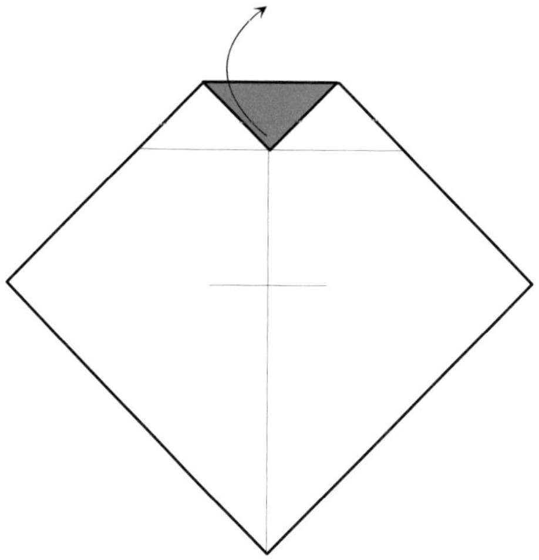

4. Unfold.

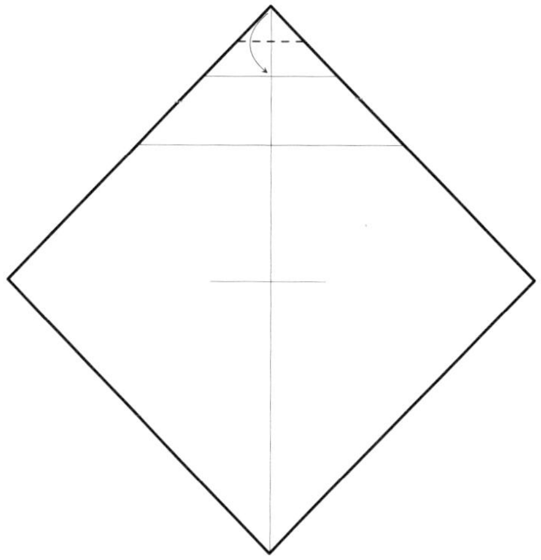

5. Valley fold the top point so it hits the crease made in step 4.

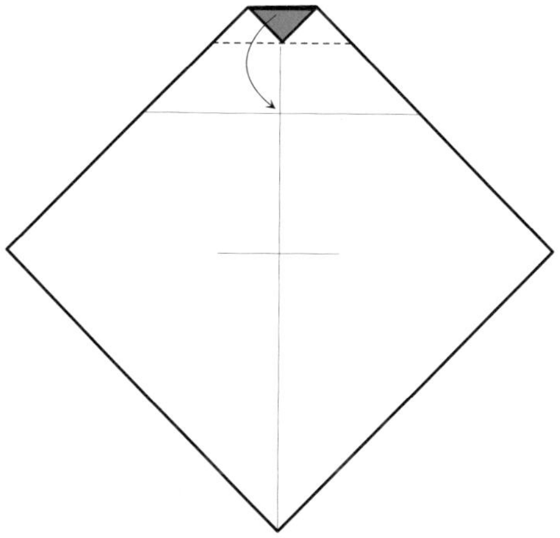

6. Valley fold along a preexisting crease.

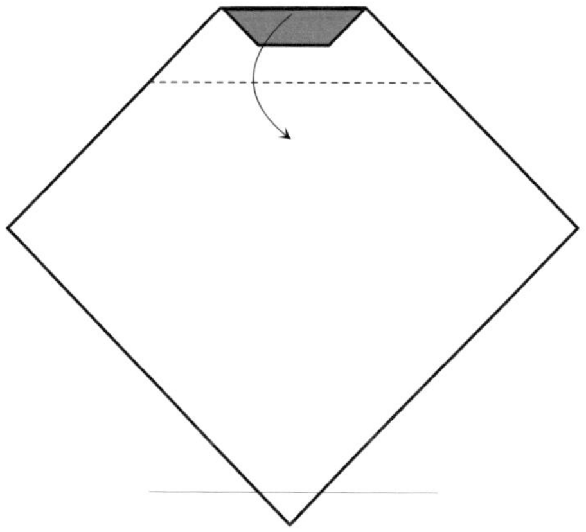

7. Valley fold again along a preexisting crease.

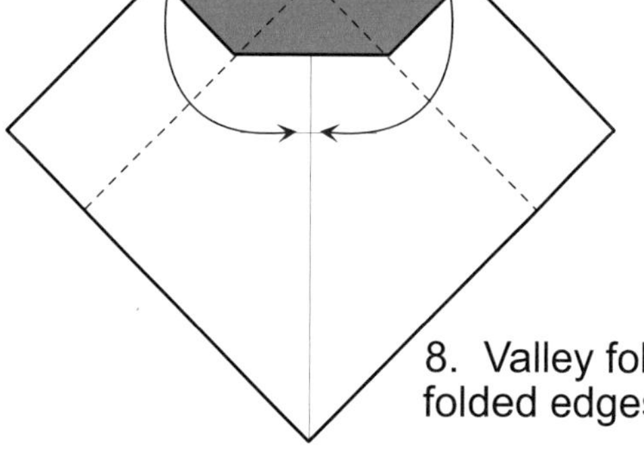

8. Valley fold so that the upper folded edges lie on the midline.

Back Piece Three Page 3

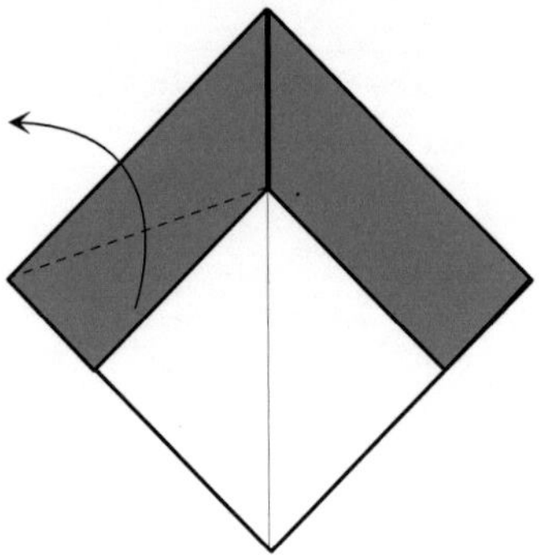

9. Valley fold the flap from corner to corner.

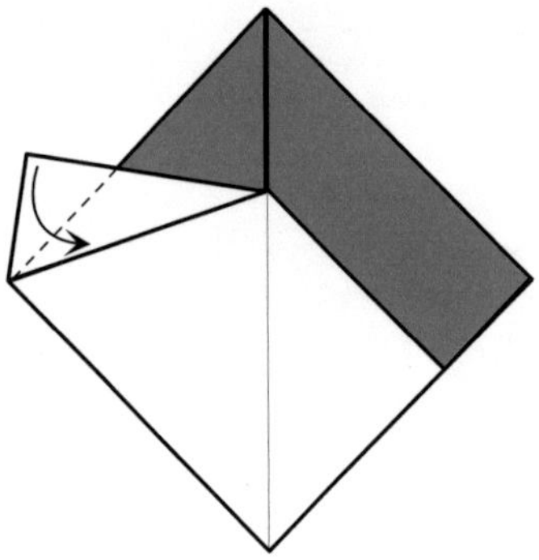

10. Valley fold back down along the hidden edge.

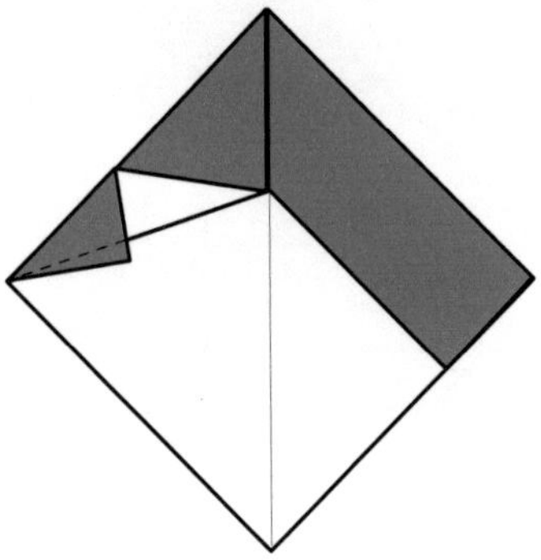

11. Valley fold back up along the hidden edge.

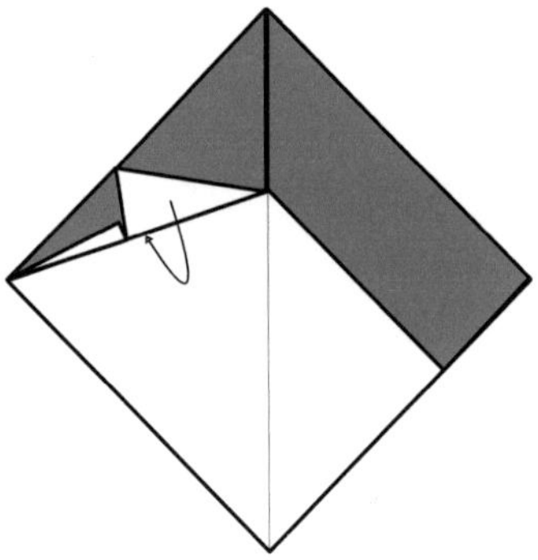

12. Mountain fold the entire flap behind.

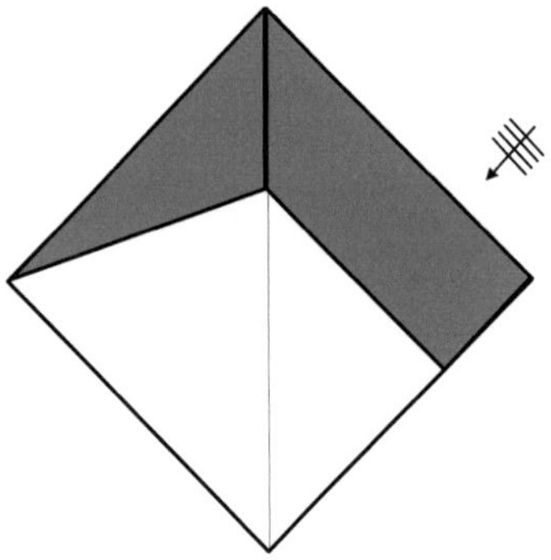

13. Repeat steps 9-12 on the other side.

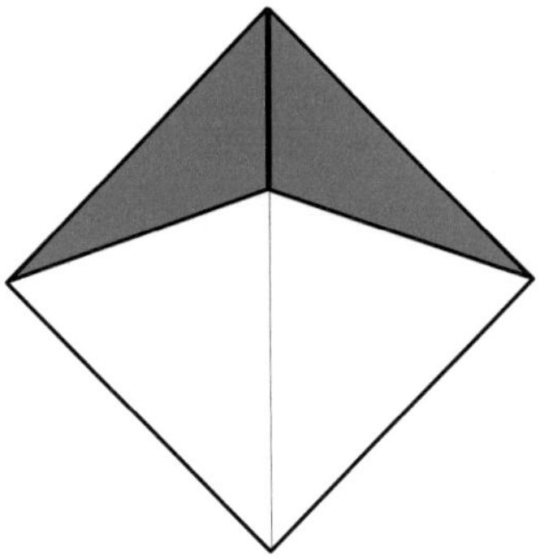

14. Folding done, time for assembly.

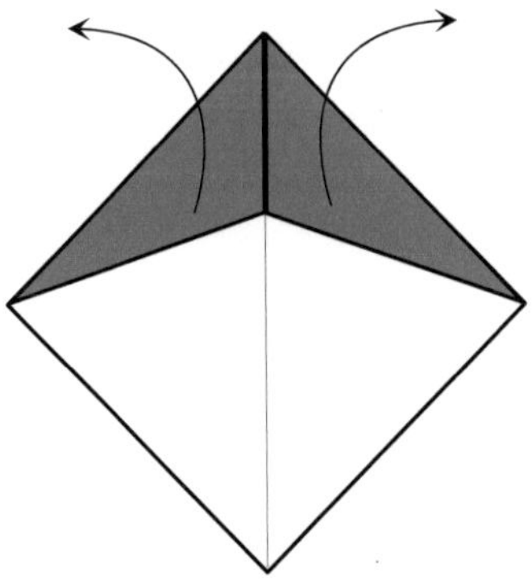

15. Unfold the two flaps.

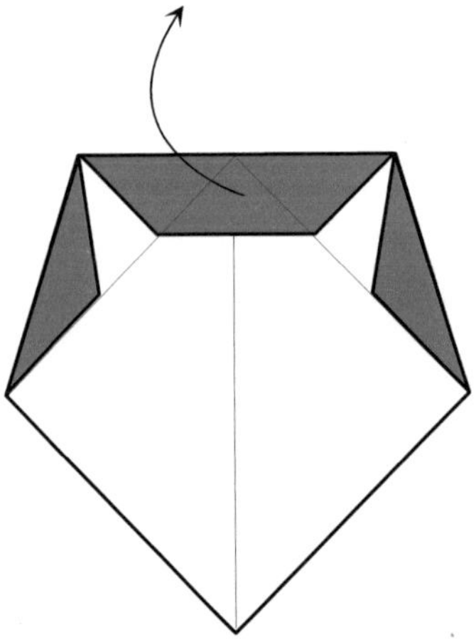

16. Unfold a flap in front.

Back Piece Three Page 4

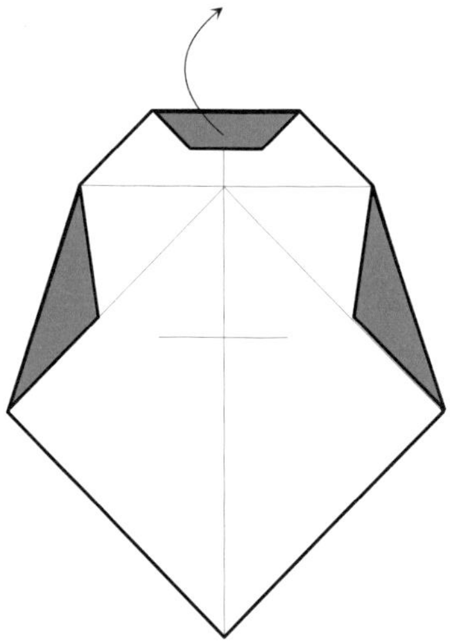

17. Unfold one more time.

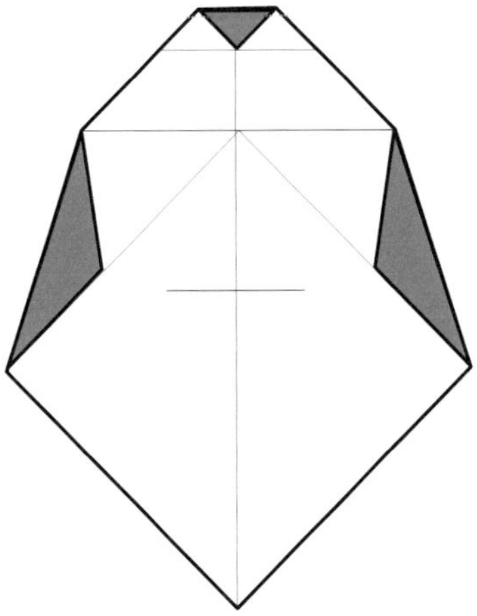

18. You're all ready for assembly. We just need to alter the front piece.

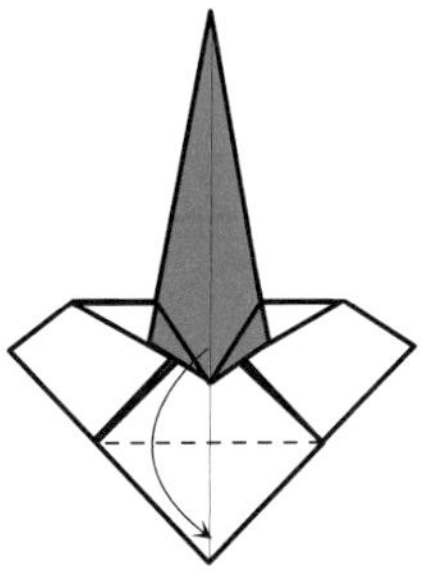

19. Valley fold the square connector flap diagonally in half.

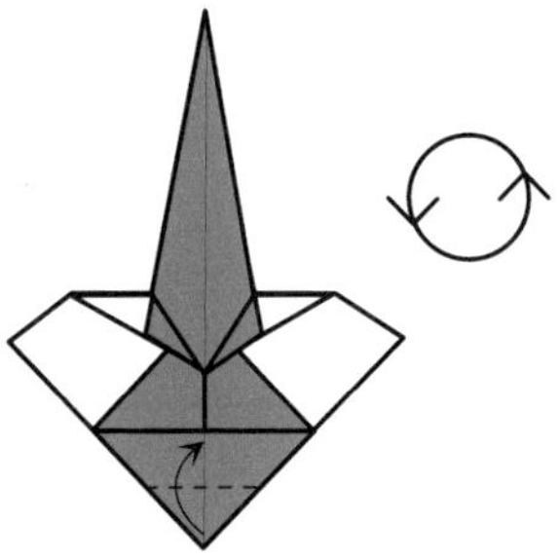

20. Valley fold the tip of the upper flap to the middle, then rotate 180 degrees.

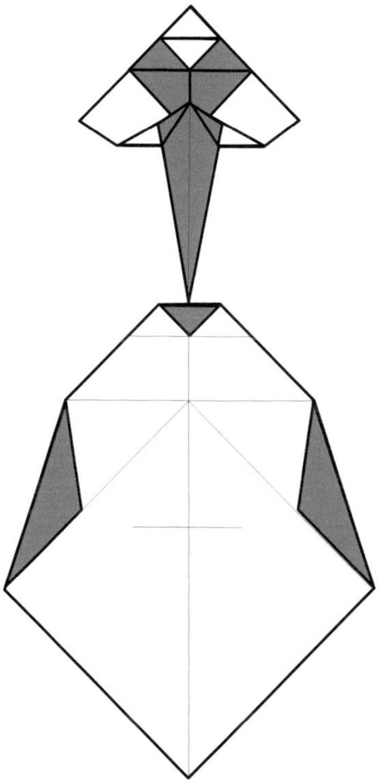

21. Line up the front and back pieces.

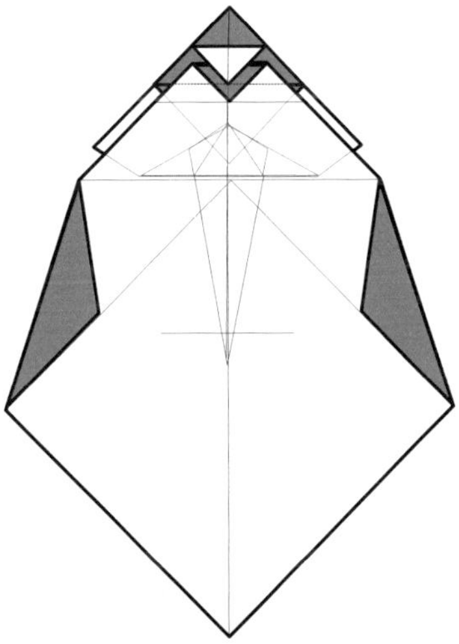

22. Superimpose the two small triangular flaps.

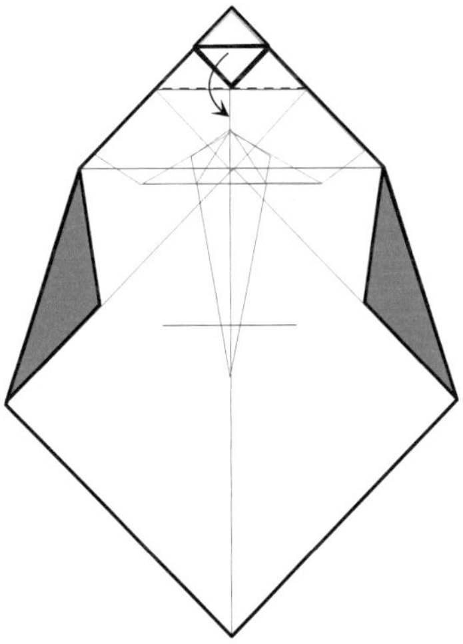

23. Valley fold both flaps together along a preexisting crease.

Back Piece Three Page 5

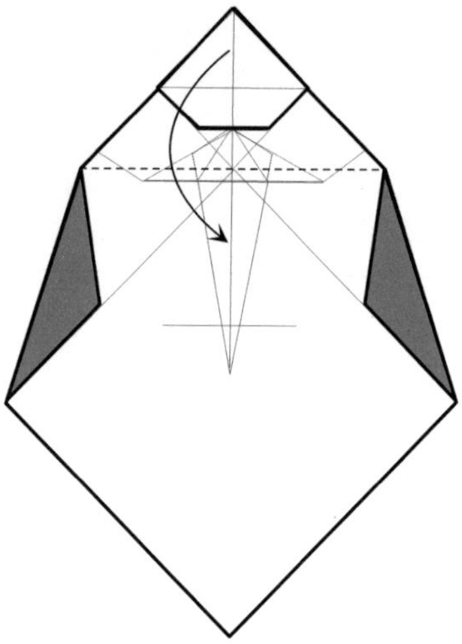

24. Valley fold alone the next preexisting crease. The front piece will flip over.

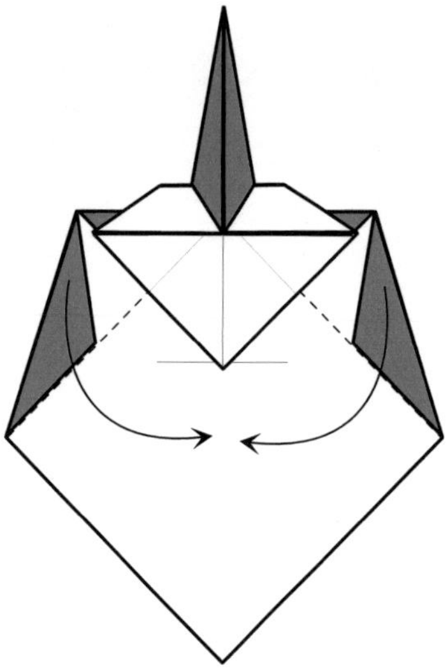

25. Reform the two lateral flaps folding along preexisting creases. These folds will lock the front together.

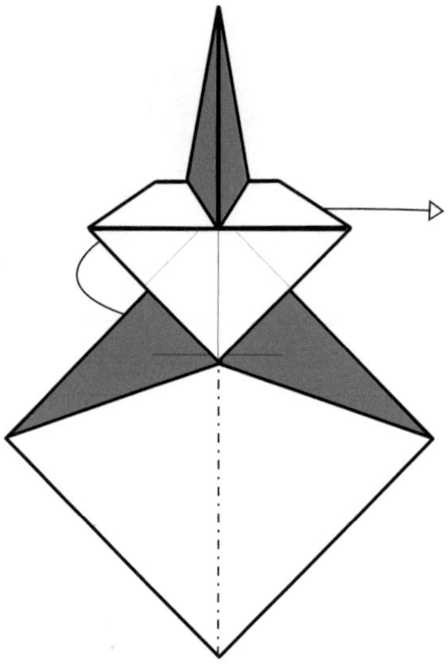

26. Mountain fold in half.

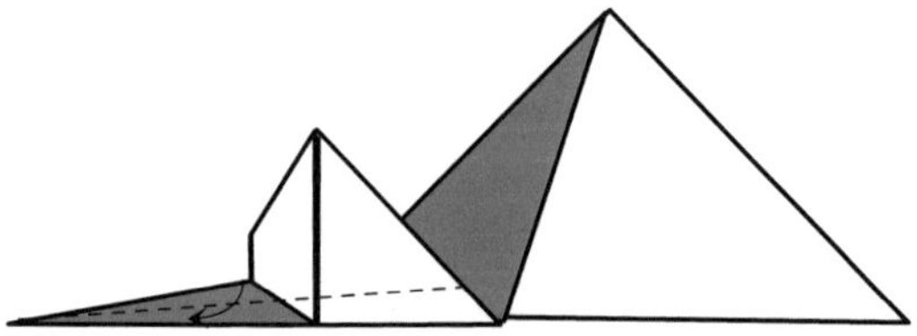

27. To fold the wings, start with front piece Gamma. Fold it to bisect the narrow angle in the nose.

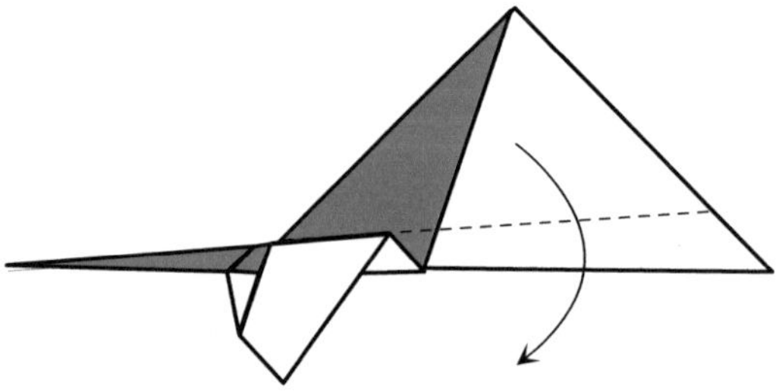

28. Follow the angle of the front wings to fold the back wings.

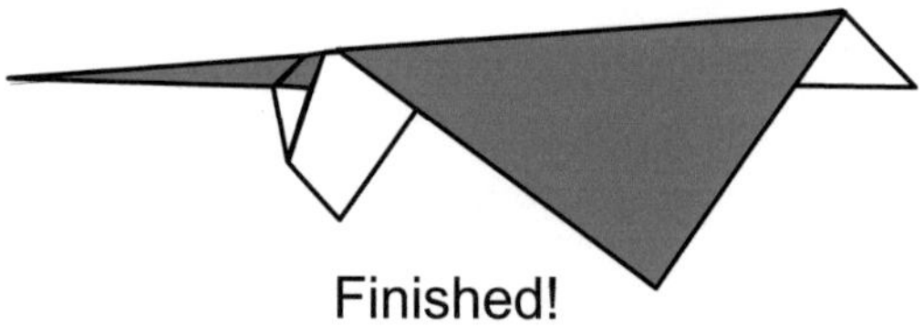

Finished!

Gamma Three

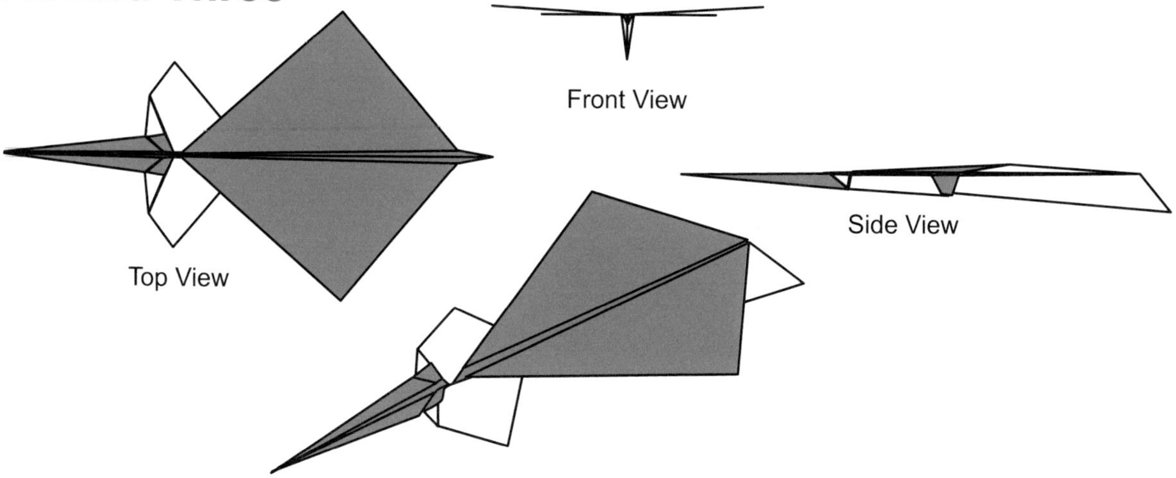

Alpha Three

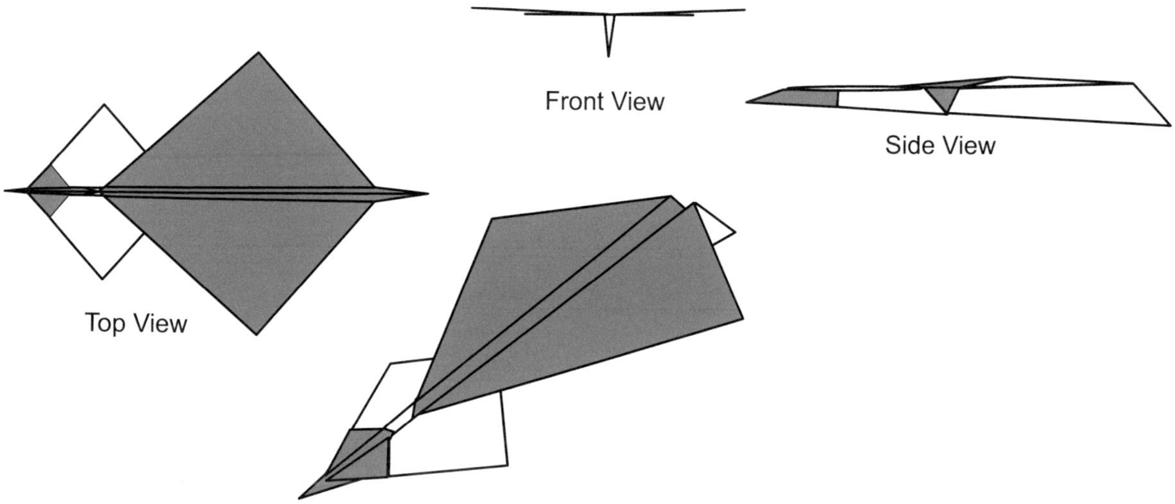

Beta Three

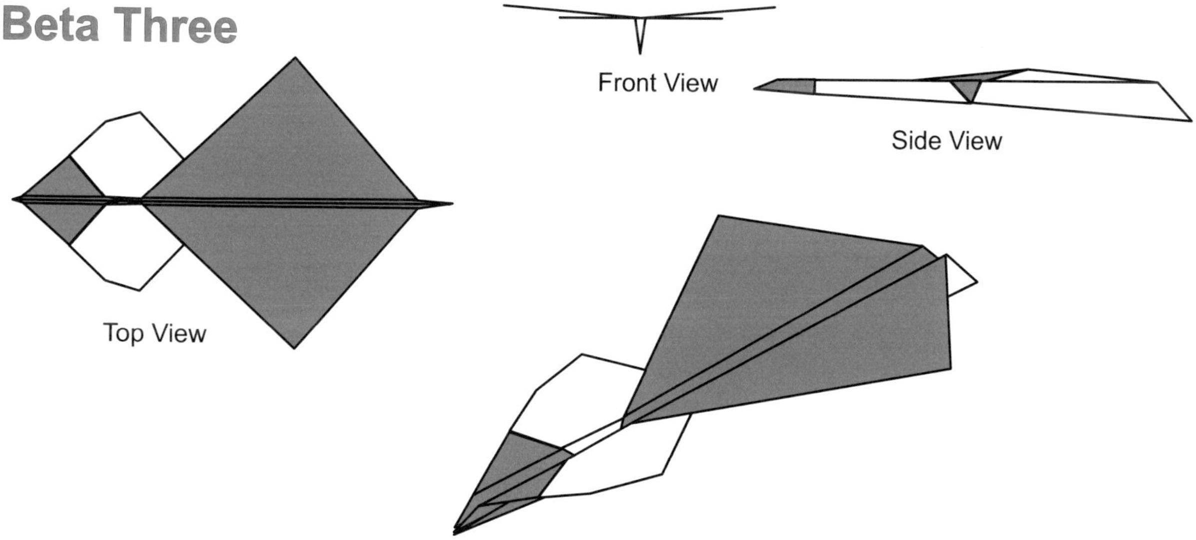

Delta Three

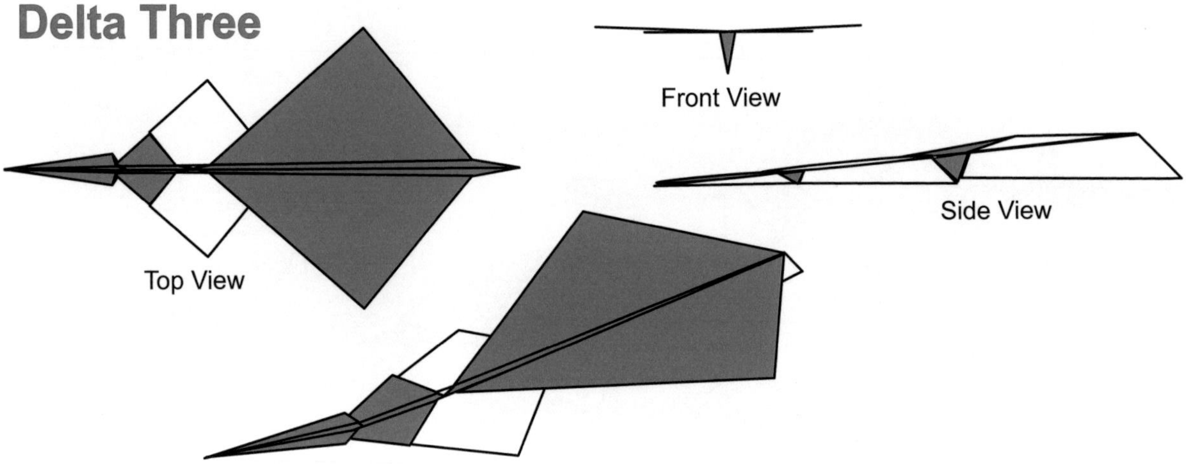

Epsilon Three

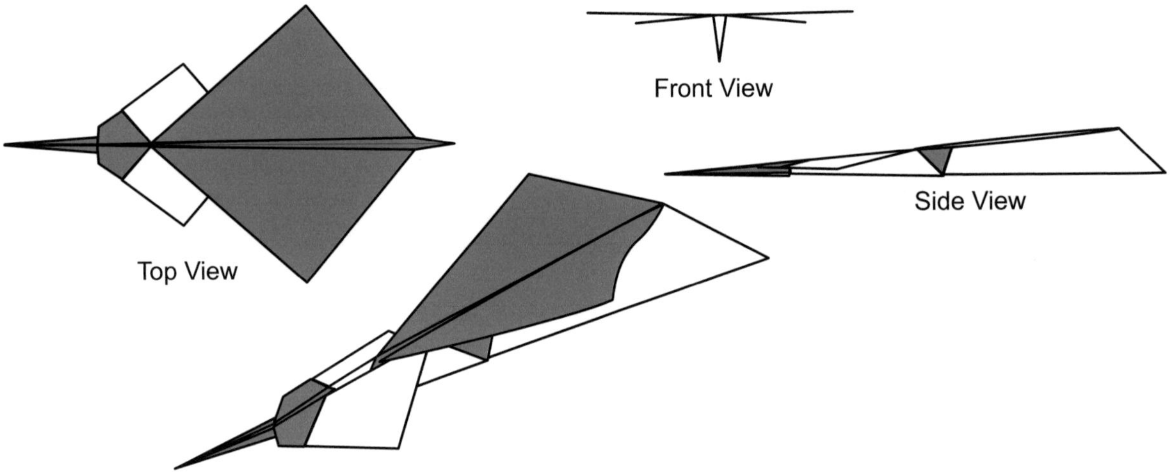

Zeta Three

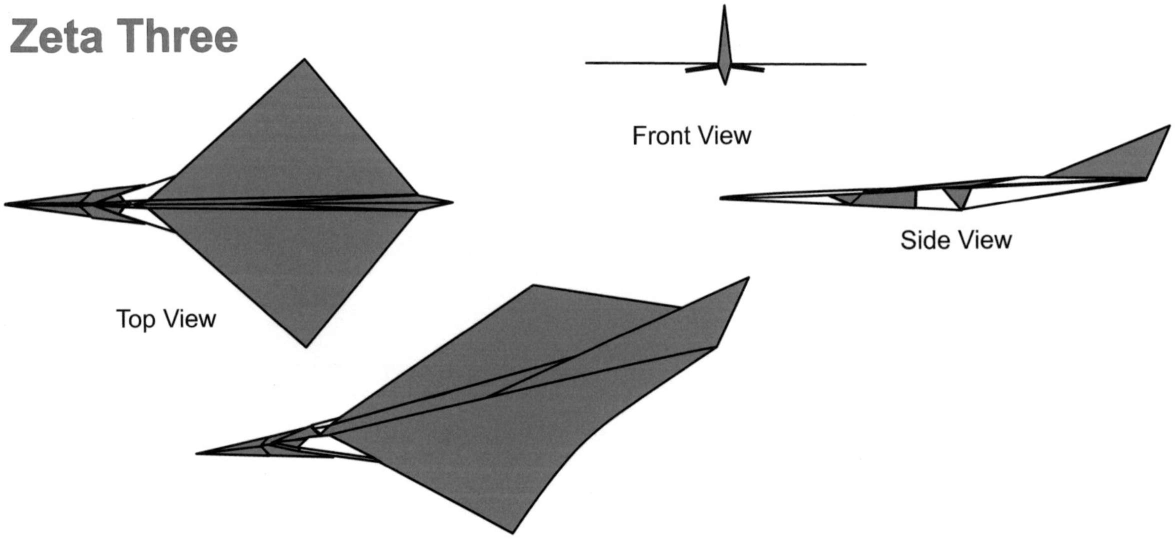

Iota Three

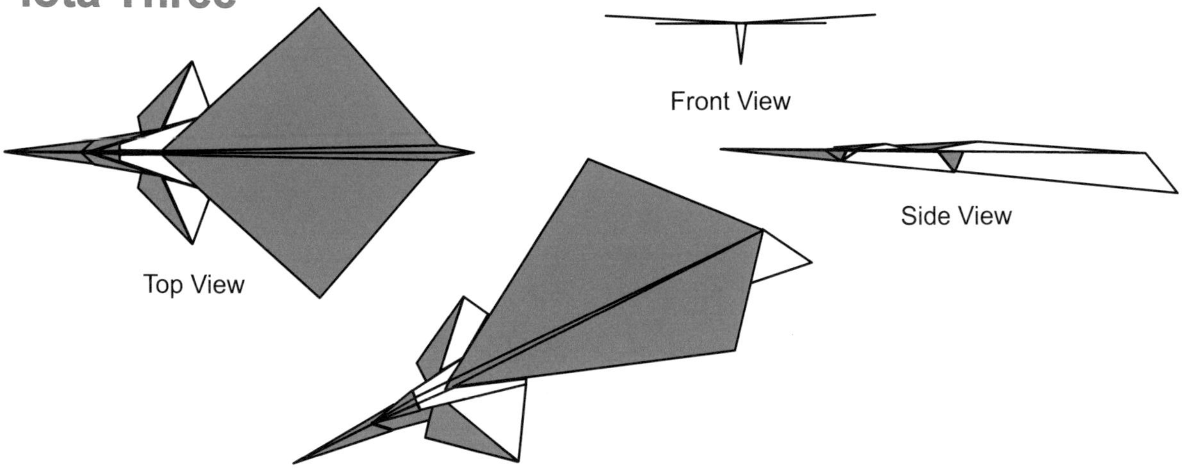

Kappa Three

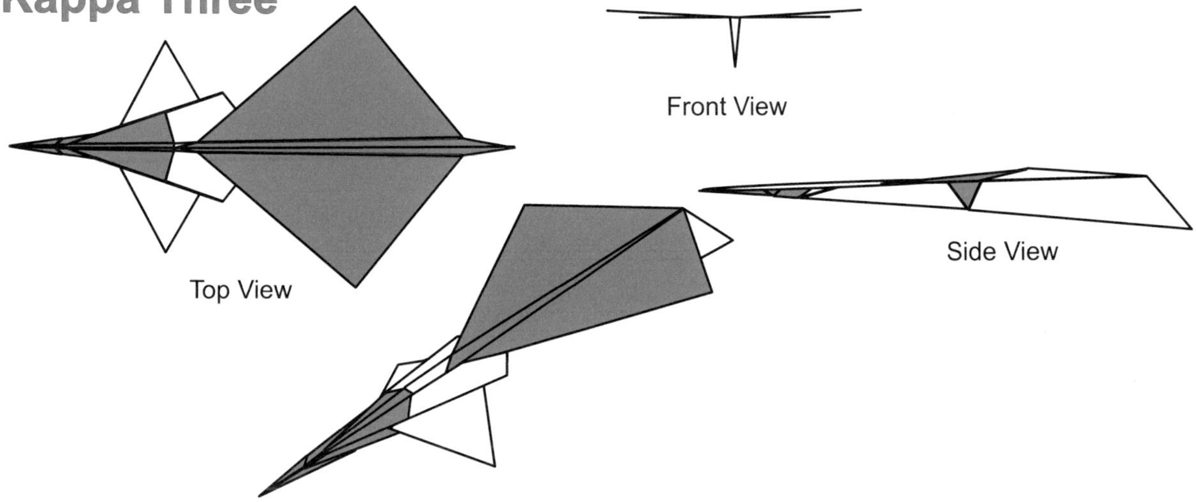

Omicron Three

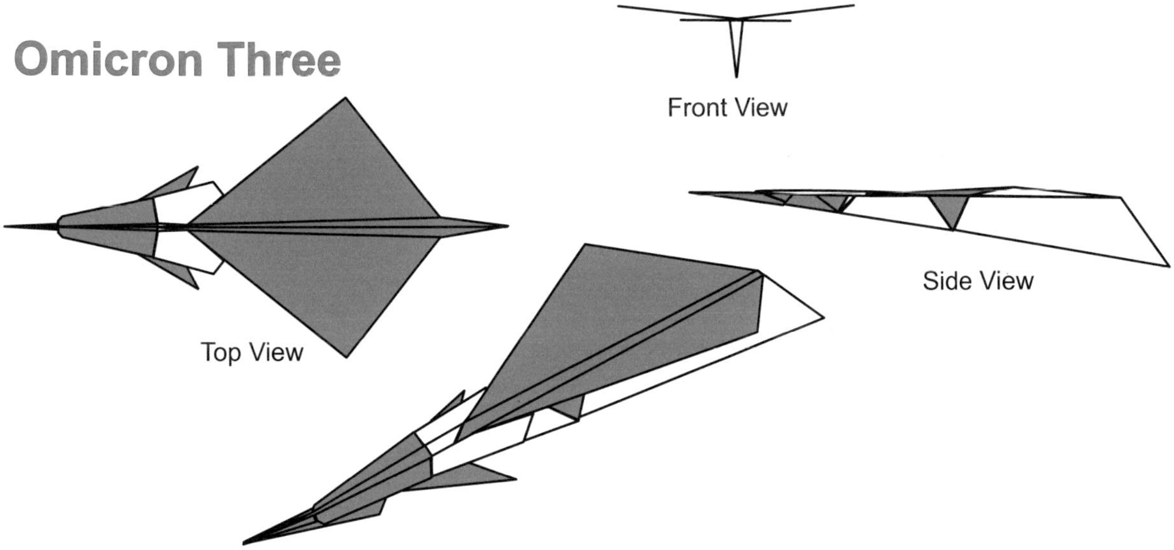

Back Piece Four Airplanes

Alpha Four

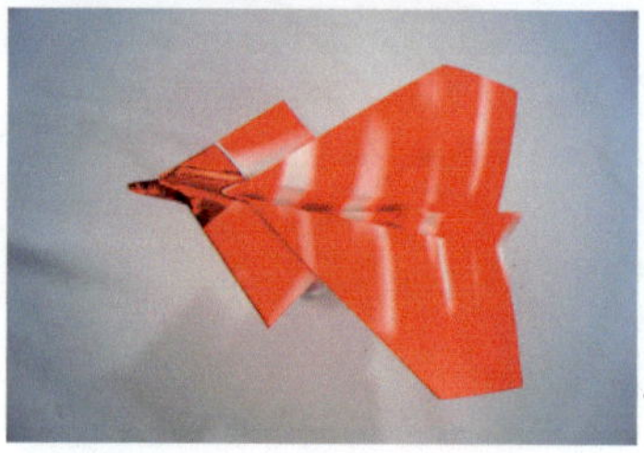

Gamma Four

Delta Four

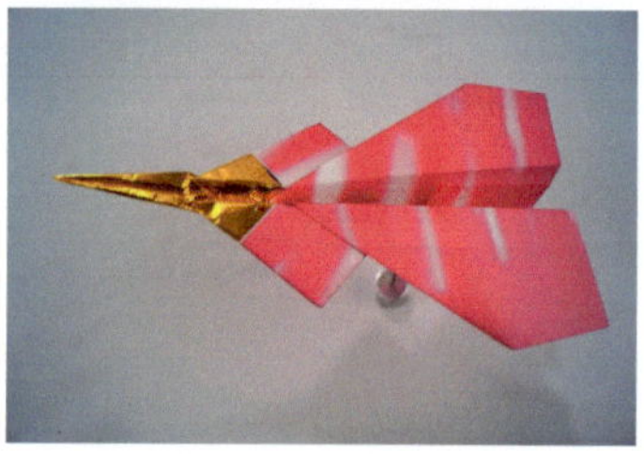

Epsilon Four

Iota Four

Zeta Four

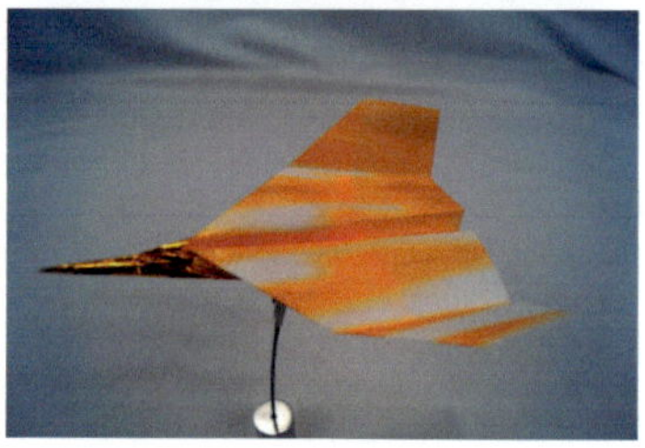

Omicron Four

Kappa Four

Lambda Four

Back Piece Four

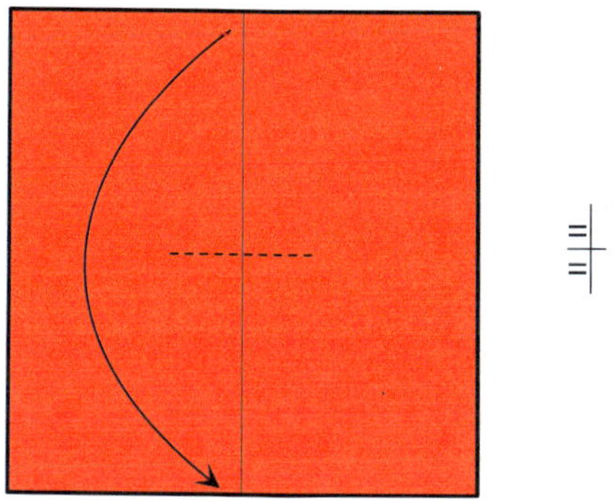

1. Valley fold the model in half lengthwise. Make a small crease and unfold.

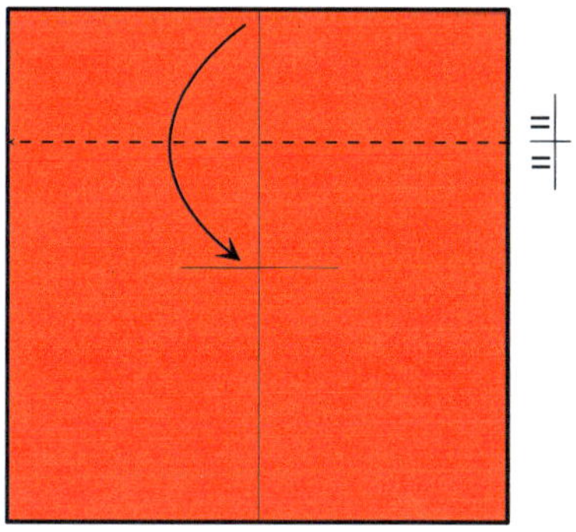

2. Valley fold so the top raw edge hits the crease made in step 1.

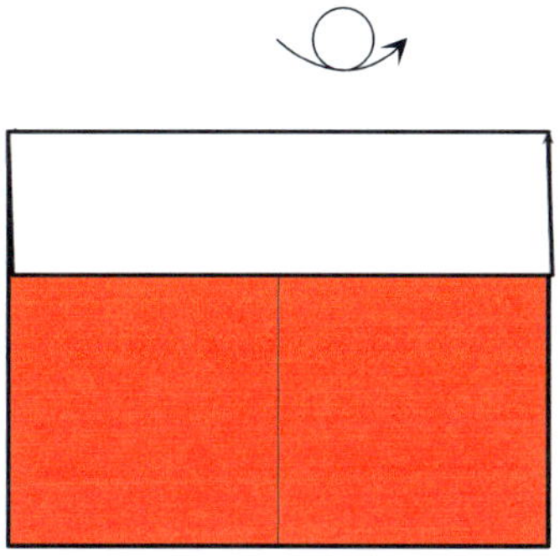

3. Turn over.

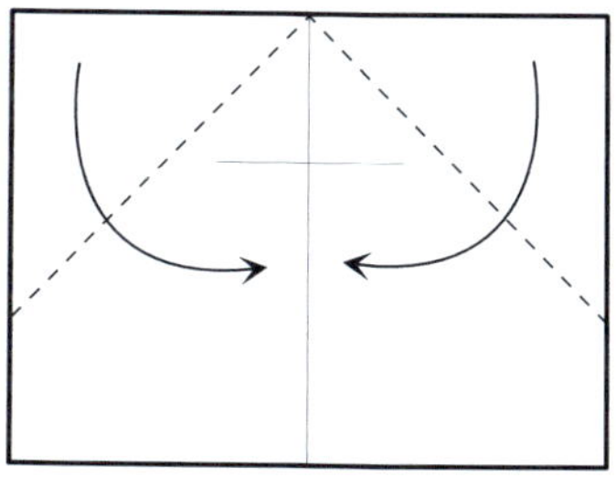

4. Valley fold both corners to the center.

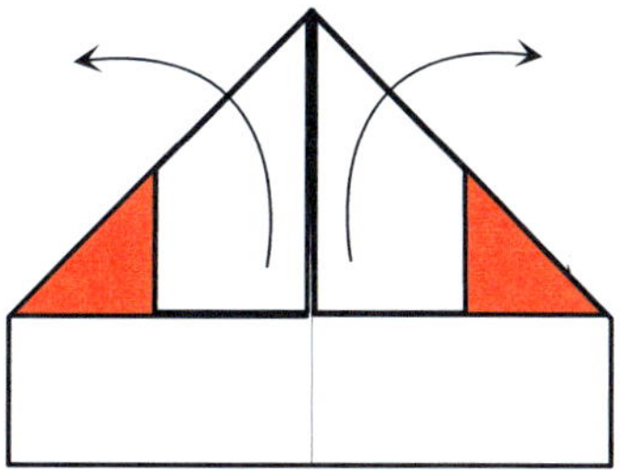

5. Unfold.

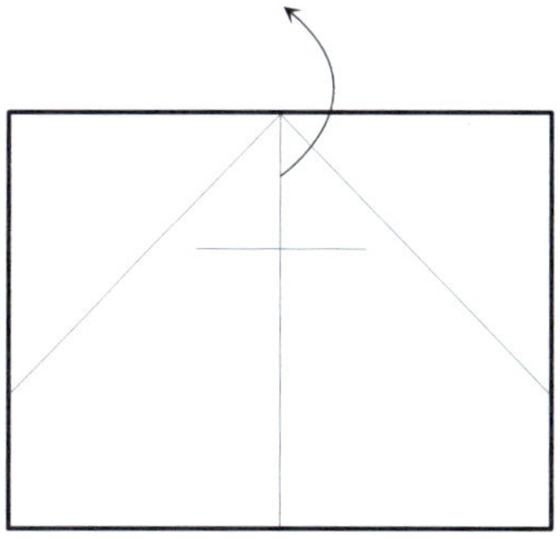

6. Unfold the top.

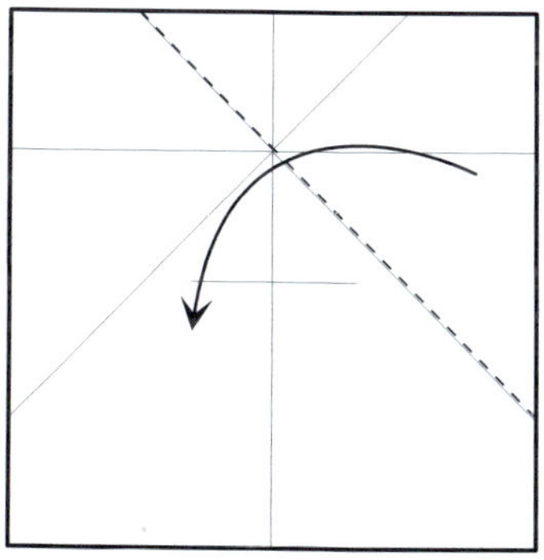

7. Valley fold along one diagonal crease.

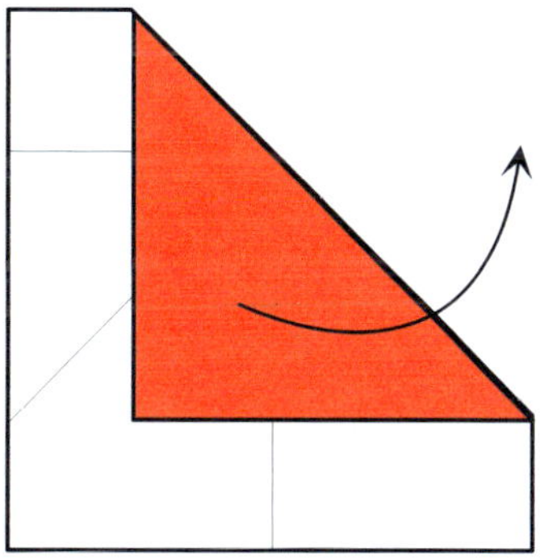

8. Unfold.

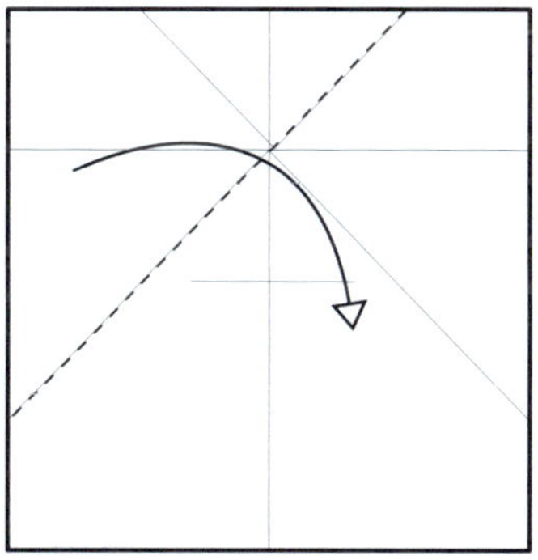

9. Repeat steps 7 and 8 on the other side.

Back Piece Four Page 2

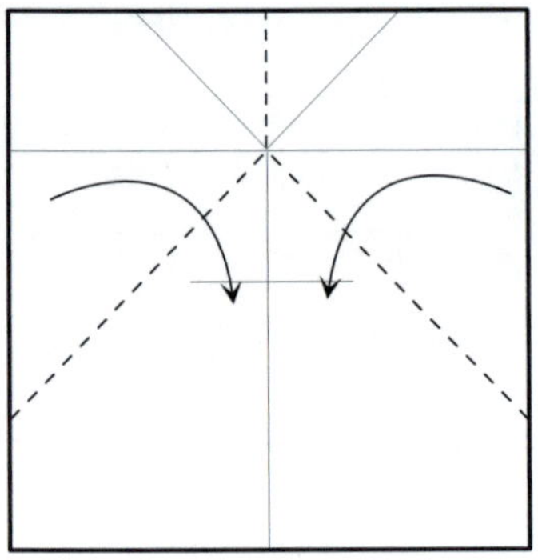

10. Rabbit ear the top down.

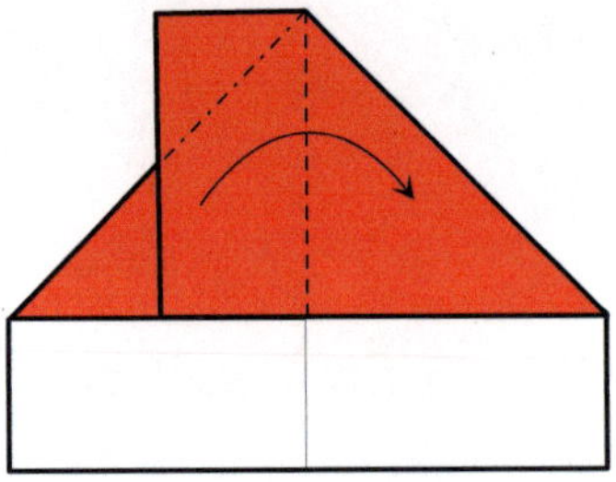

11. Valley fold the top layer over, squash folding the top.

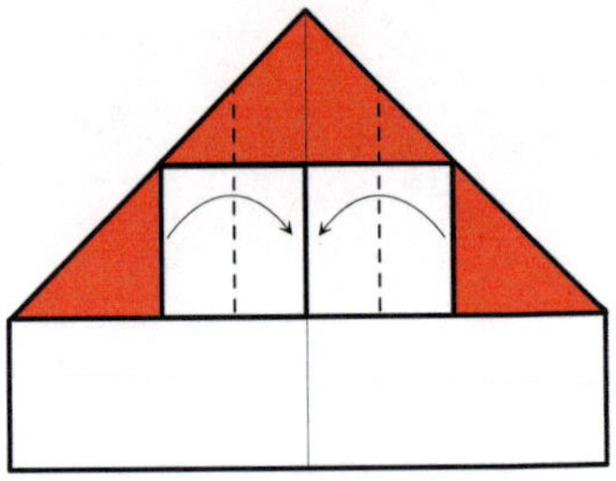

12. Valley fold the two short flaps into the middle.

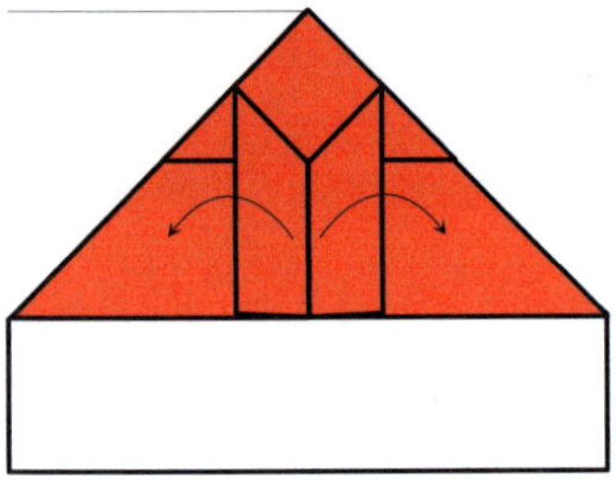

13. Unfold.

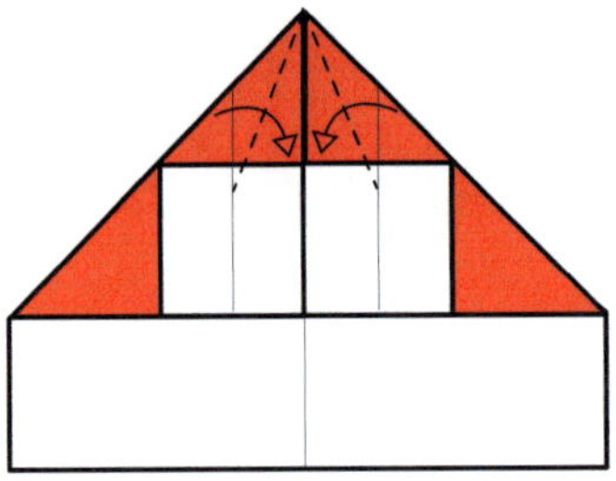

14. Valley fold the angle bisectors at the front. Fold up to the vertical creases.

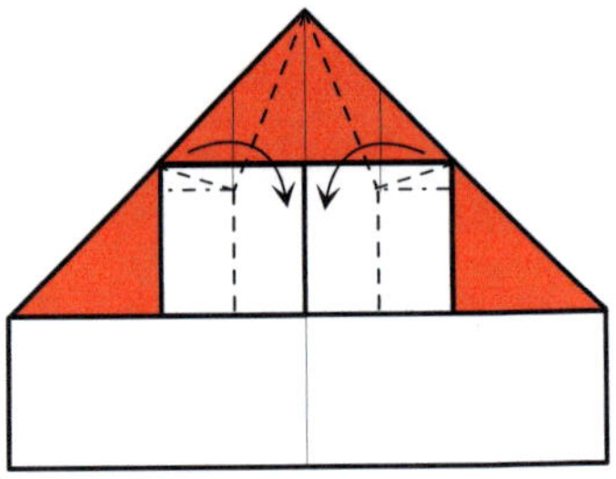

15. Rabbit ear fold the flaps in following the creases made in the previous two steps.

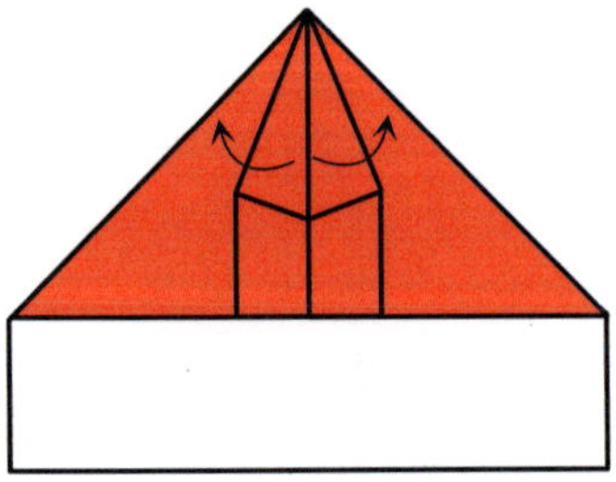

16. Unfold.

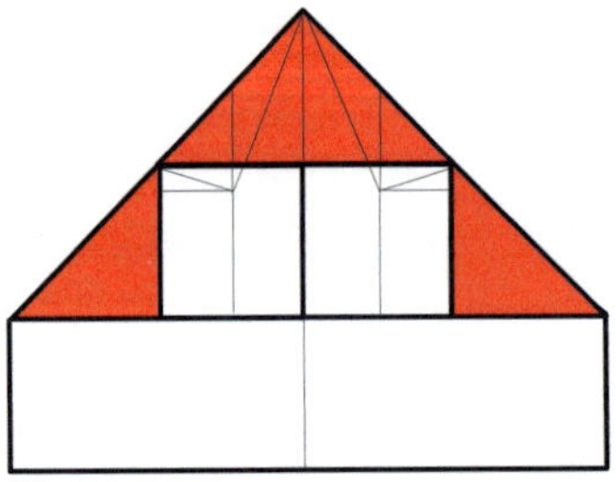

17. Ready for assembly.

Back Piece Four Page 3

18. The Front Piece goes on with the join on the back.

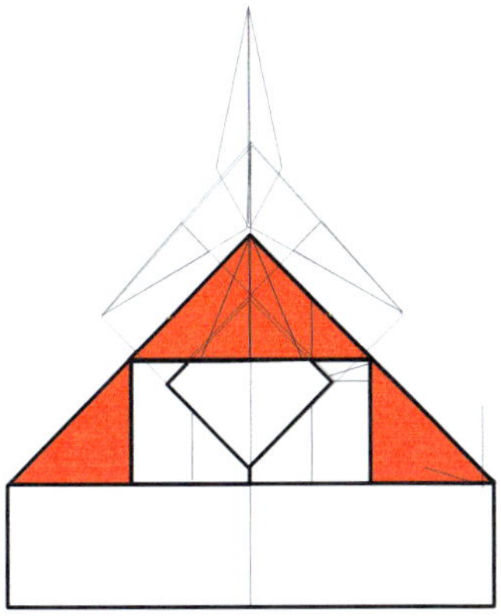

19. The loose flap of the join goes inside the front flap of the back piece.

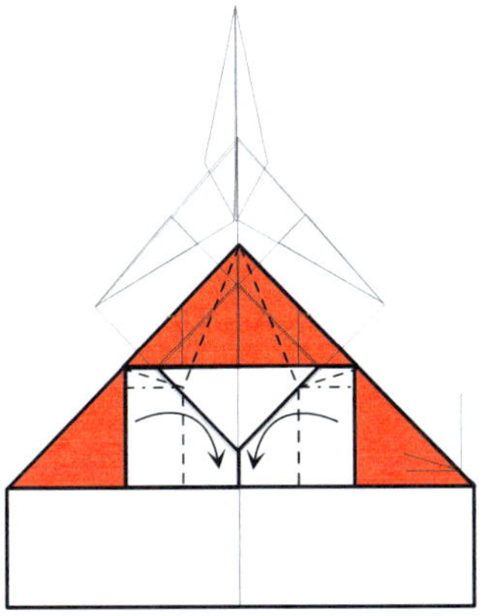

20. Refold the rabbit ears around the join on the back of the front piece, enclosing it.

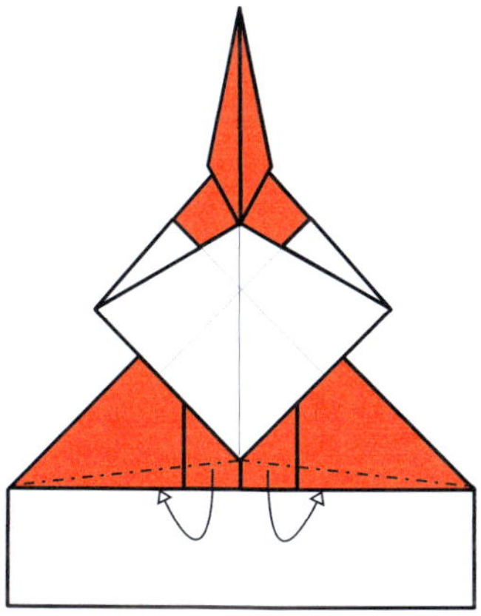

21. Mountain fold the loose flaps in the back behind to secure the hind end of the front piece.

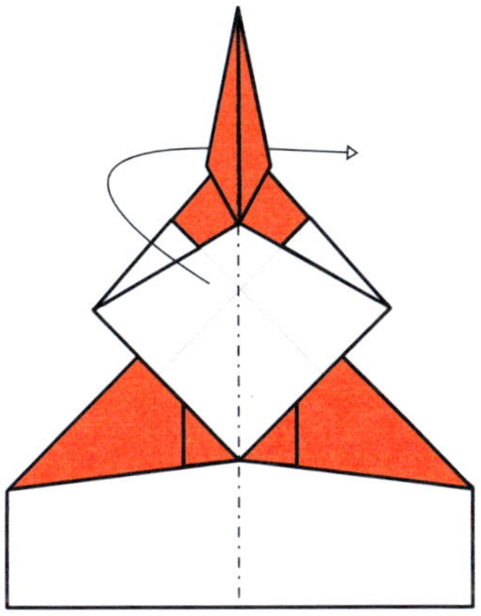

22. Mountain fold in half.

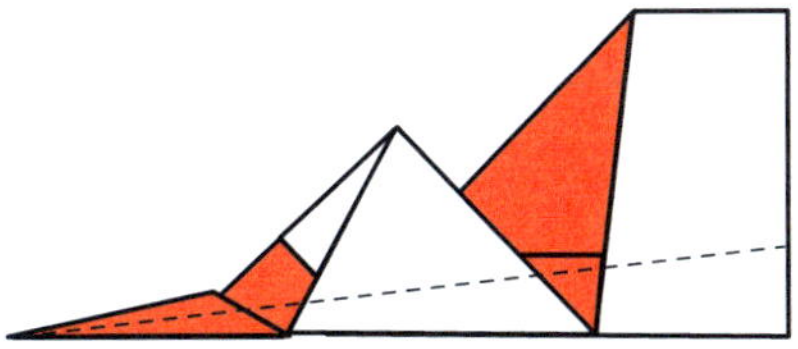

23. For Front Piece Delta the wings are folded so they bisect the front. Fold the front piece, and then follow the angle on the back piece.

Front View

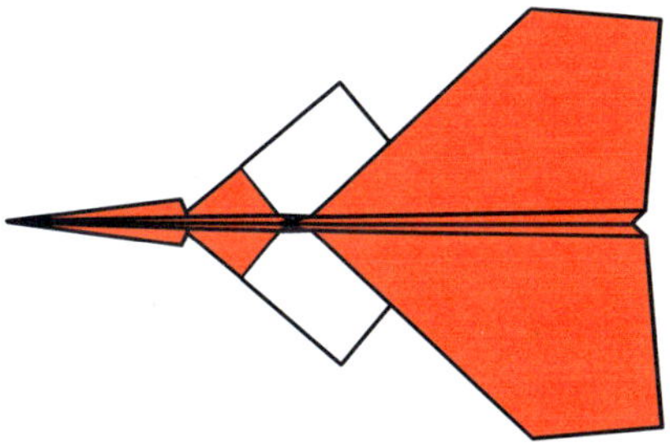

Top View

Side View

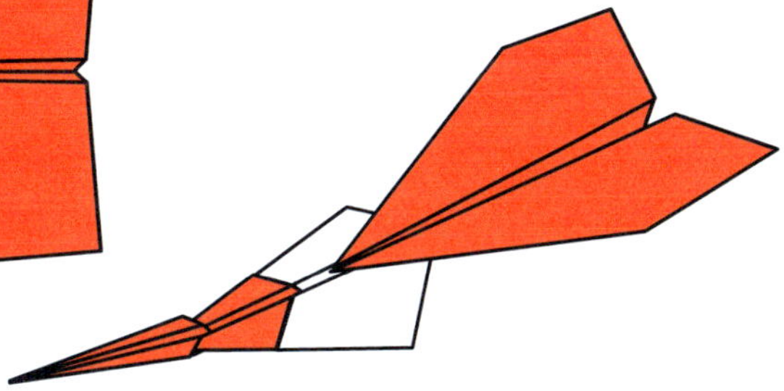

Alpha Four

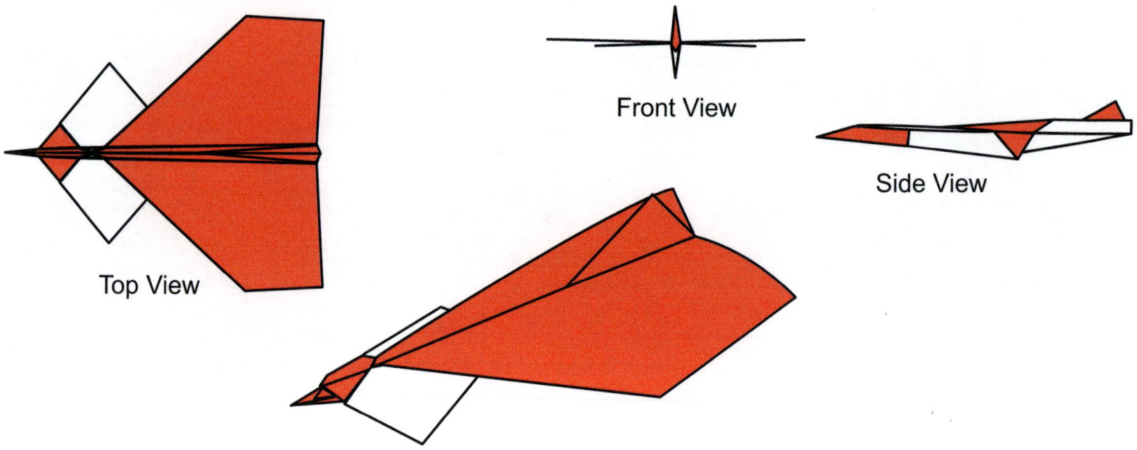

Epsilon Four

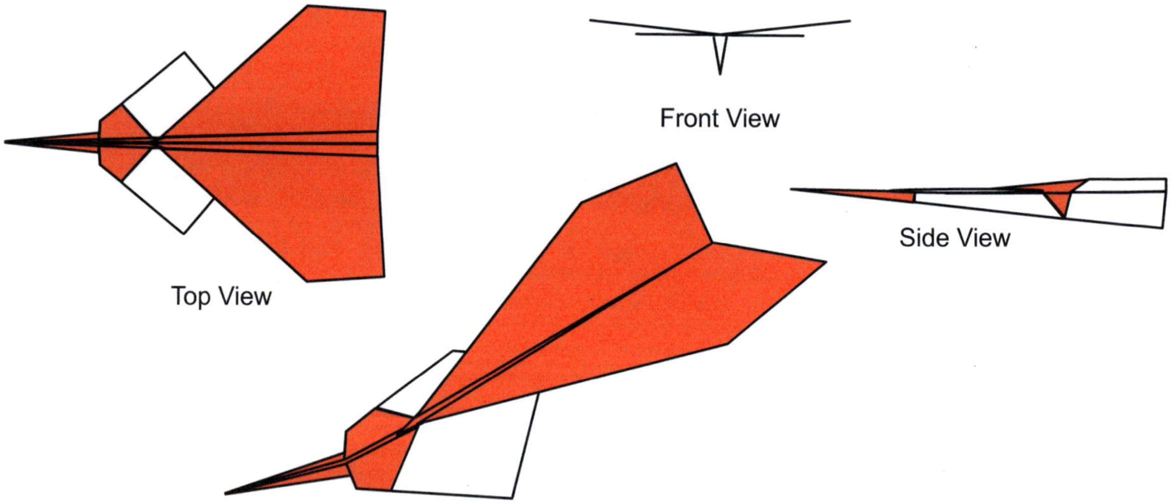

Zeta Four

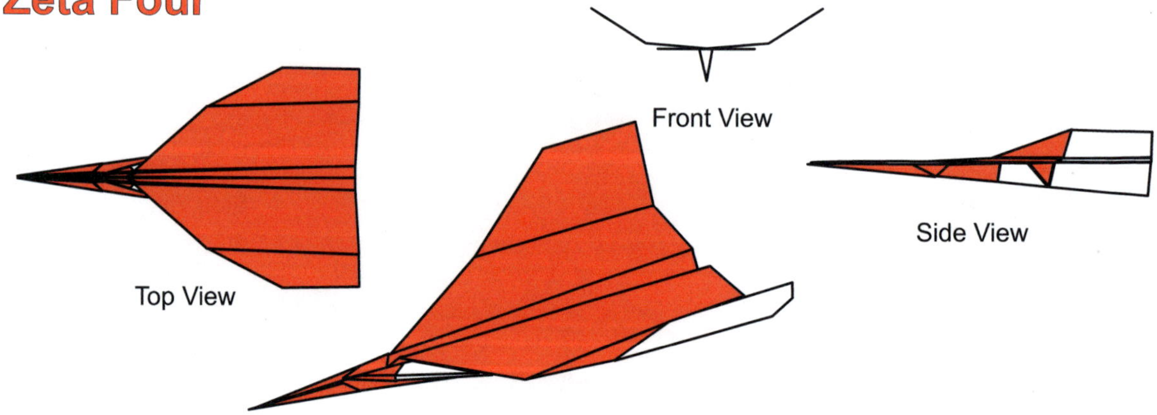

Gamma Four

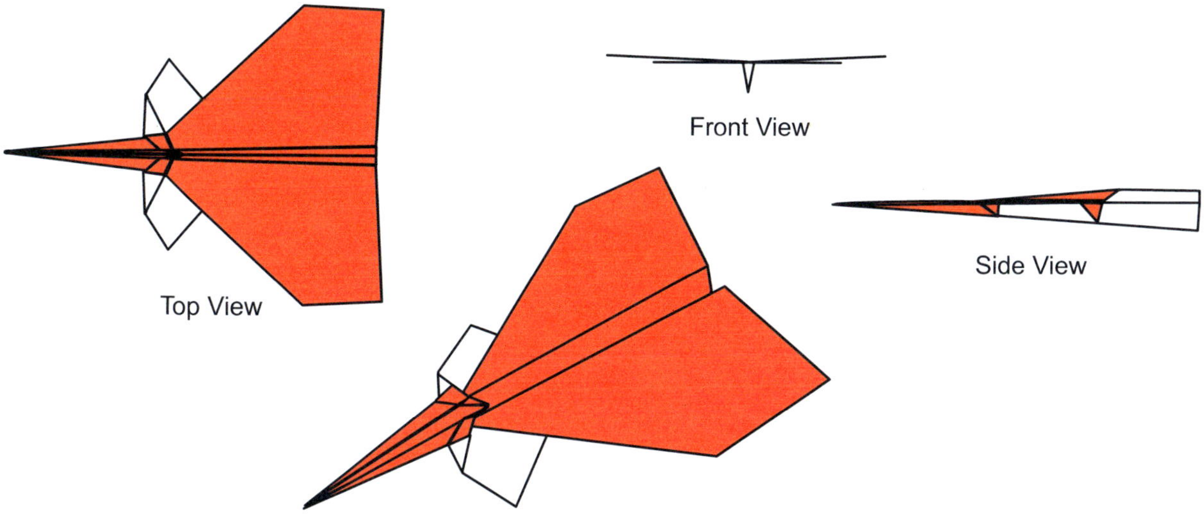

Iota Four

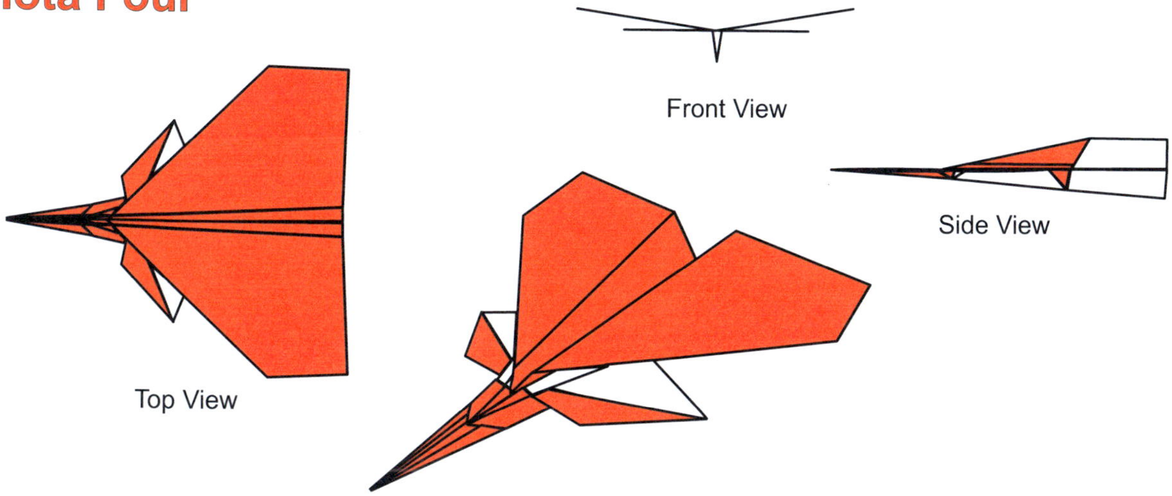

Kappa Four

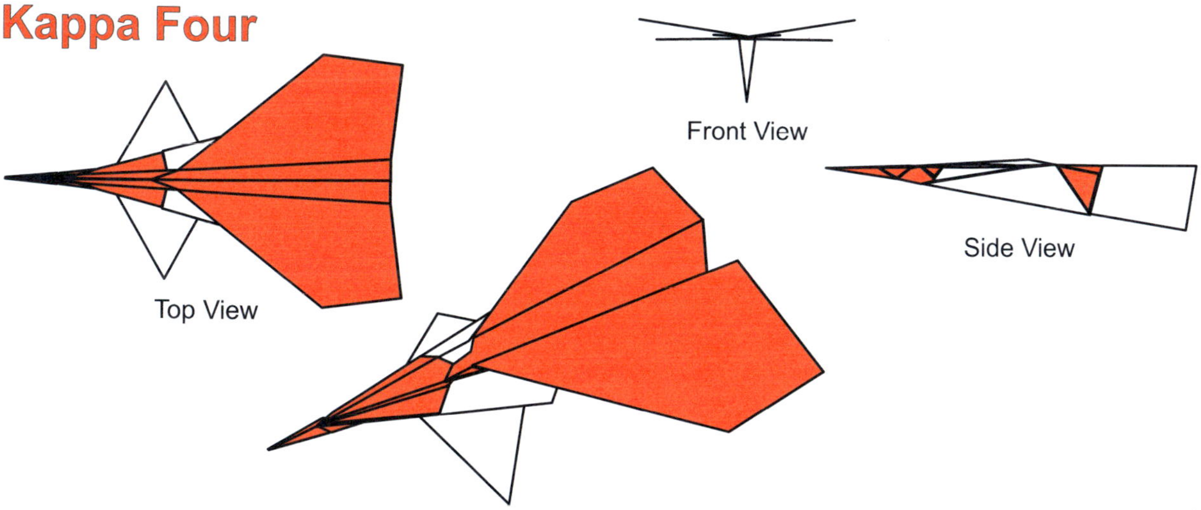

Lambda Four

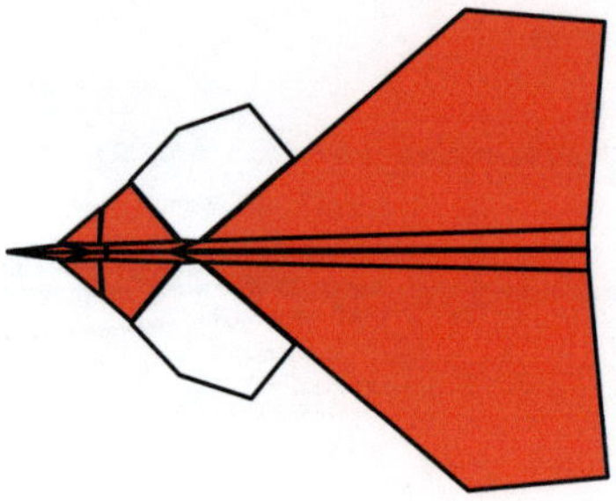

Top View

Front View

Side View

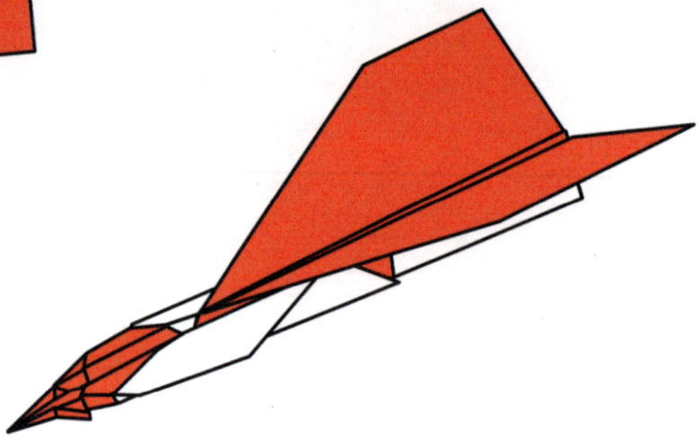

Omicron Four

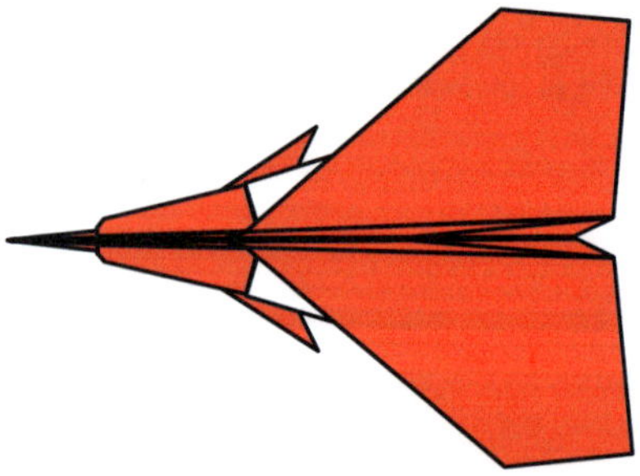

Top View

Front View

Side View

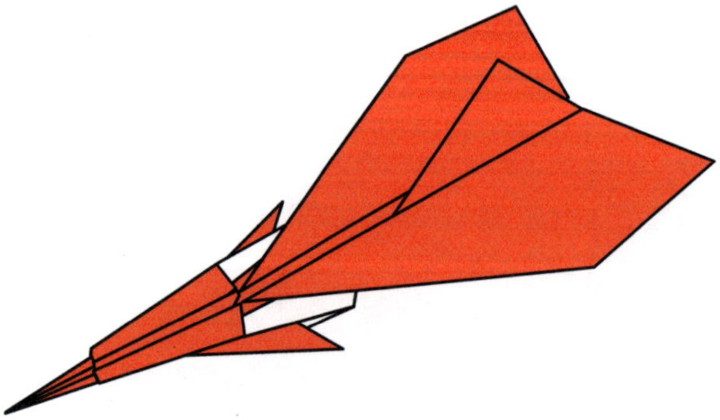

Back Piece Five Airplanes

Alpha Five

Beta Five

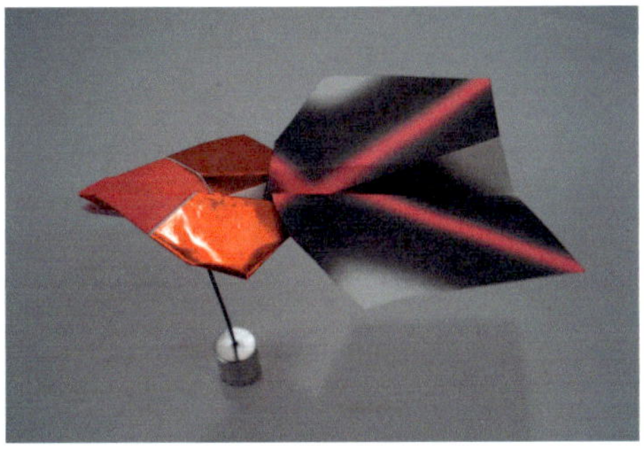

Gamma Five

Zeta Five

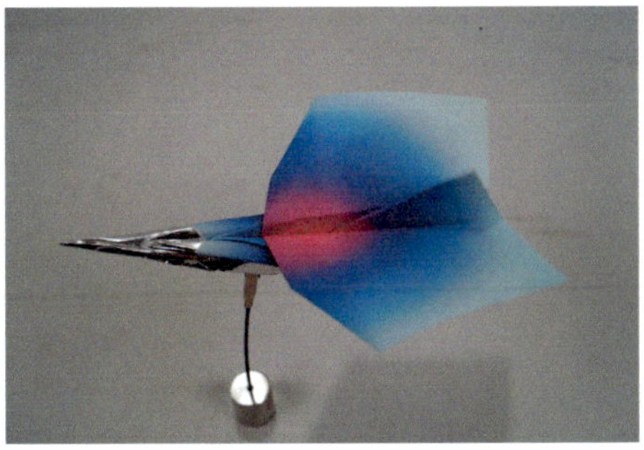

Iota Five

Kappa Five

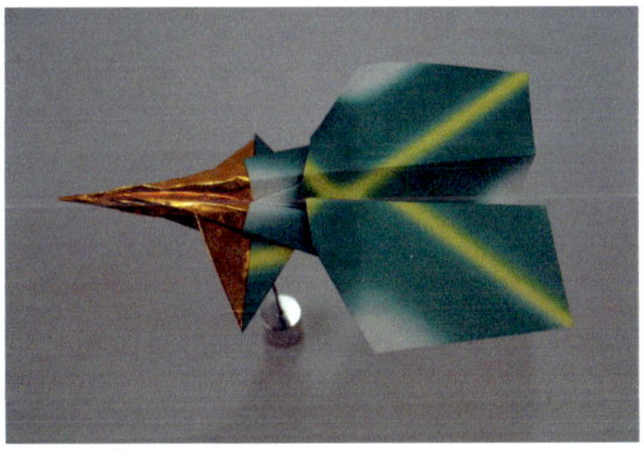

Omicron Five

Lambda Five

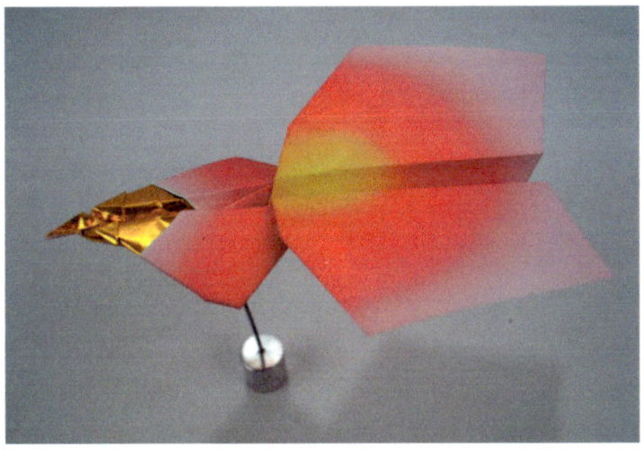

Back Piece Five

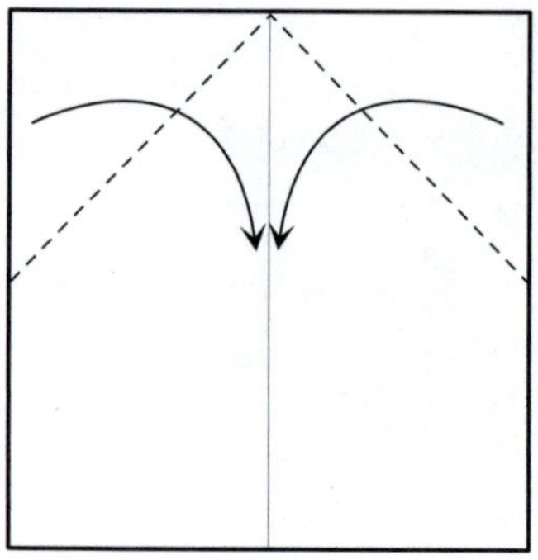

1. Valley fold the two corners into the middle.

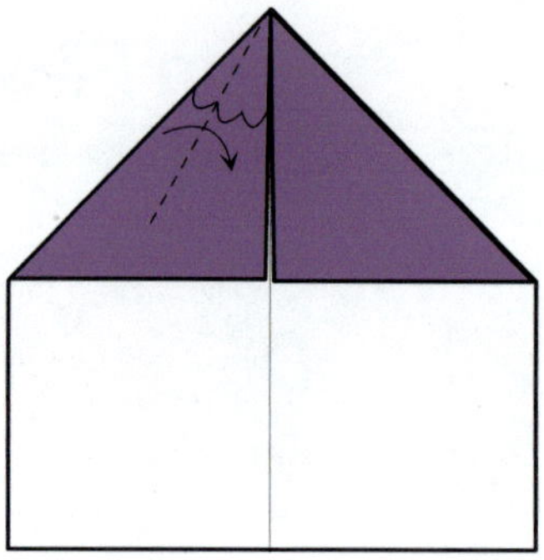

2. Valley fold one third of the angle in. No good reference for this, estimate best you can.

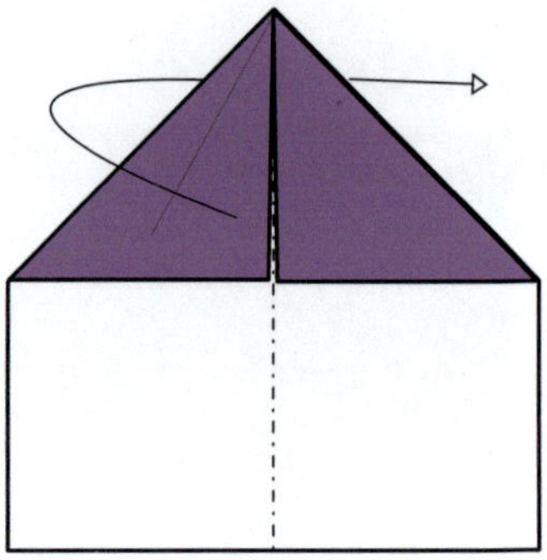

3. Mountain fold in half.

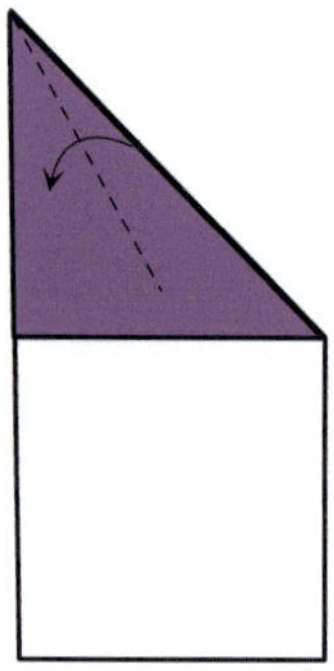

4. Use the first crease you made to make the second on the other side. You want them to be the same.

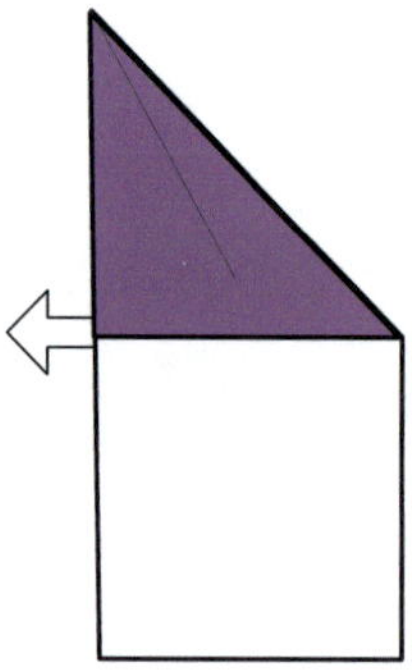

5. Open back up.

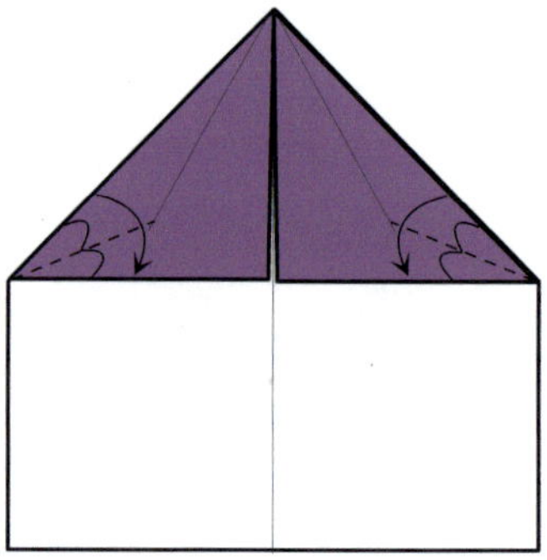

6. Valley fold bisectors of the lower angles.

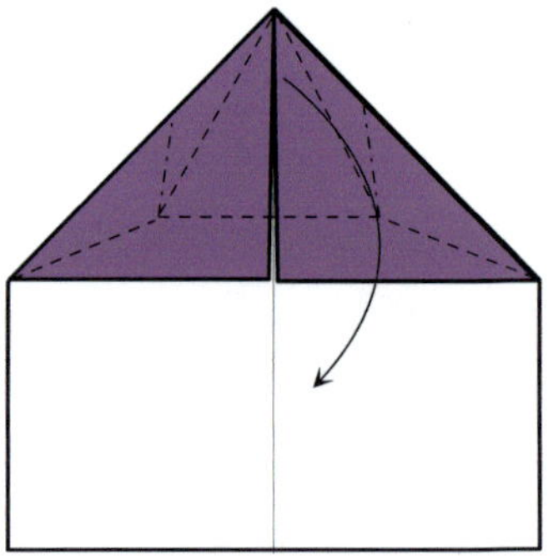

7. Collapse along the creases you've made.

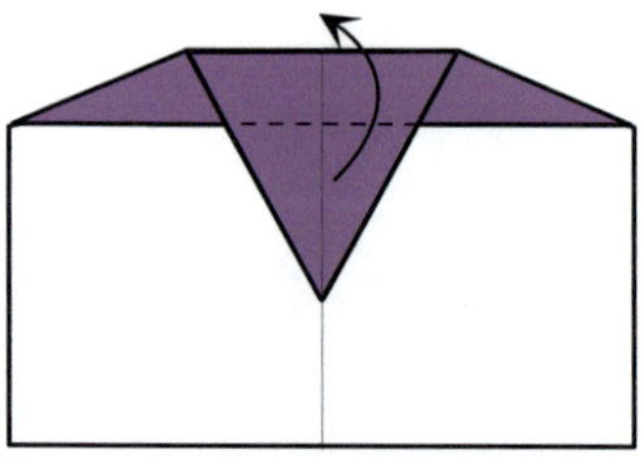

8. Valley fold the triangular flap up as far as it goes without resistance.

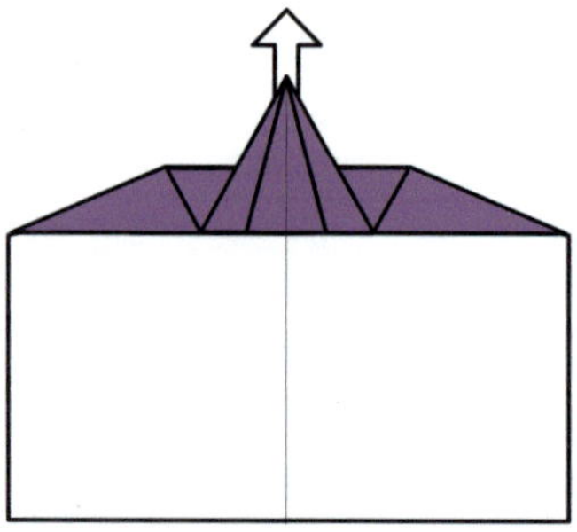

9. Unfold to step 7.

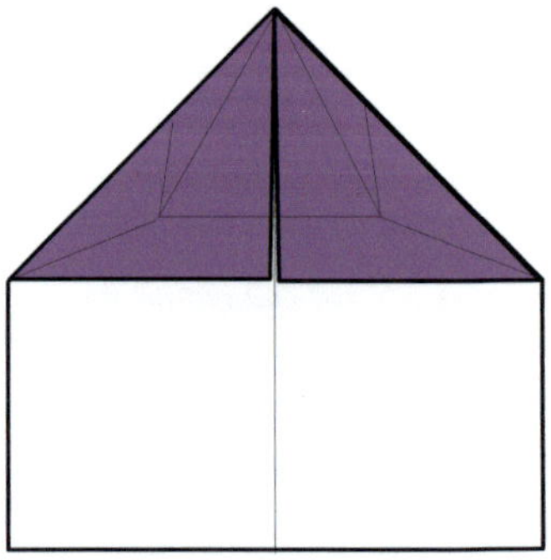

10. Time for assembly.

Back Piece Five Page 2

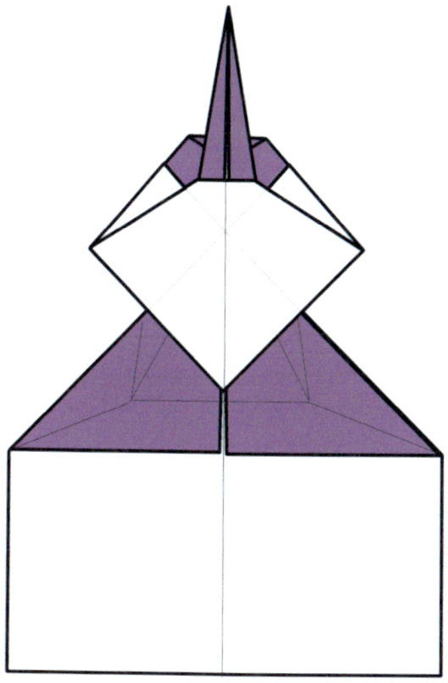

11. The front piece fits in the nose.

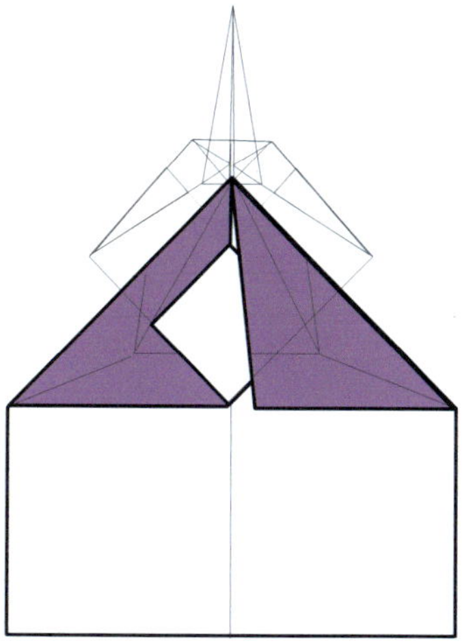

12. Slide the front piece join underneath the flaps of the back piece.

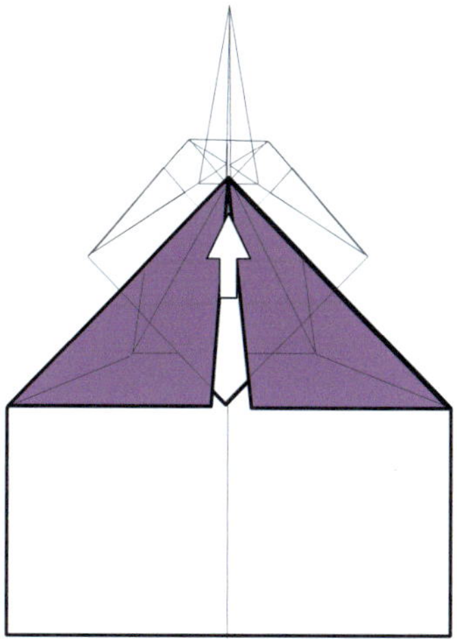

13. Then slide the front piece forward as far as it will go.

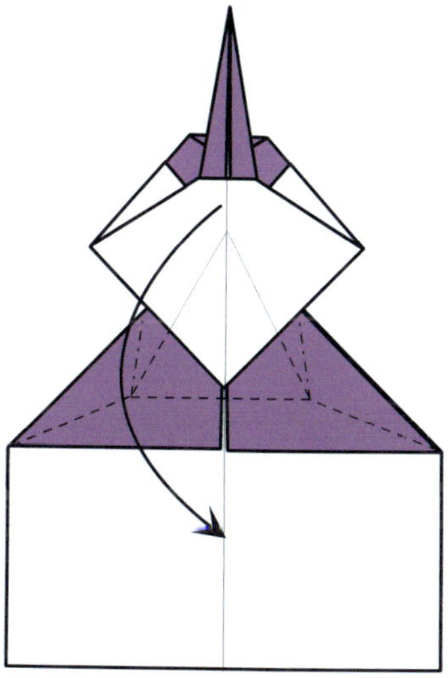

14. Assemble the front of the back piece, inserting some flaps in between the front and back pieces.

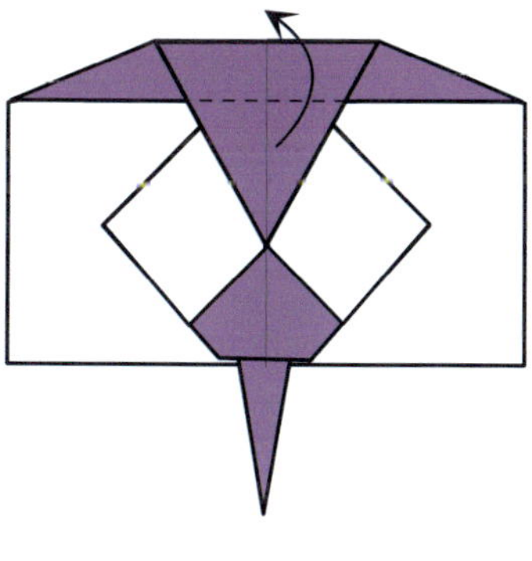

15. Valley fold the front back up along a preexisting crease.

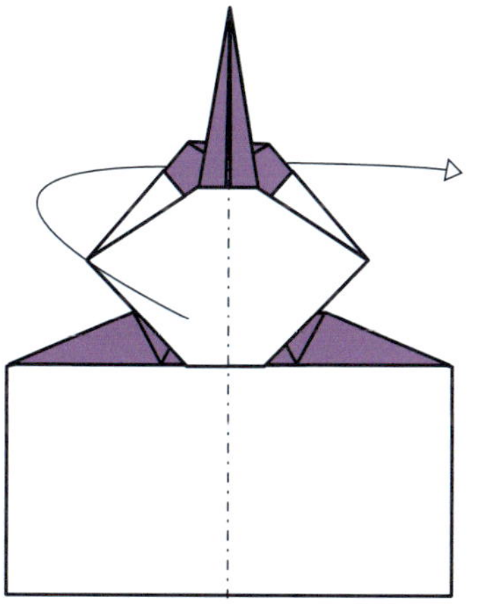

16. Mountain fold in half.

Back Piece Five Page 3

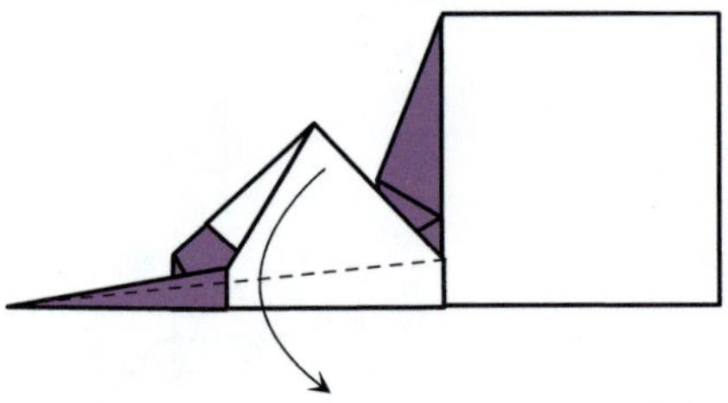

17. For front piece Epsilon, fold the front wings down where they intersect the back wings.

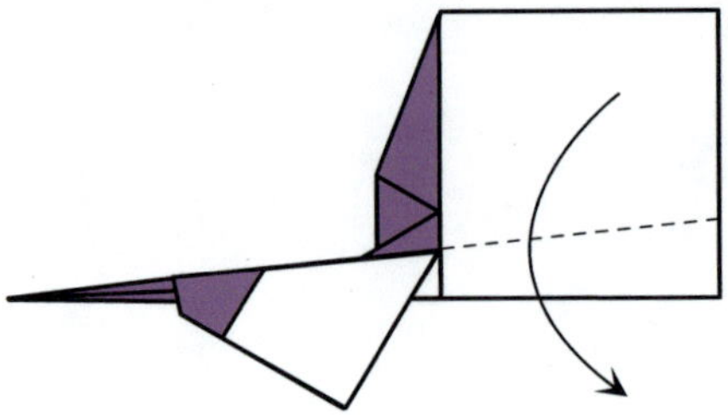

19. Follow the angle of the forward wings to fold the rear wings. All of the aircraft with back piece 5 are a bit nose heavy. All will need a tail or elevators.

Epsilon Five

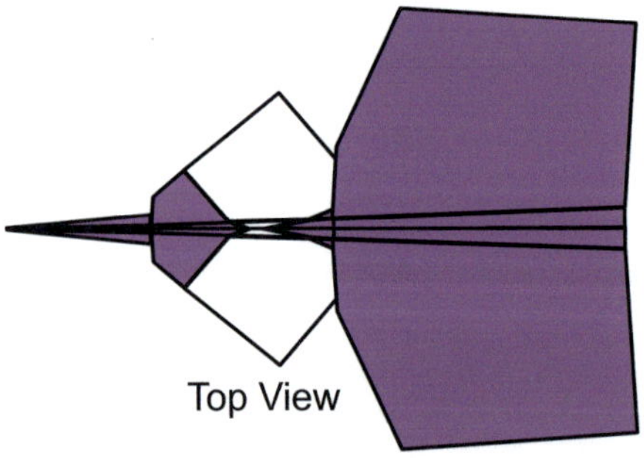

Top View

Front View

Side View

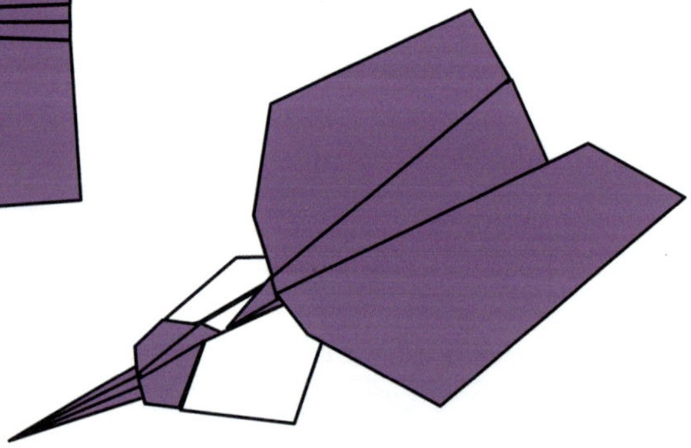

Beta Five

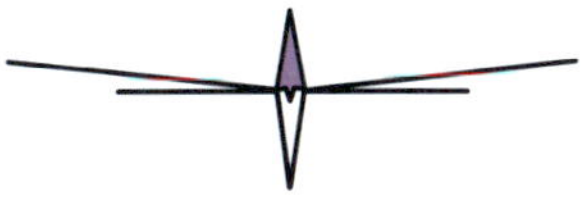

Front View

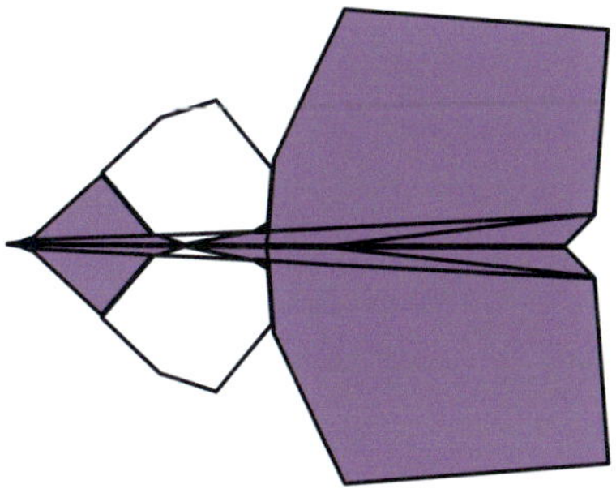

Top View

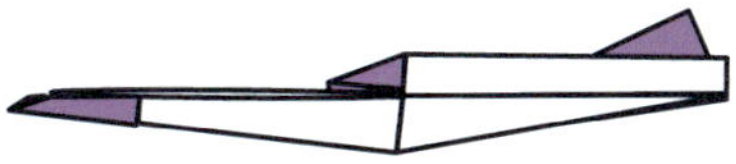

Side View

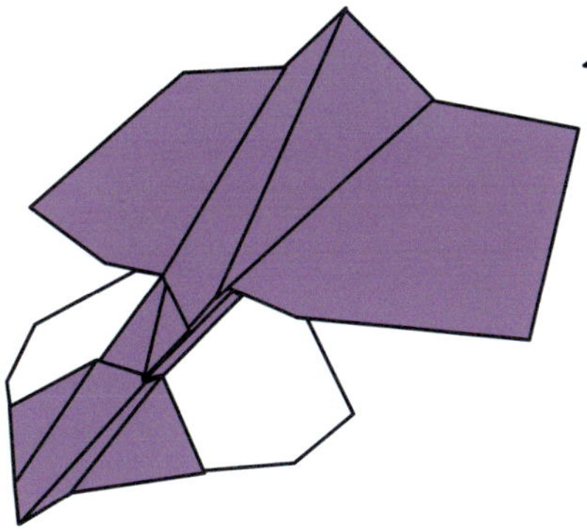

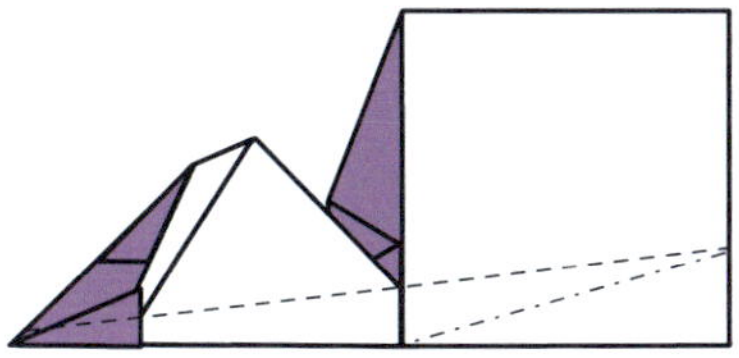

Delta Five

Front View

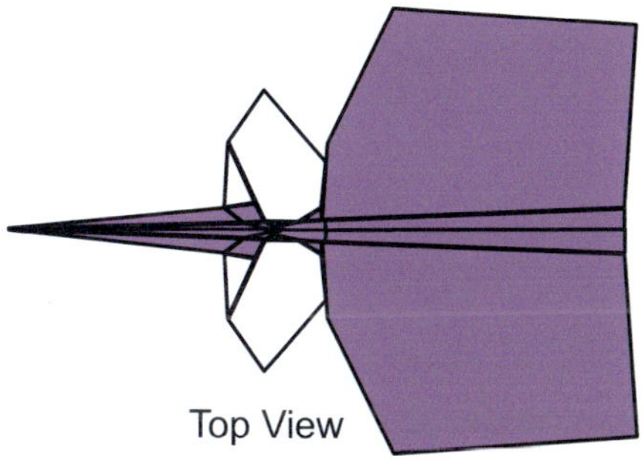

Top View

Side View

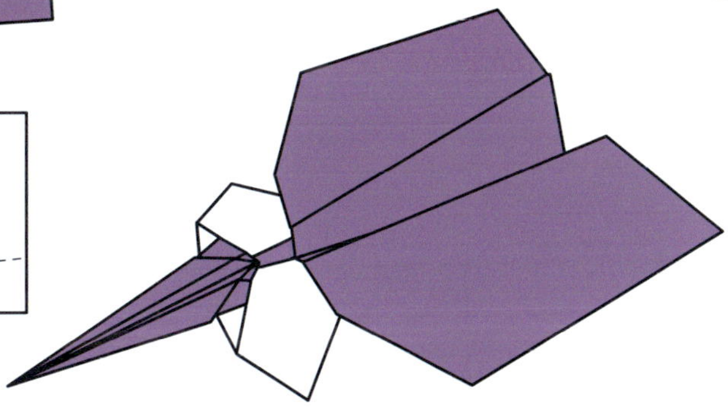

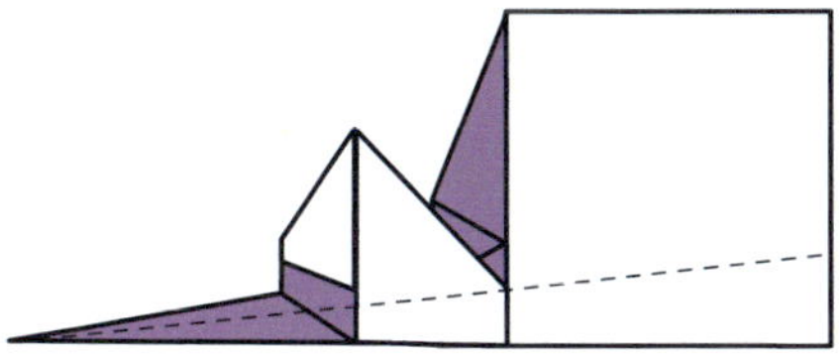

Iota Five

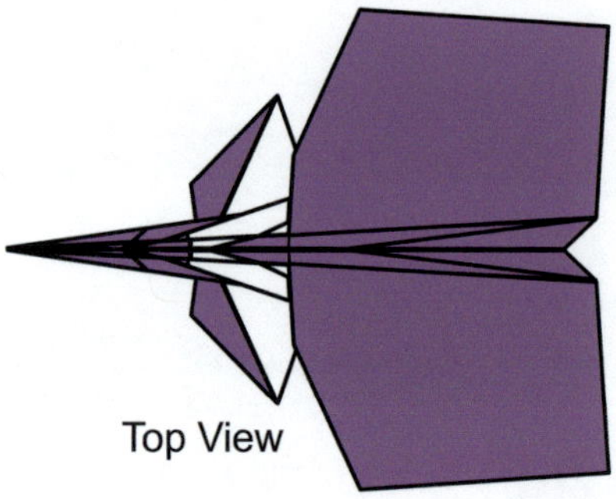

Top View

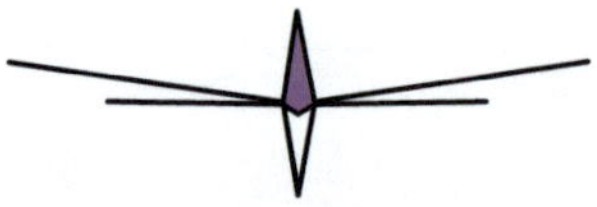

Front View

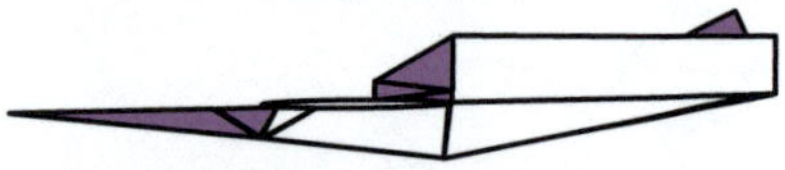

Side View

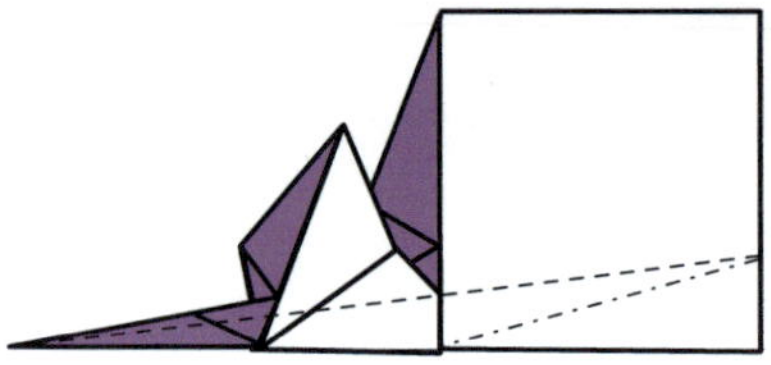

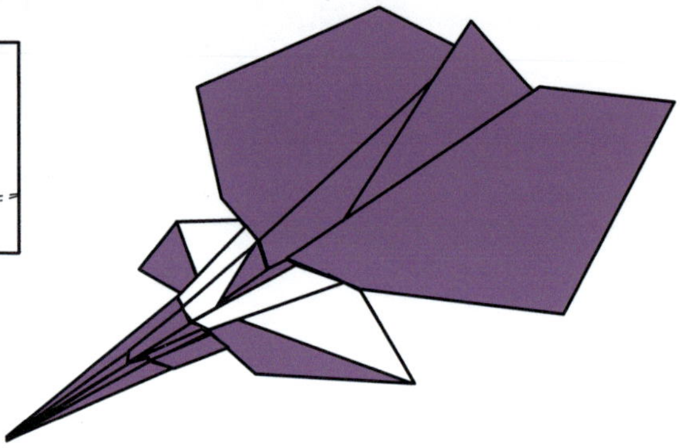

Kappa Five

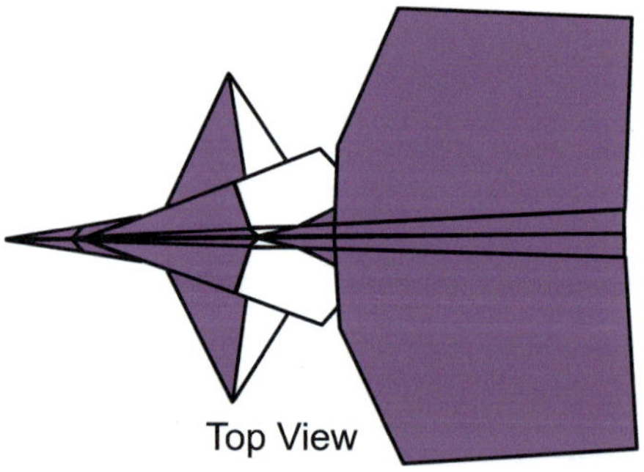

Top View

Front View

Side View

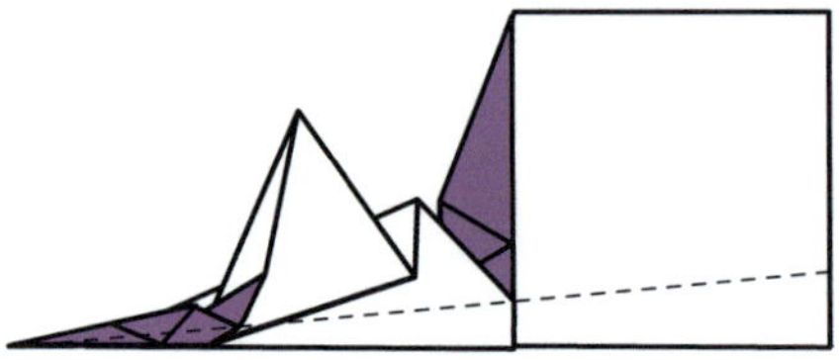

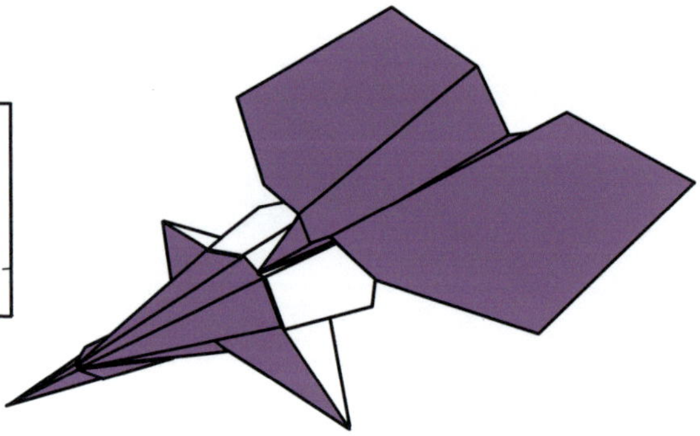

Zeta Five

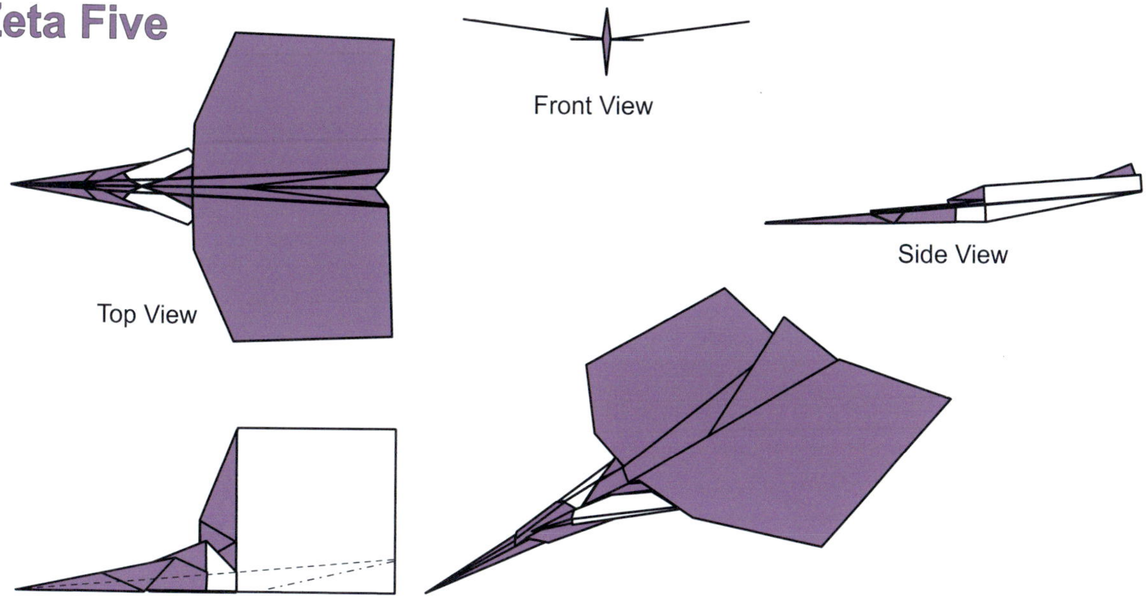

Alpha Five

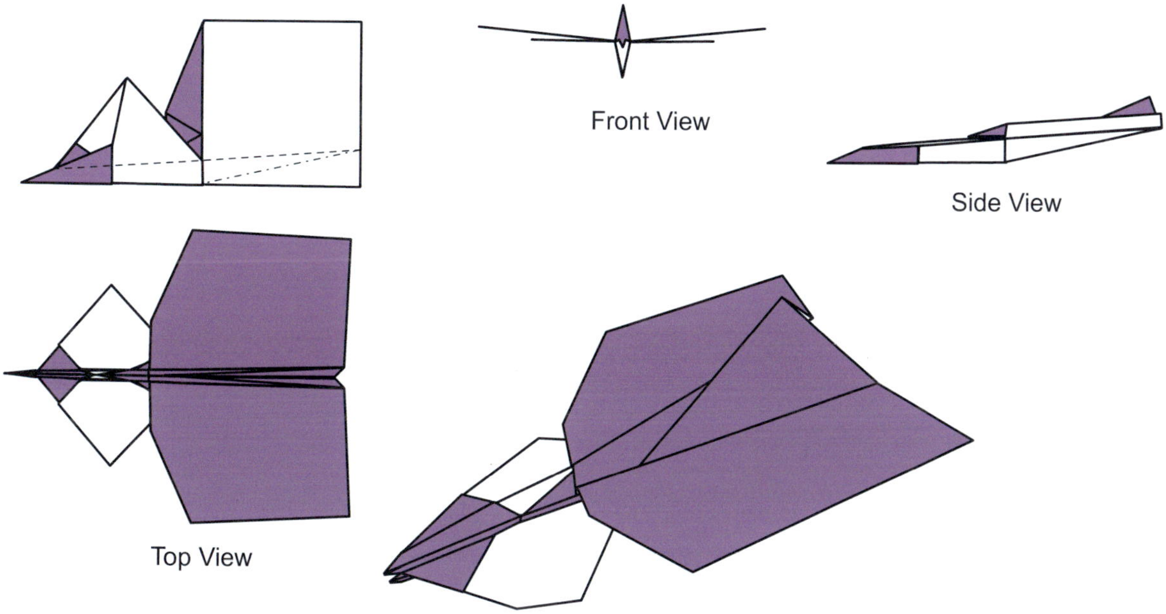

Lambda Five

Front View

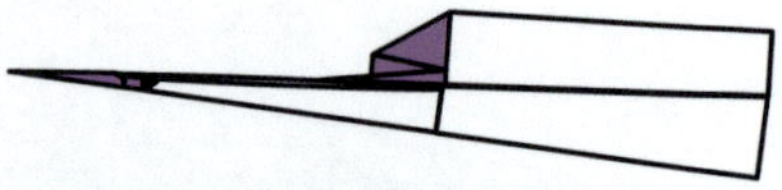

Side View

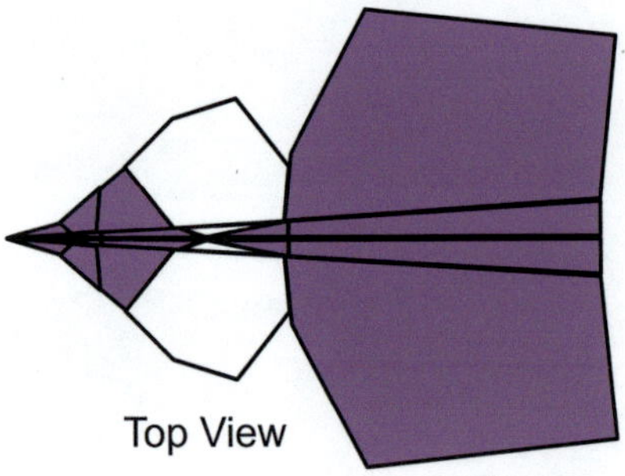

Top View

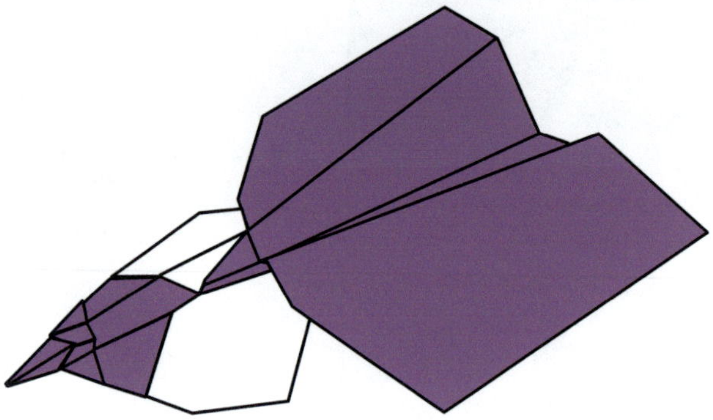

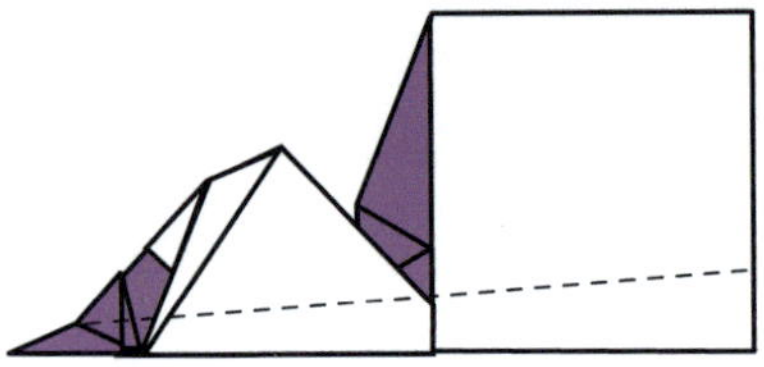

Omicron Five

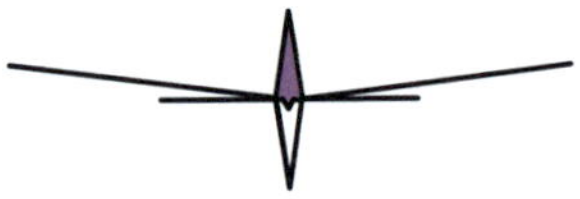

Front View

Side View

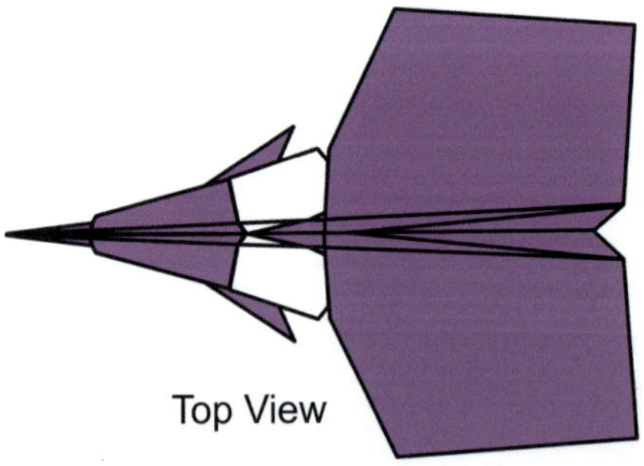

Top View

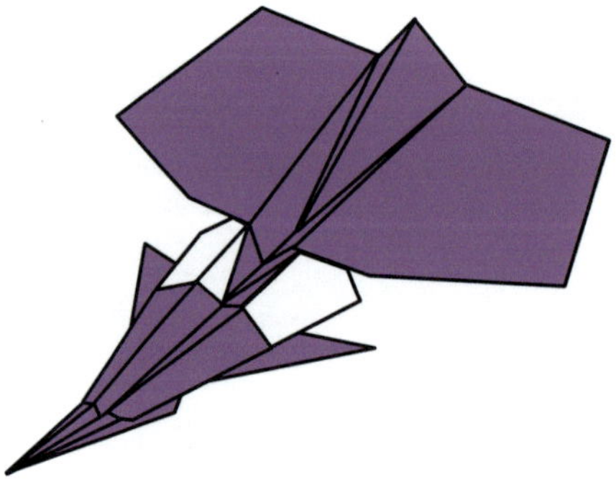

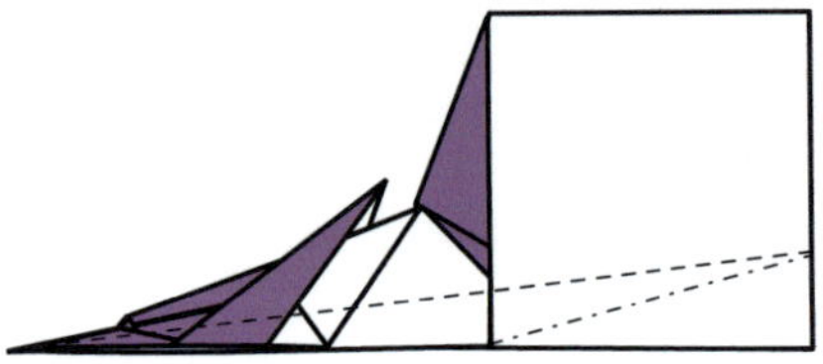

Back Piece Six Airplanes

Alpha Six

Beta Six

Gamma Six

Epsilon Six

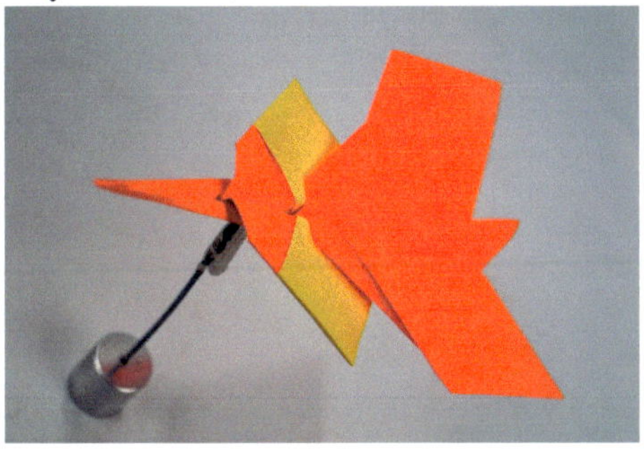

Kappa Six

Iota Six

Omicron Five

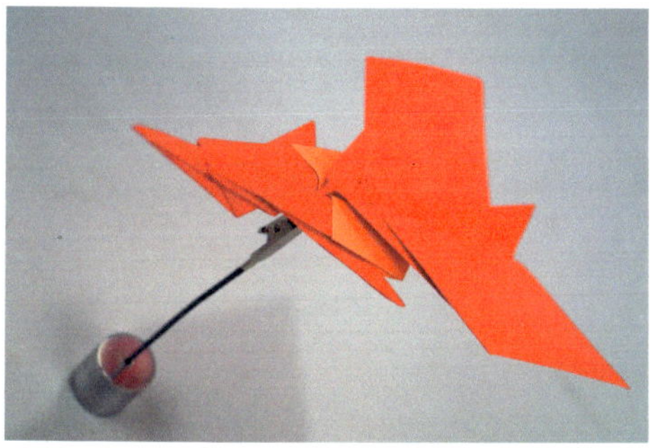

Lambda Six

Back Piece Six

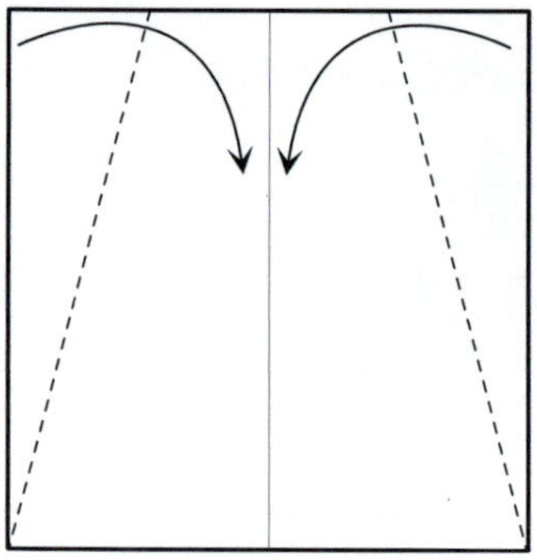

1. Valley fold each corner diagonally so that the corner hits the midline. Make certain both sides match.

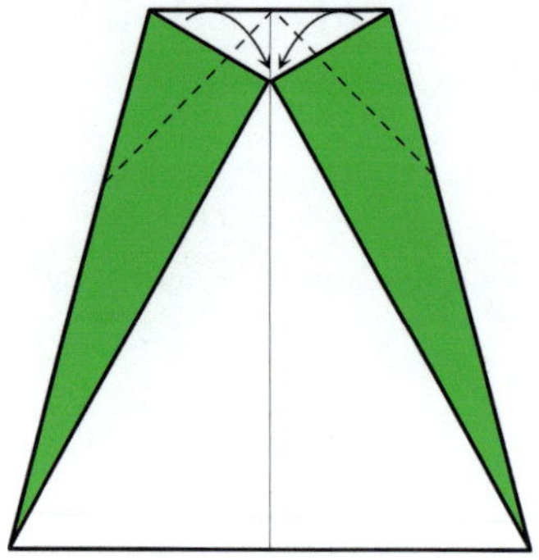

2. Valley fold in the front so the front edges lie on the center.

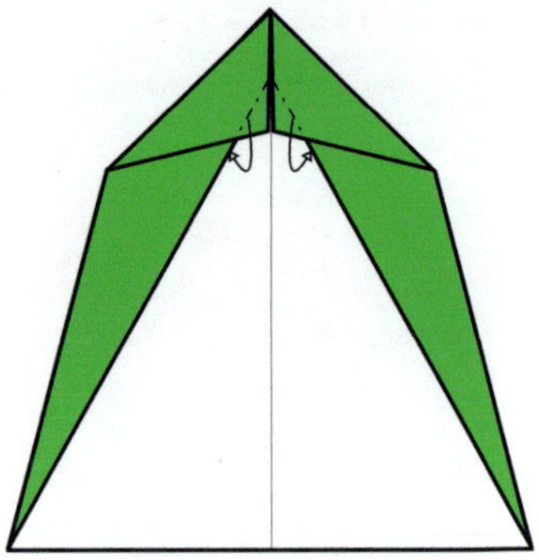

3. Mountain fold the small corners behind where they overlap with the layers below.

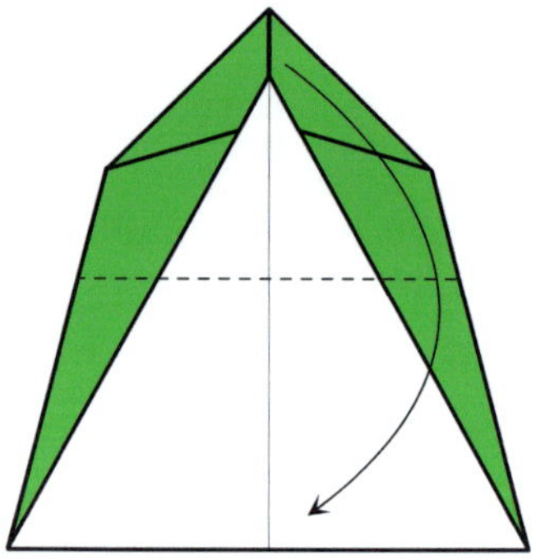

4. Valley fold in half lengthwise.

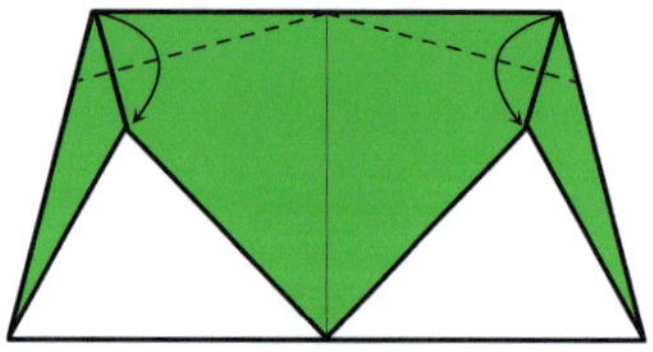

5. Valley fold the corners so they lie on the intersection of the layers.

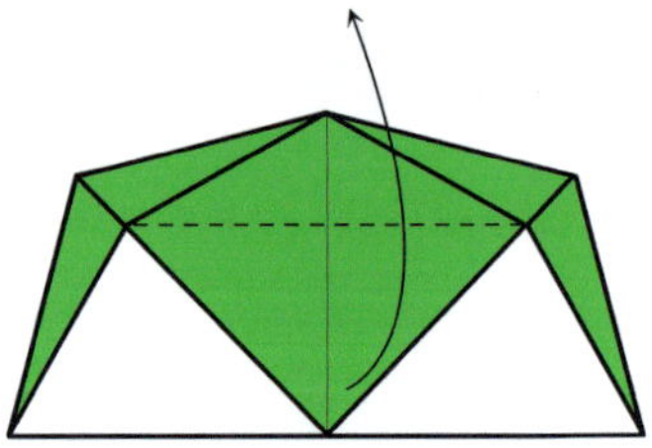

6. Valley fold the front point up as far as it will go without resistance.

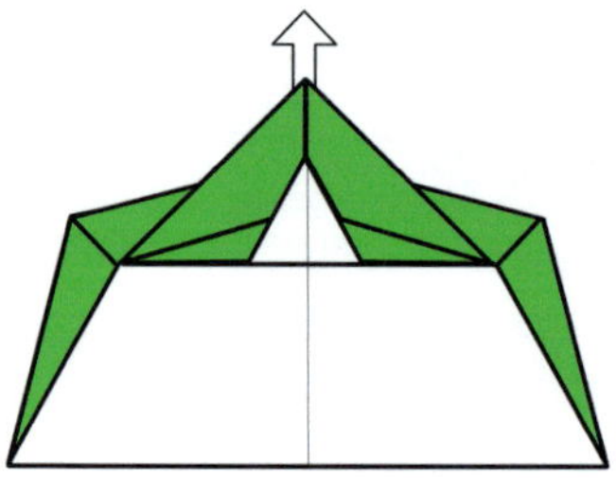

7. Unfold to step 4.

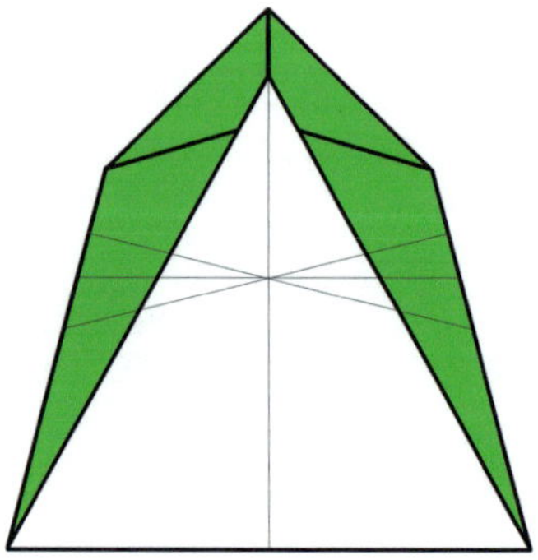

8. Assembly time.

Back Piece Six Page 2

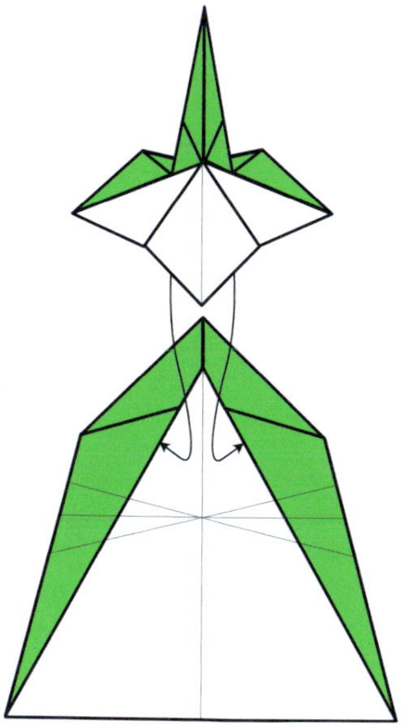

9. To assemble, place the join of the front piece inside the layers of the back piece.

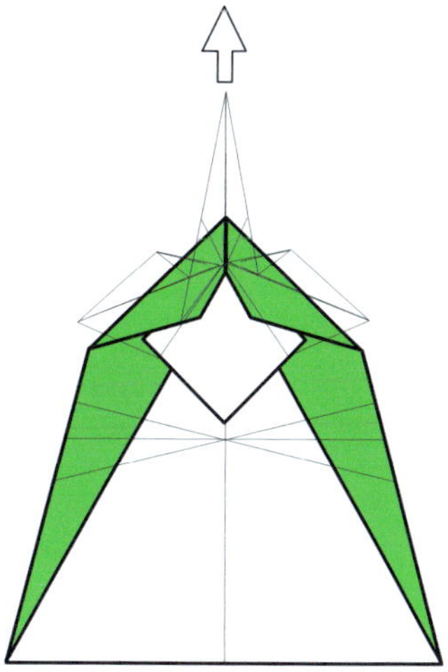

10. Slide the front piece all the way to the front of the back piece.

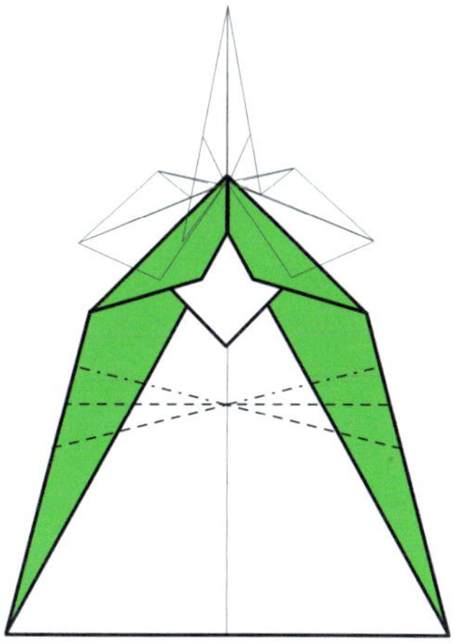

11. Reform back piece 6 around the front piece.

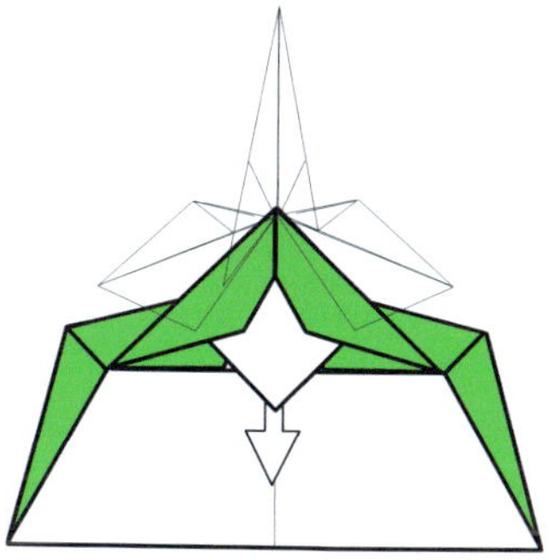

12. Slide the front piece back as far as it will go.

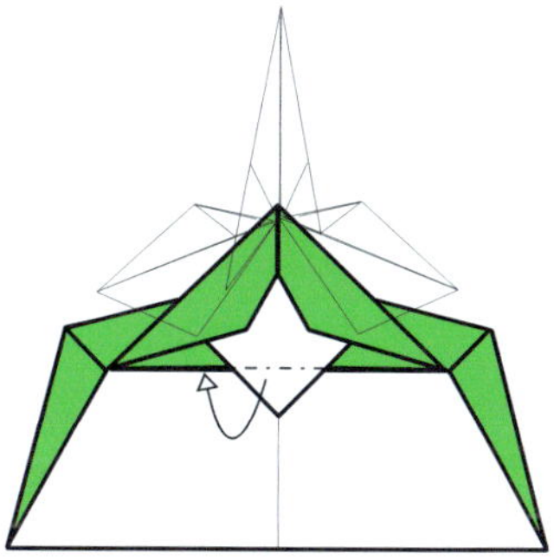

13. Mountain fold the front piece back where it overlaps.

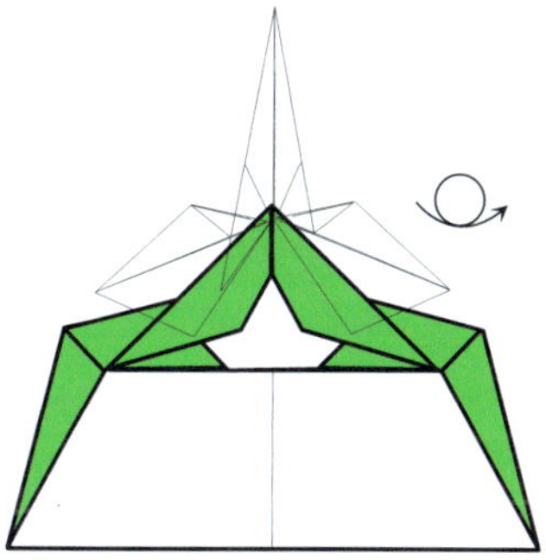

14. Turn over.

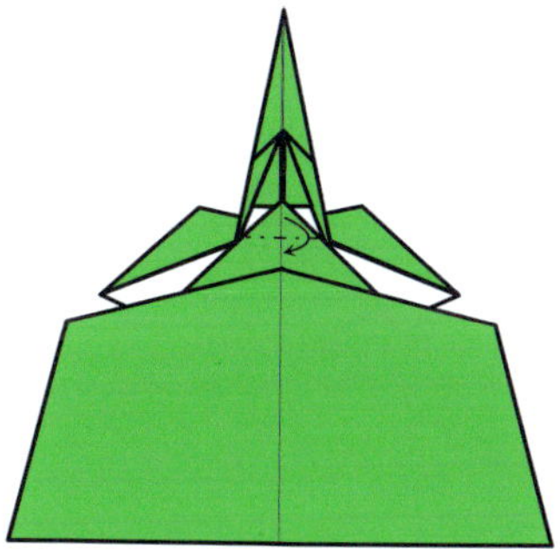

15. Valley fold the tip of the back piece to lock in the front piece.

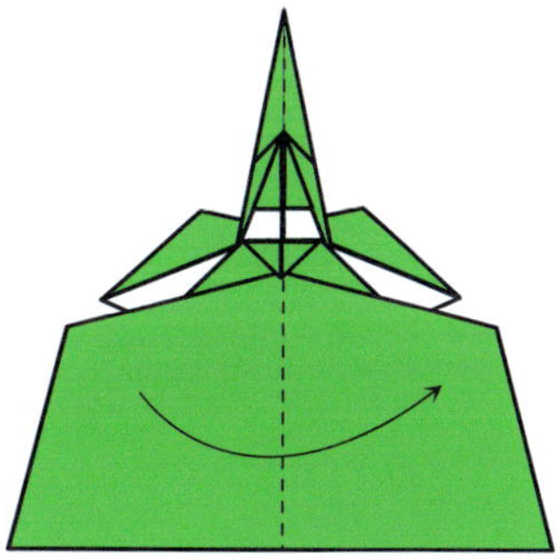

16. Valley fold in half.

Iota Six

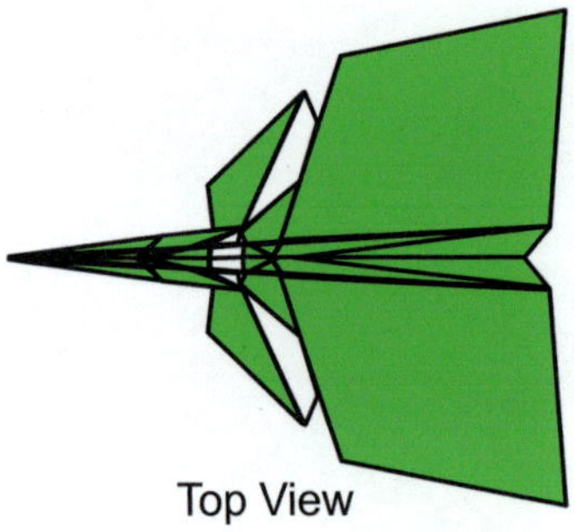

Top View

Front View

Side View

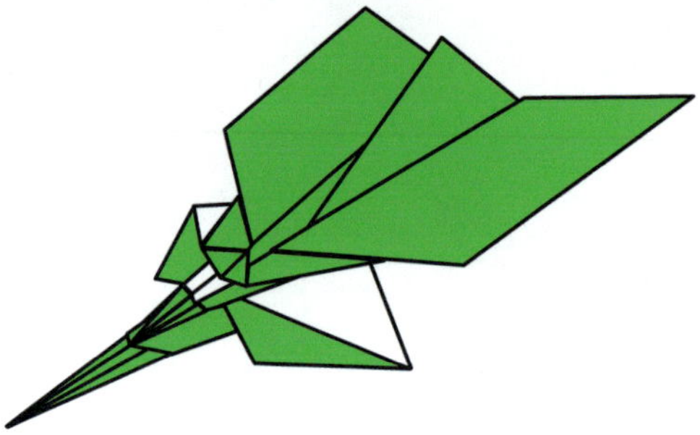

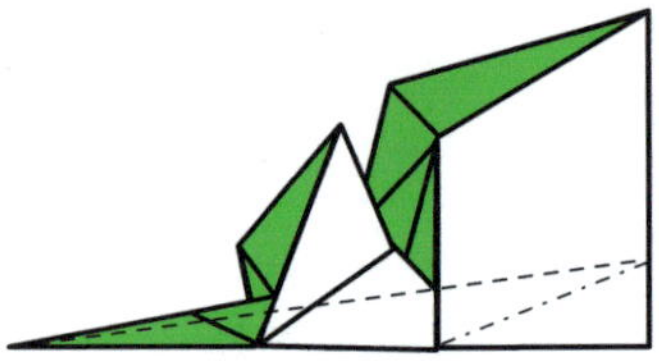

Valley fold the front wings where they meet the back. Fold both sets of wings at the same angle.

Alpha Six

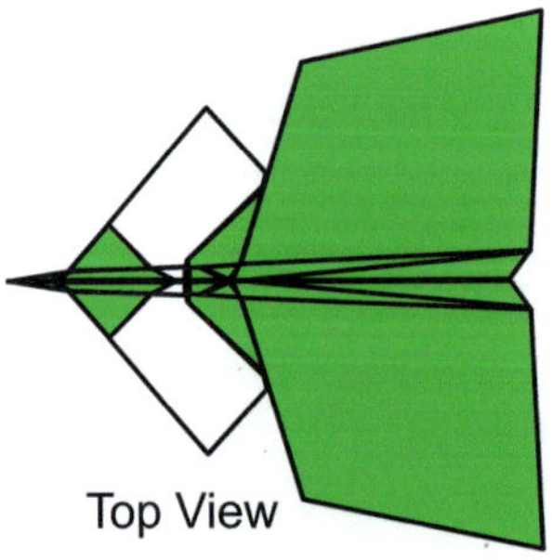

Top View

Front View

Side View

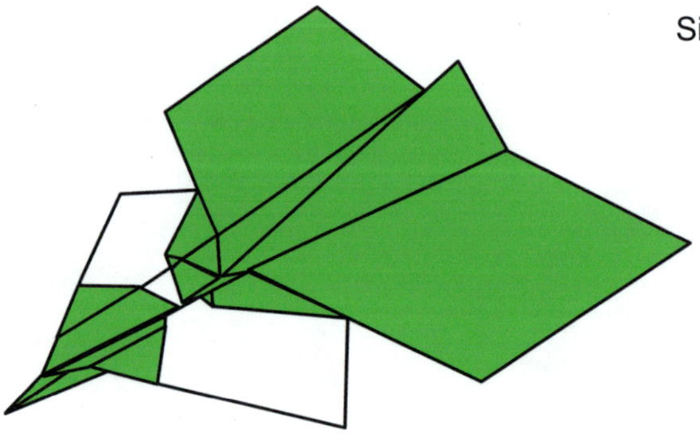

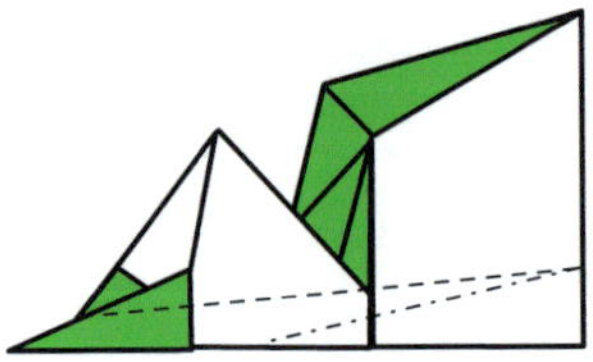

Beta Six

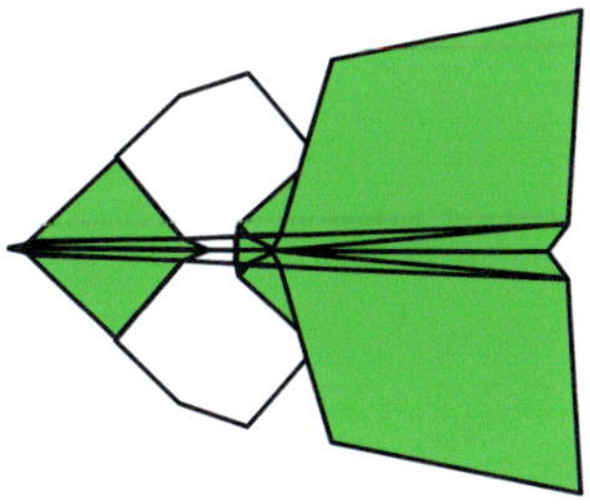

Top View

Front View

Side View

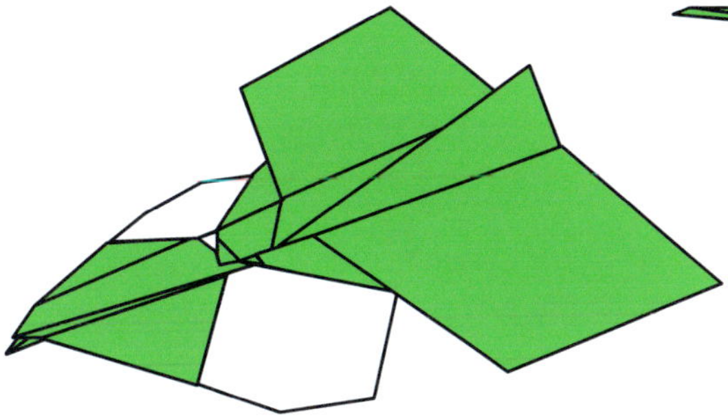

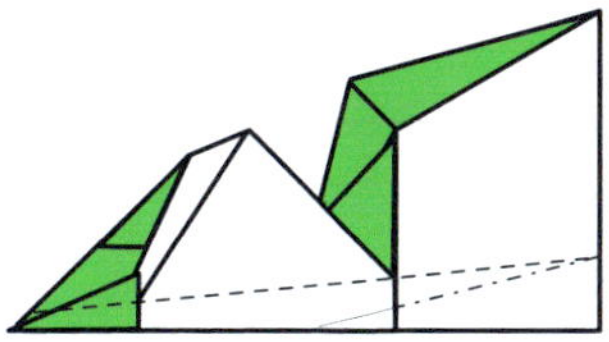

Gamma Six

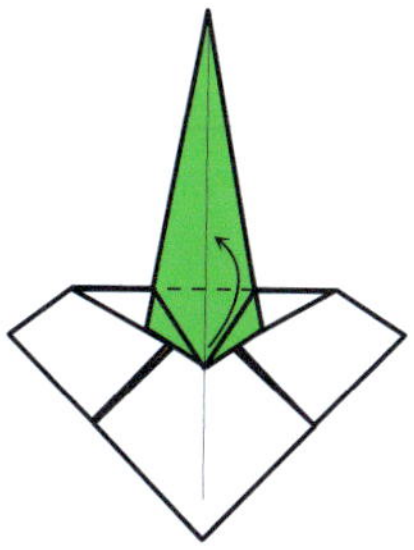

1'. Valley fold the thick flap back up for assembly.

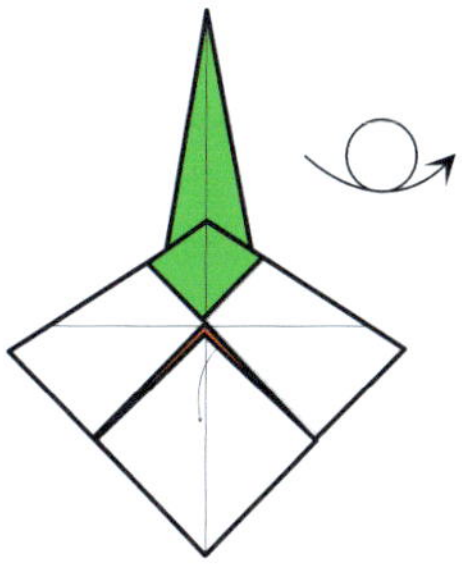

2'. Turn over.

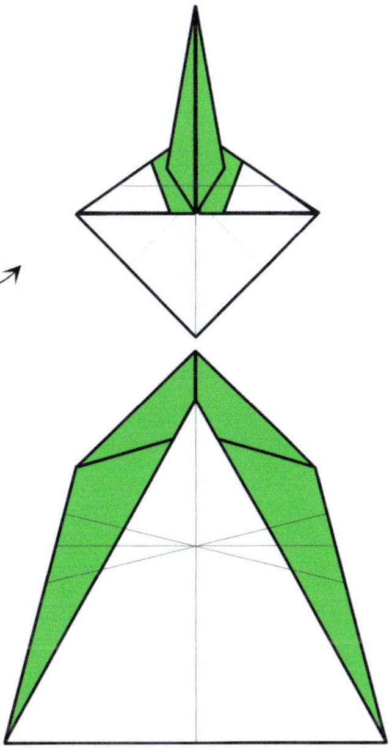

3'.Assemble with the back piece as in the other models.

4'.Valley fold the front of the back piece down as in the other models.

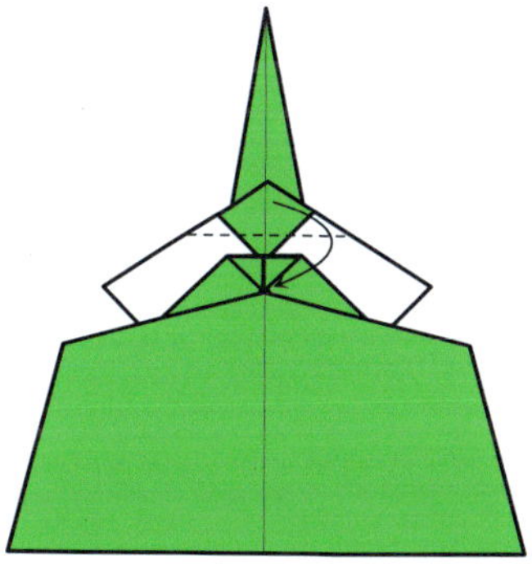

5'.Valley fold the front of the back piece back down.

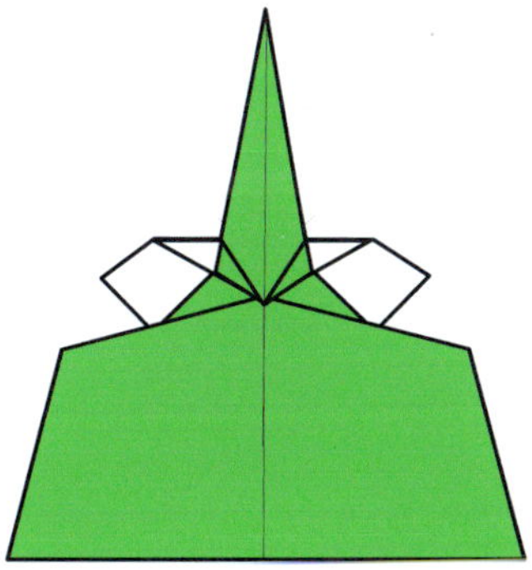

Gamma Six

Front View

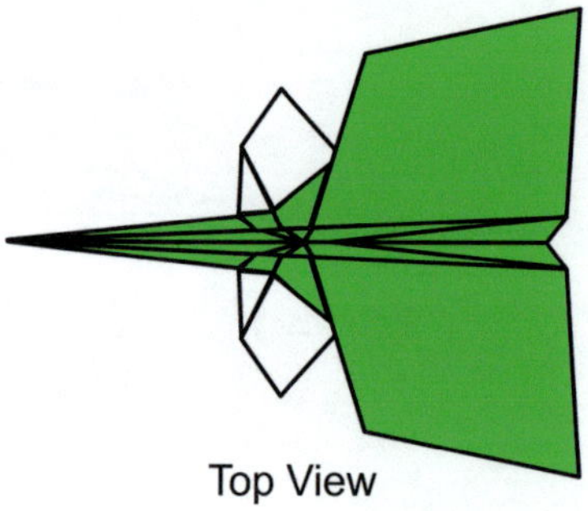

Top View

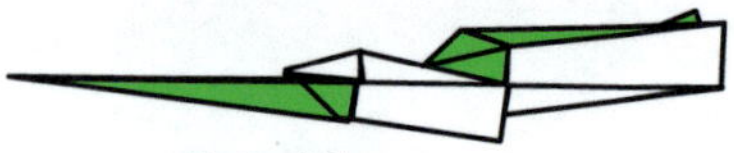

Side View

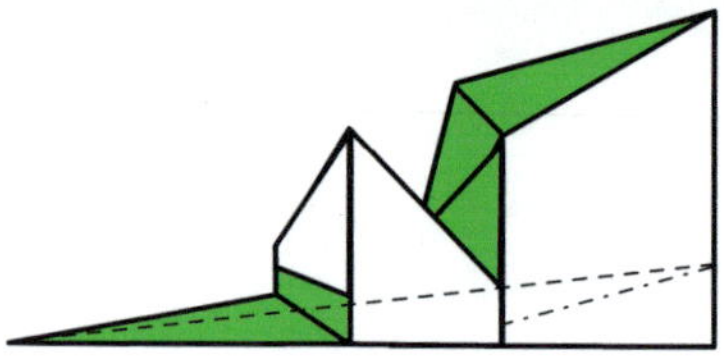

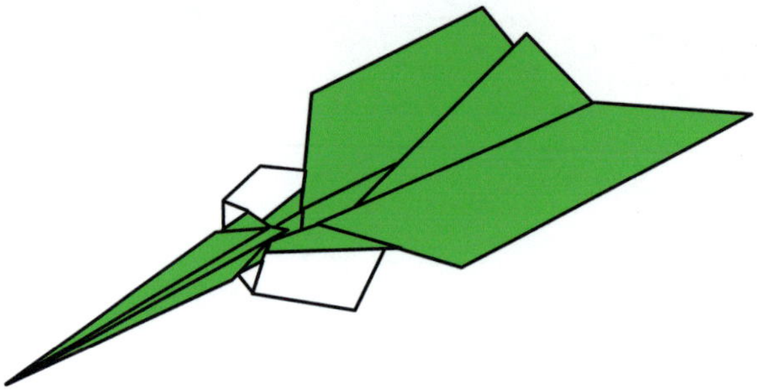

Epsilon Six

Front View

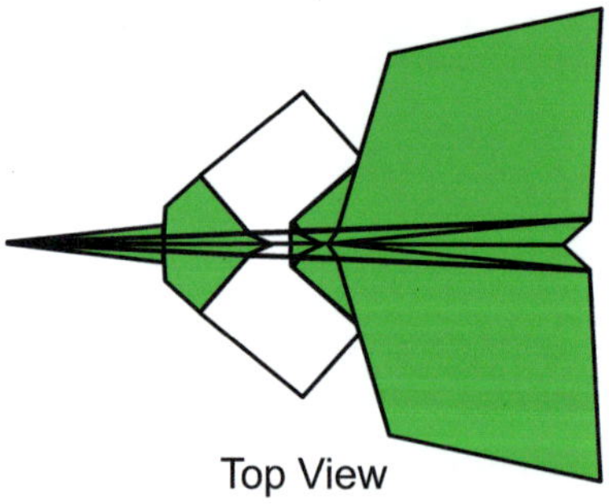

Top View

Side View

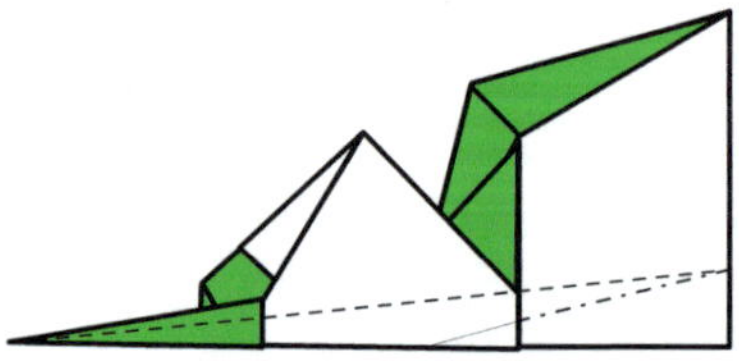

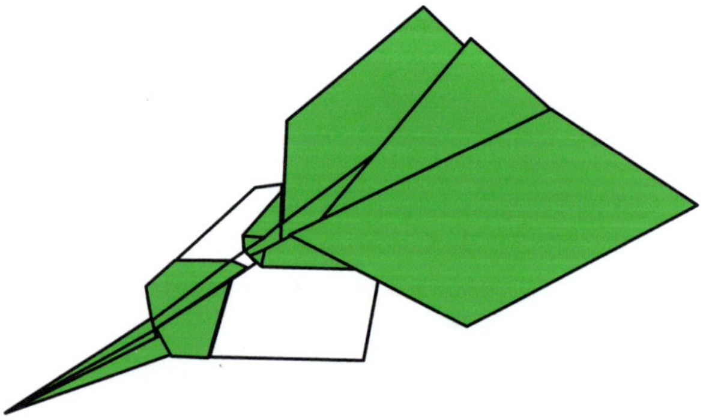

Kappa Six

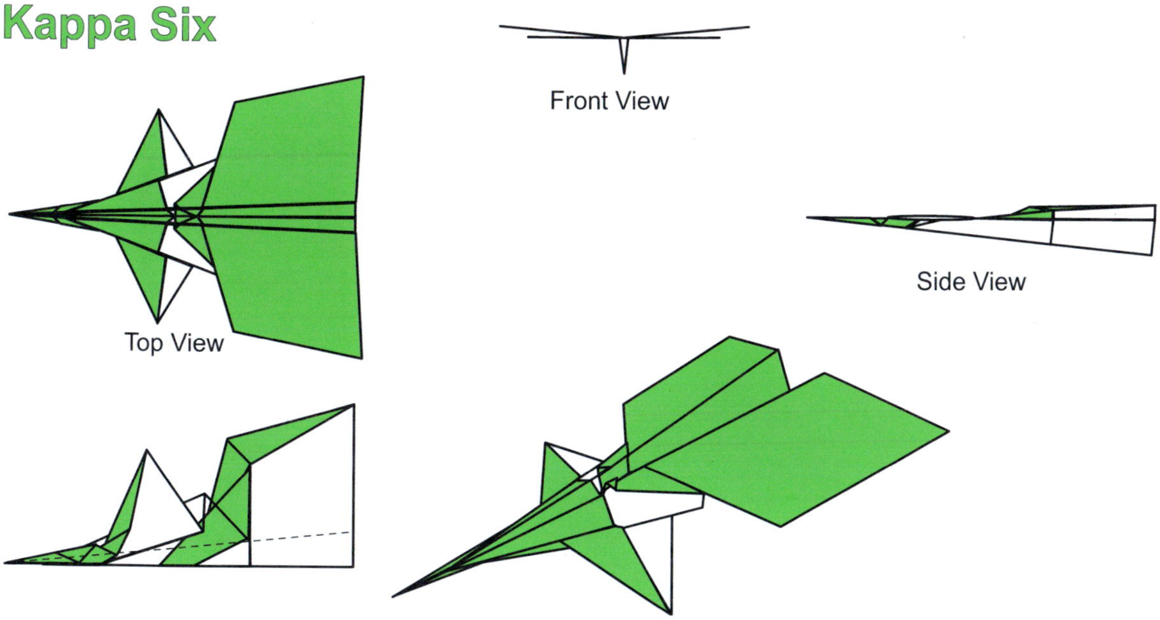

Lambda Six

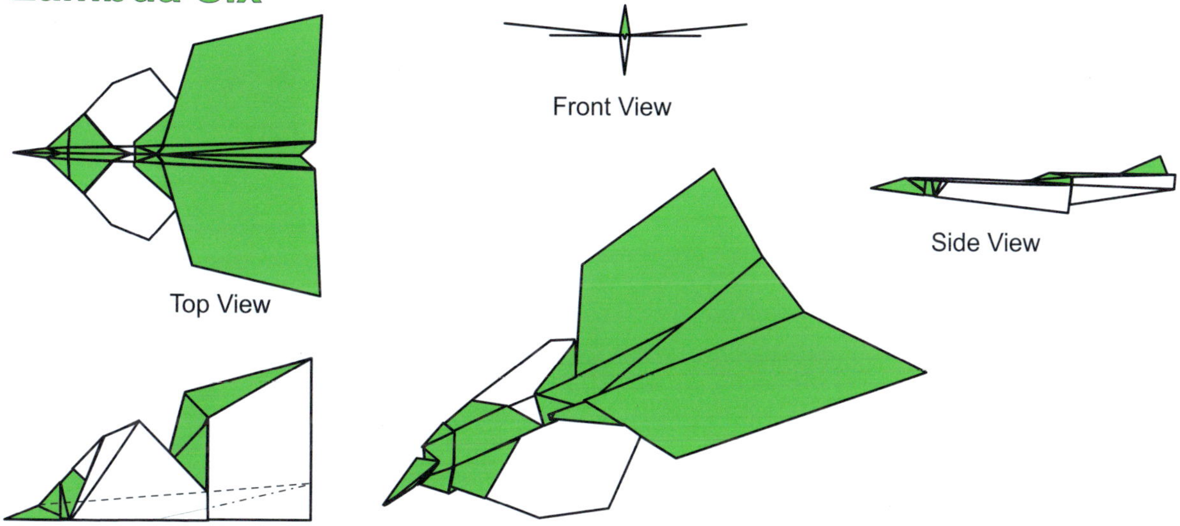

Back Piece Seven Airplanes

Alpha Seven

Beta Seven

Gamma Seven

Delta Seven

Epsilon Seven

Kappa Seven

Omicron Seven

Lambda Seven

Back Piece Seven

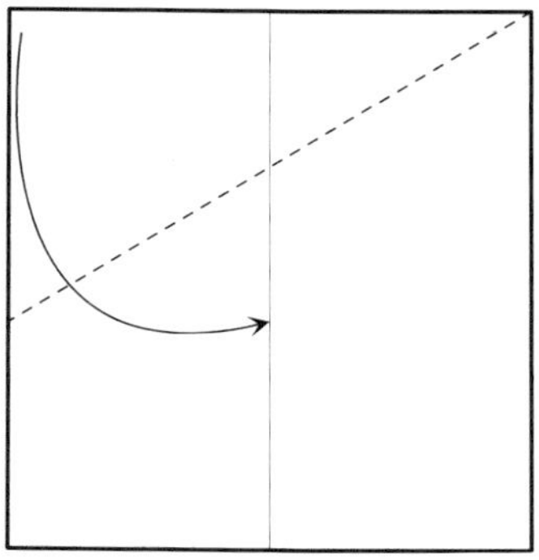

1. Valley fold from one corner so the adjacent corner lies on the midline.

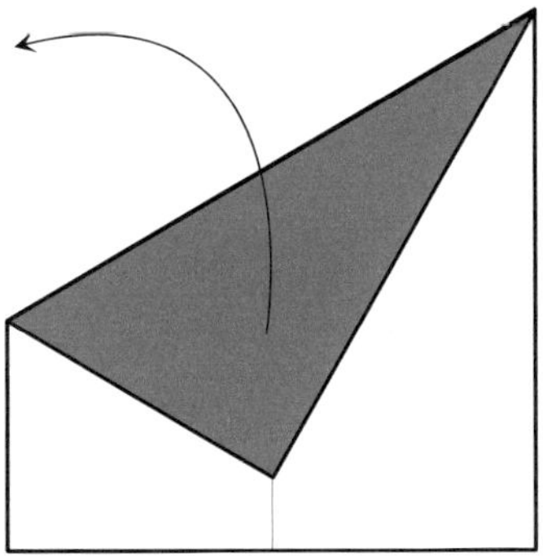

2. Unfold.

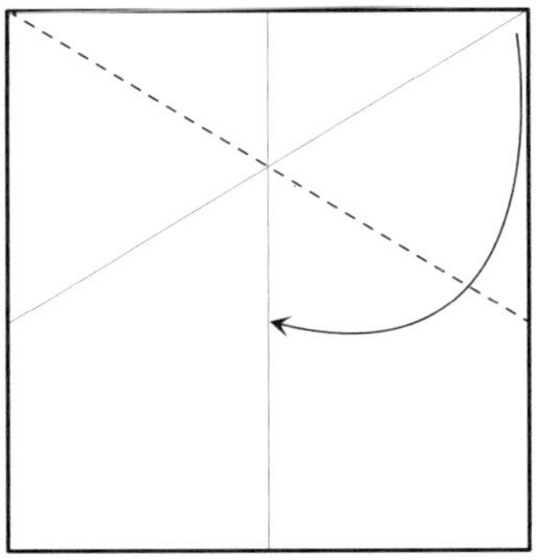

3. Repeat steps 1 and 2 on the other side.

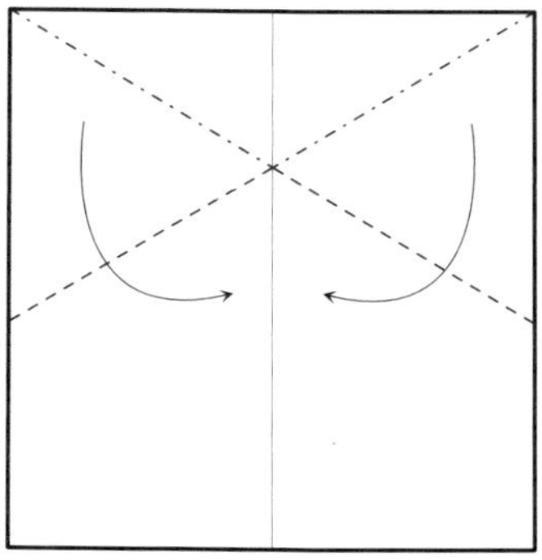

4. Rabbit ear the top down using the creases made in the previous steps.

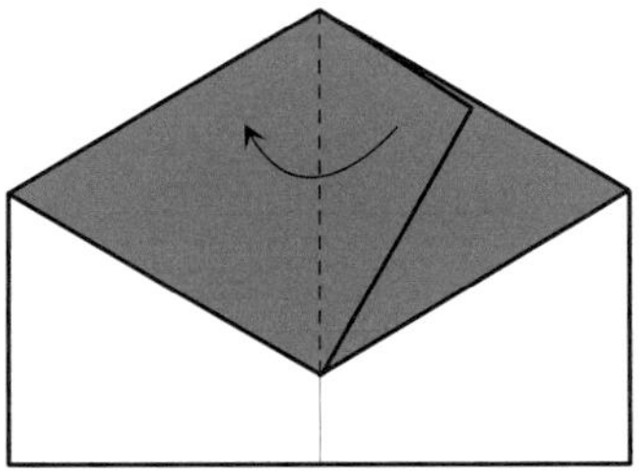

5. Valley fold the rabbit ear to the other side and crease well.

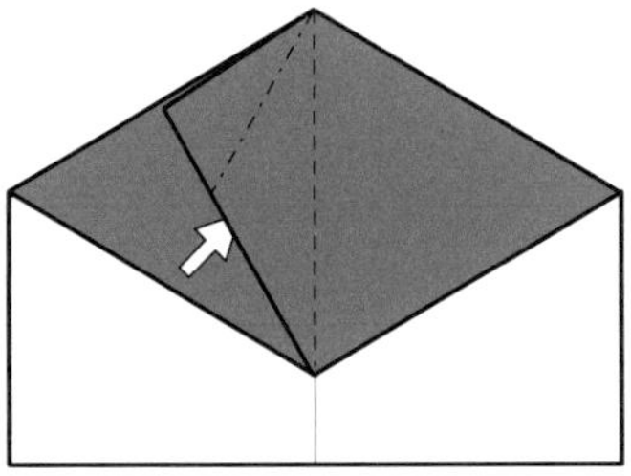

6. Squash fold the rabbit ear flap.

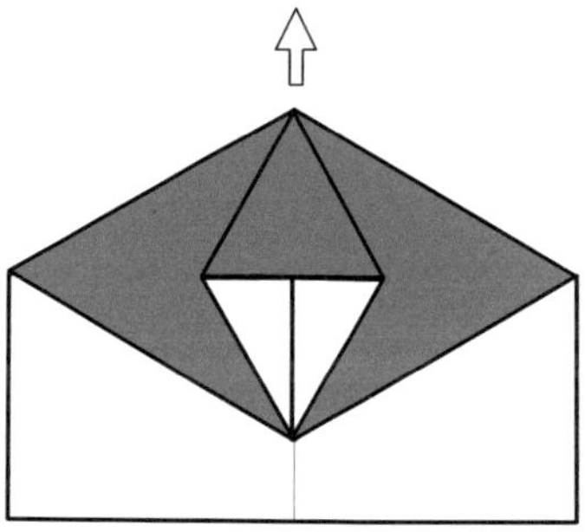

7. Unfold completely.

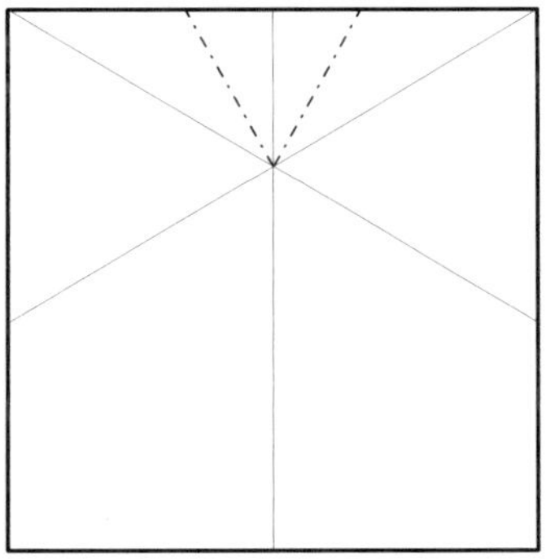

8. Turn the two inner creases into mountain folds.

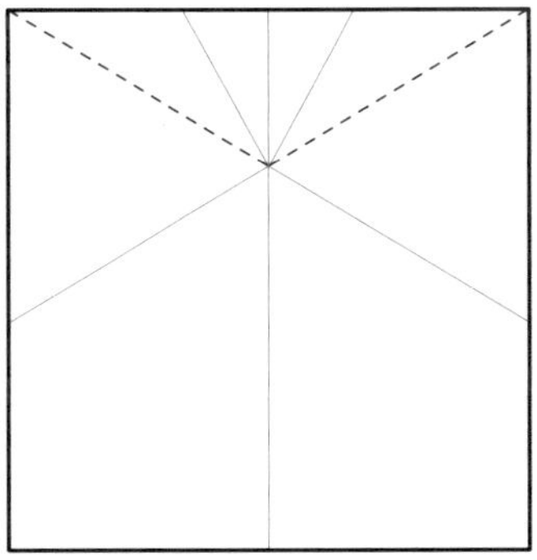

9. Turn the two outer folds into valley folds.

Back Piece Seven Page 2

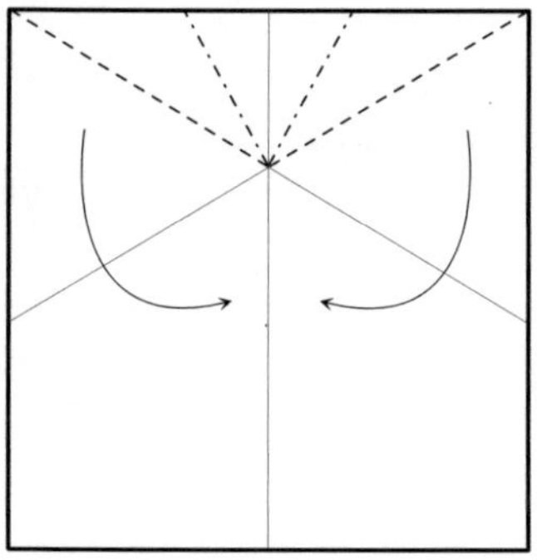

10. Fold the front down sending the squash folded flaps to the inside.

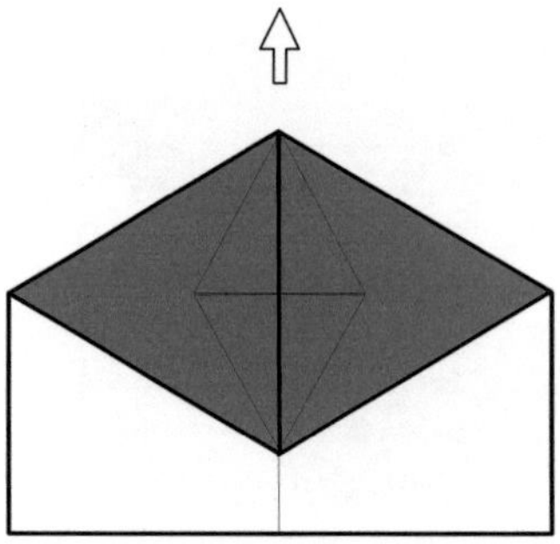

11. Open back up.

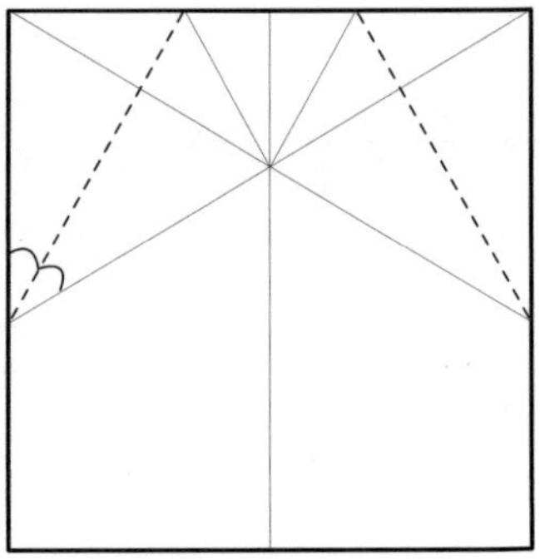

12. Crease the valley folds shown.

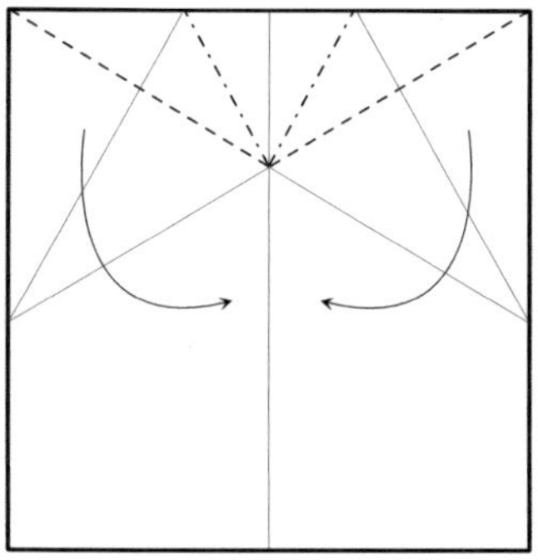

13. Reform the front.

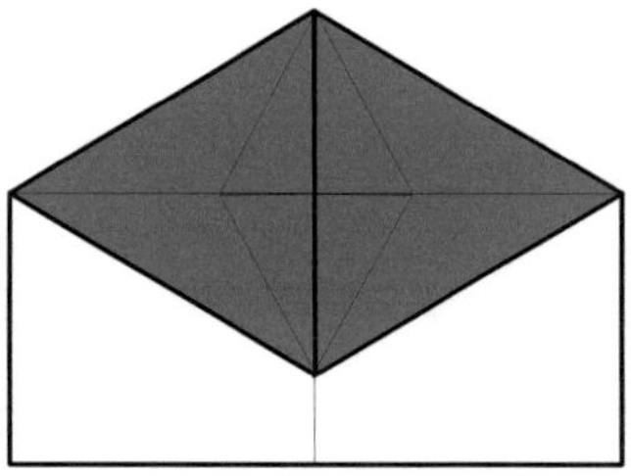

14. Ready for assembly.

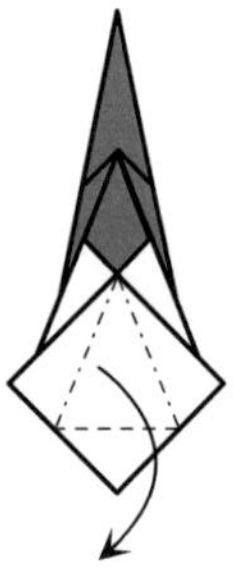

15. Petal fold the join of front piece Zeta.

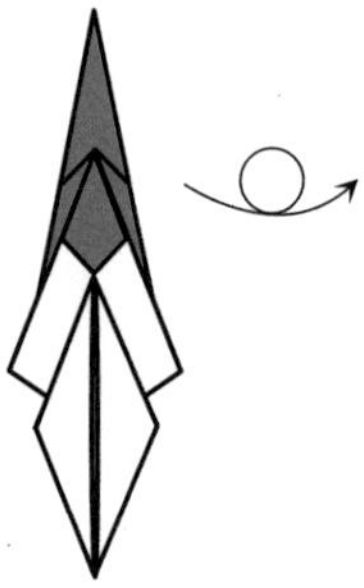

17. Turn over.

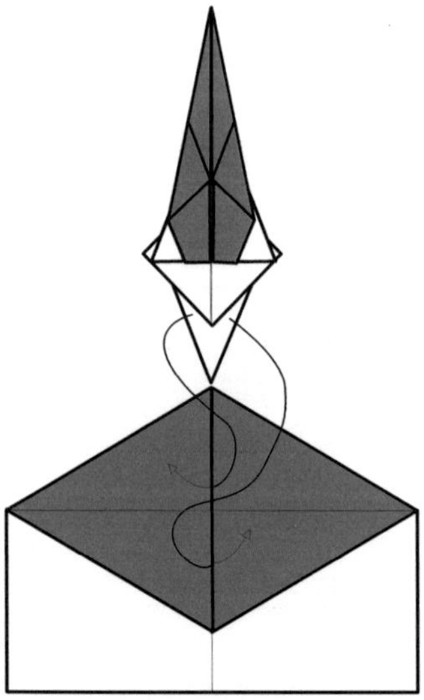

18. Slide the flaps of the join into the pockets in the back piece.

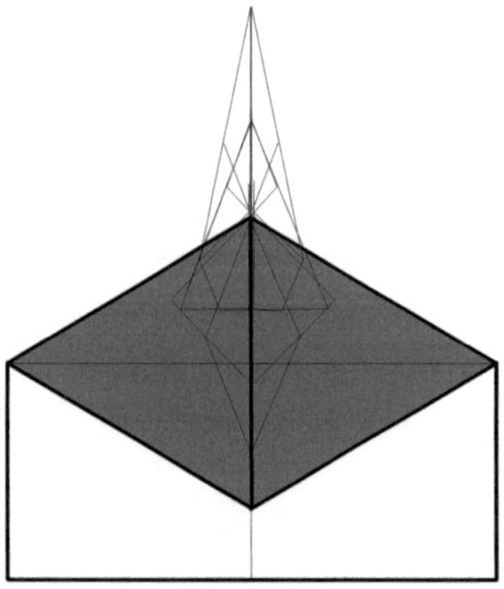

19. Move the front piece as far forward as it will go.

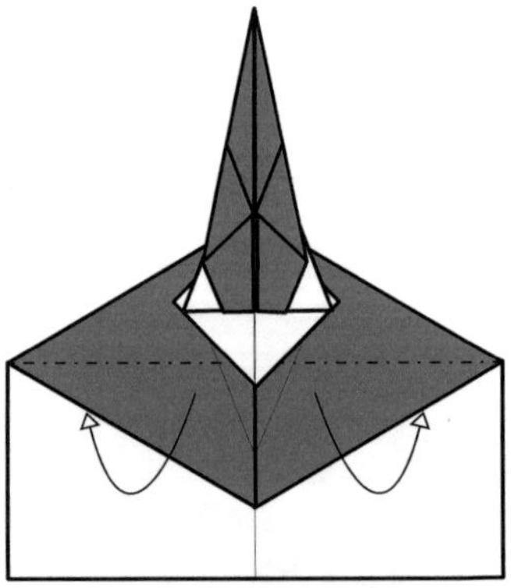

20. Mountain fold the flaps behind on preexisting creases. Part of the front piece goes with them.

Back Piece Seven Page 3

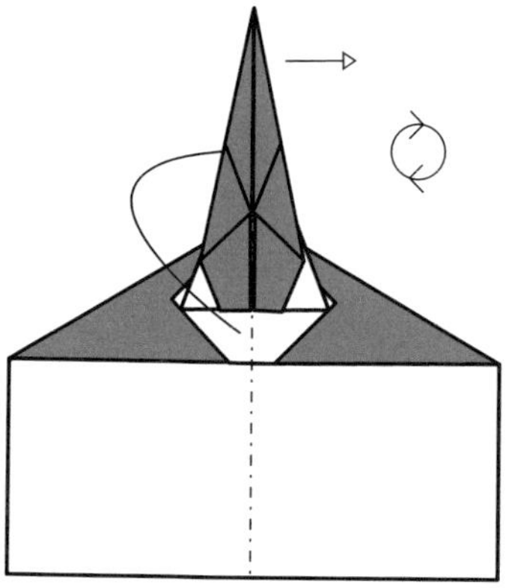

21. Mountain fold in half and rotate.

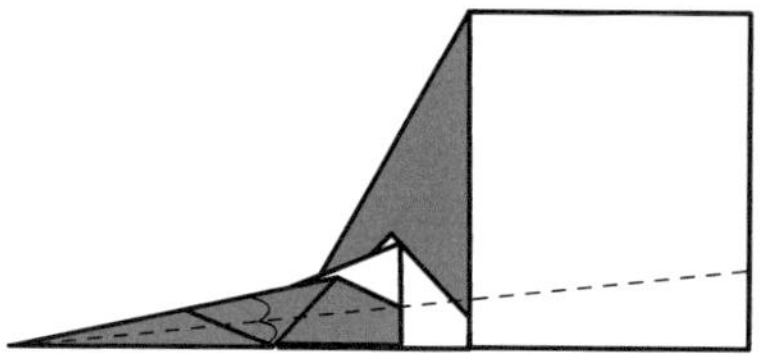

The wings of Front Piece Zeta are usually folded to bisect the angle at the front. The rear wings are then folded at the same angle.

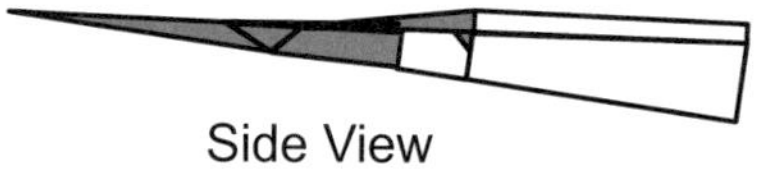

Side View

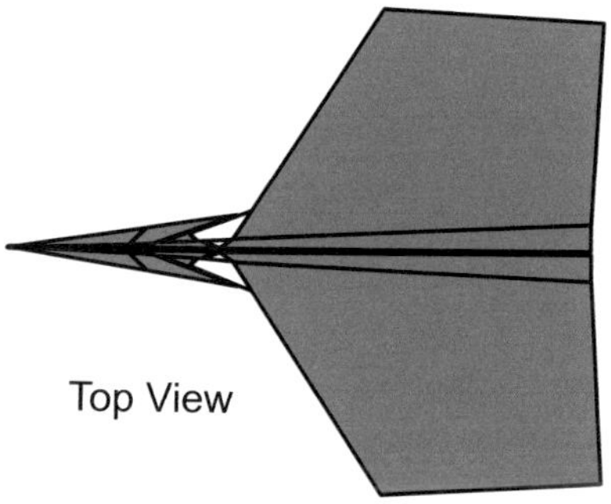

Top View

Front View

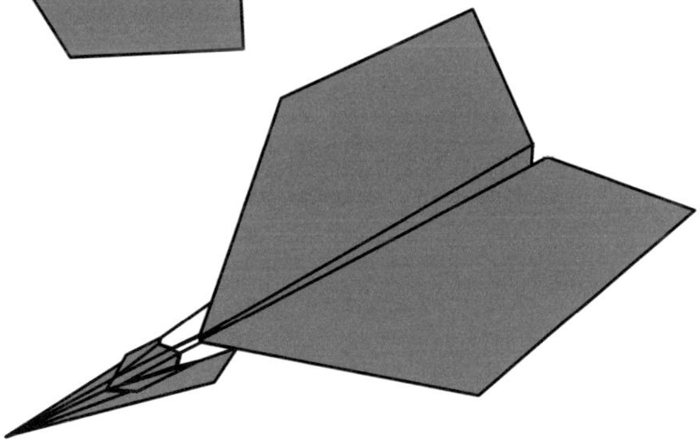

Beta Seven

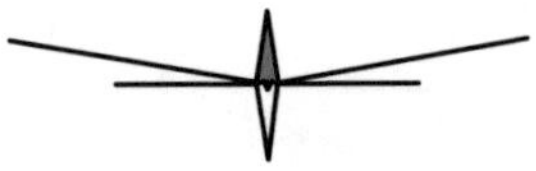

Front View

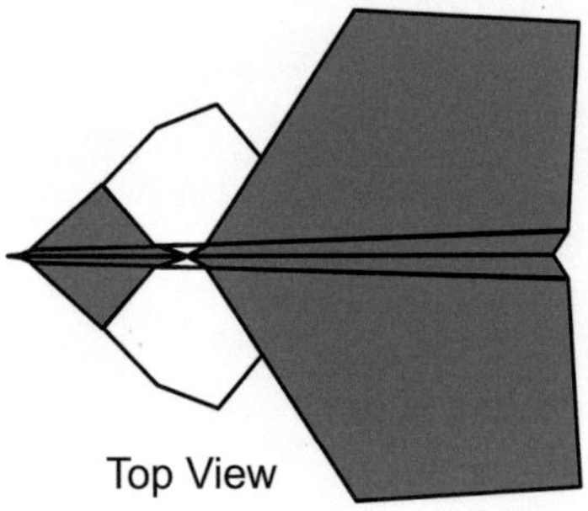

Top View

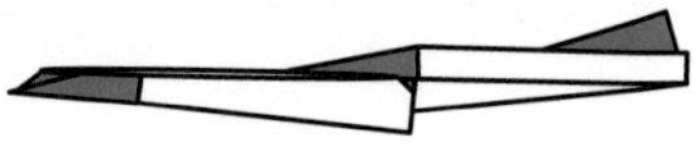

Side View

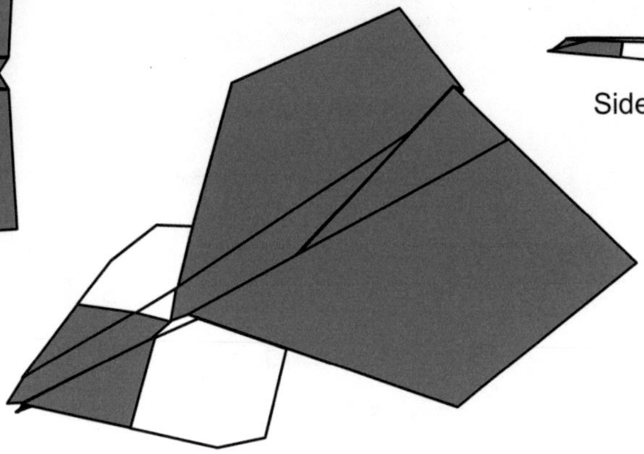

Gamma Seven

Front View

Side View

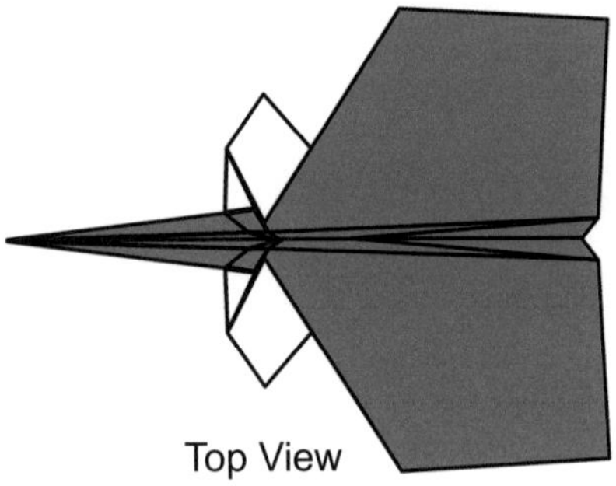

Top View

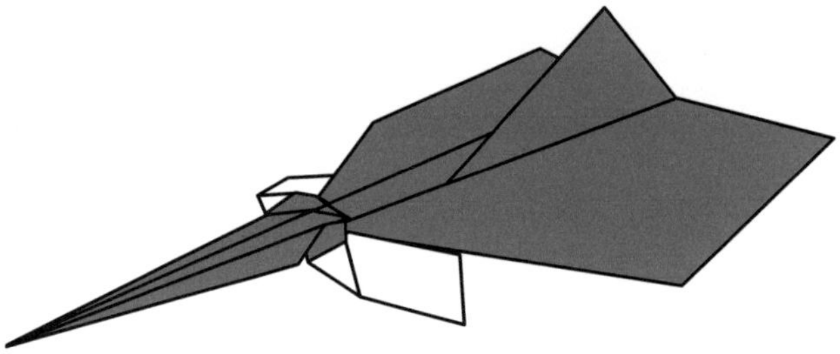

Delta Seven

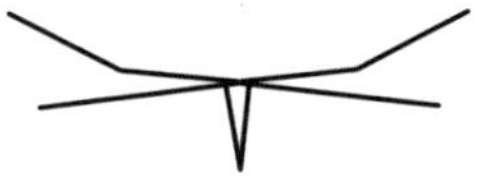

Front View

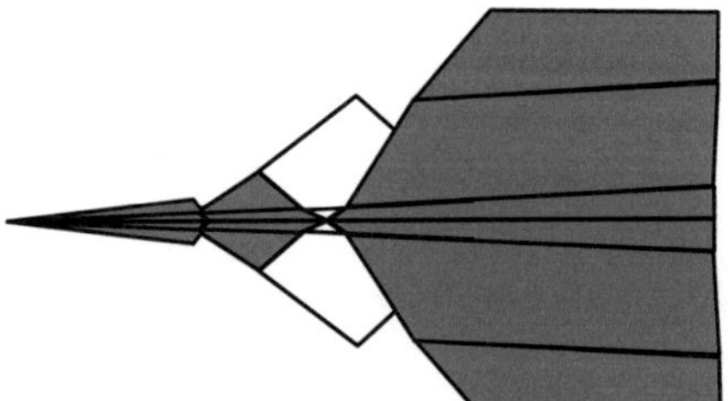

Top View

Side View

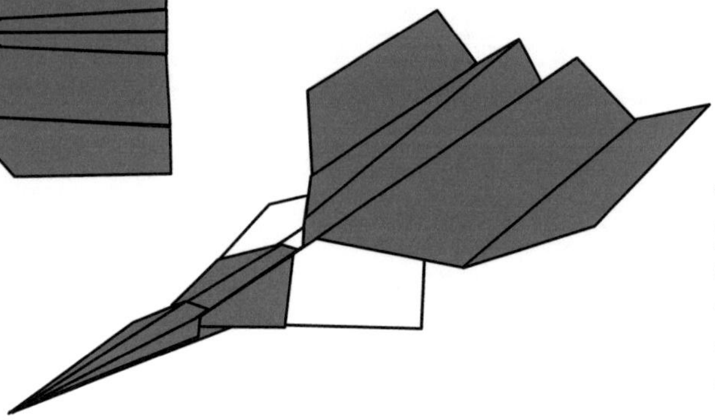

Vertical stabilizers can be folded on the back wings by lining up their outer edges with the center of the wings. It isn't perfect, but works out just fine.

Epsilon Seven

Front View

Side View

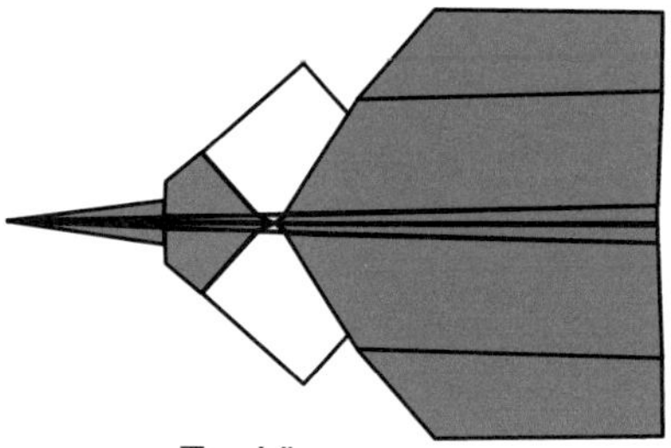

Top View

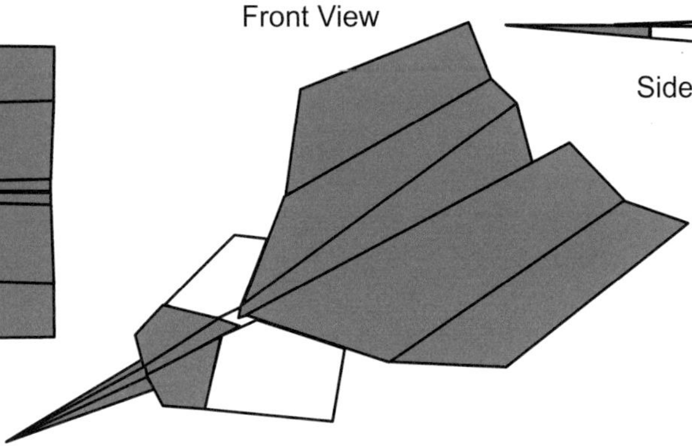

Kappa Seven

Front View

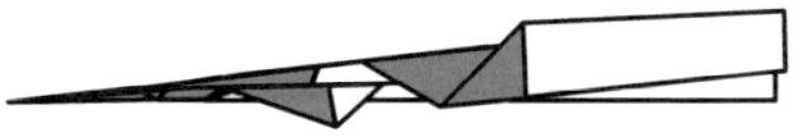

Side View

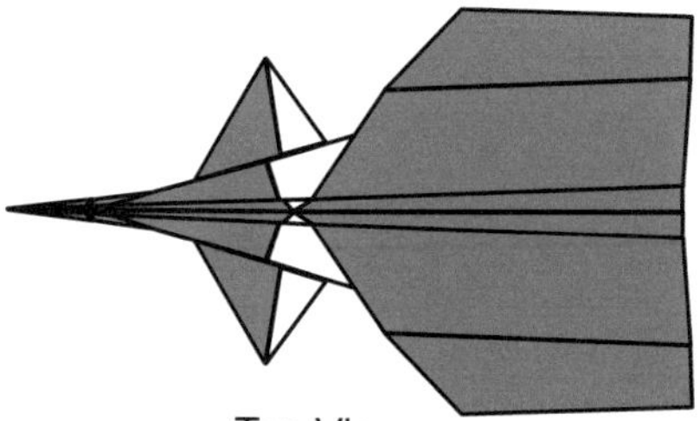

Top View

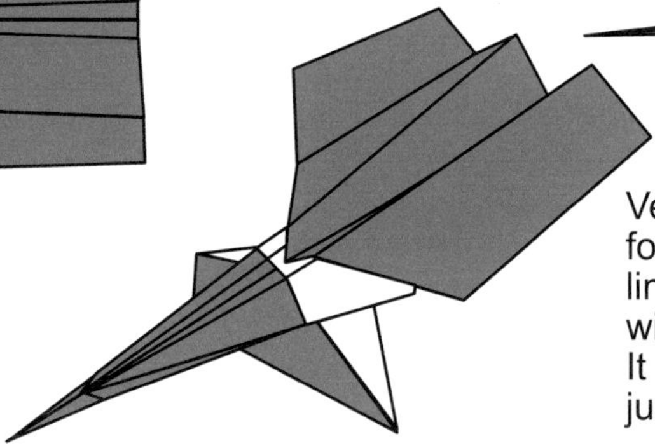

Vertical stabilizers can be folded on the back wings by lining up their outer edges with the center of the wings. It isn't perfect, but works out just fine.

Lambda Seven

Front View

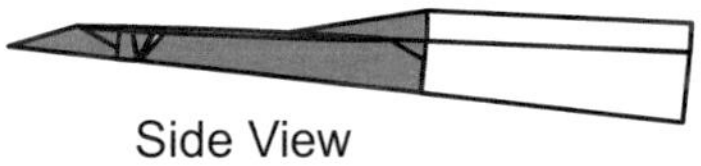

Side View

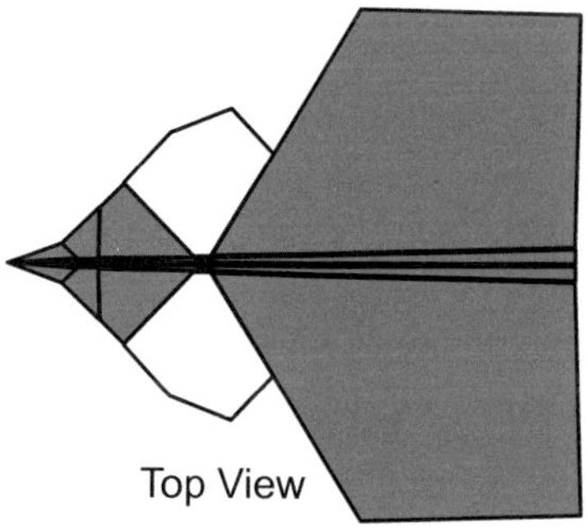

Top View

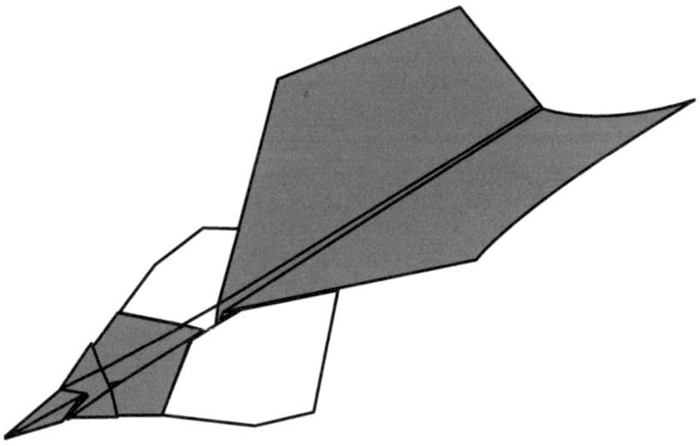

Alpha Seven

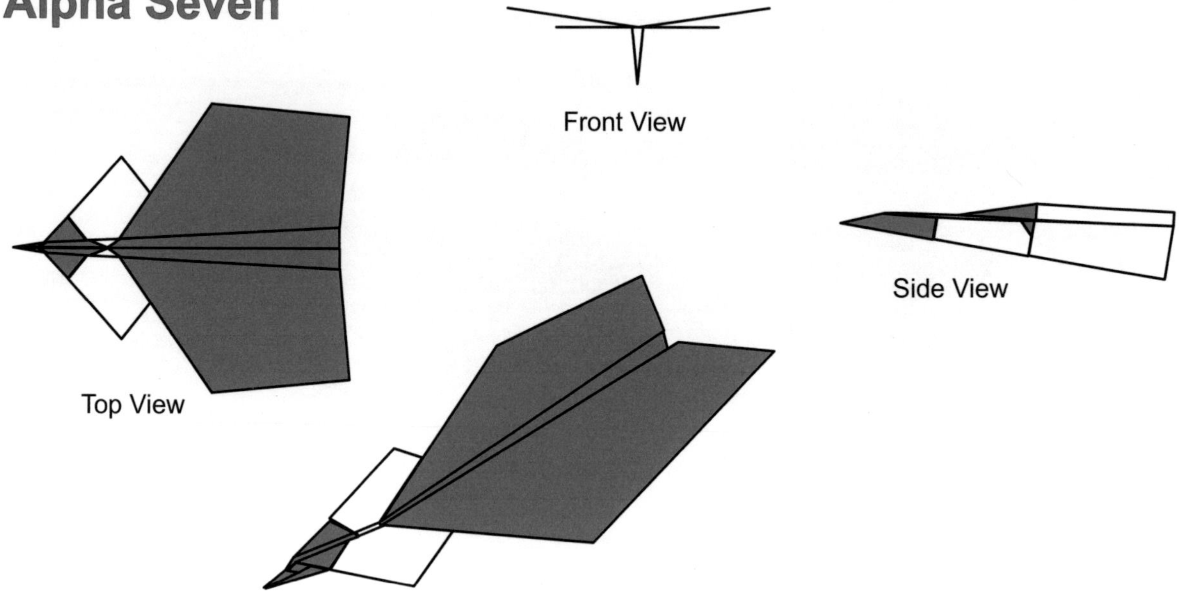

Omicron Seven

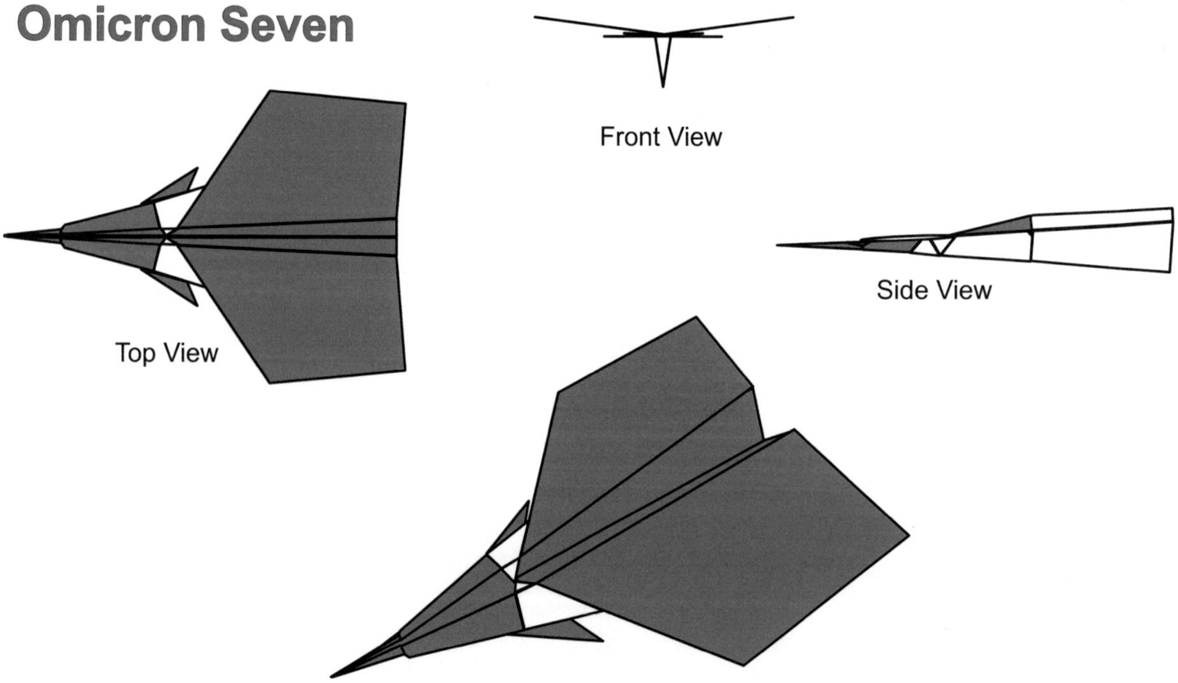

Back Piece Eight Airplanes

Alpha Eight

Beta Eight

Gamma Eight

Epsilon Eight

Delta Eight

Zeta Eight

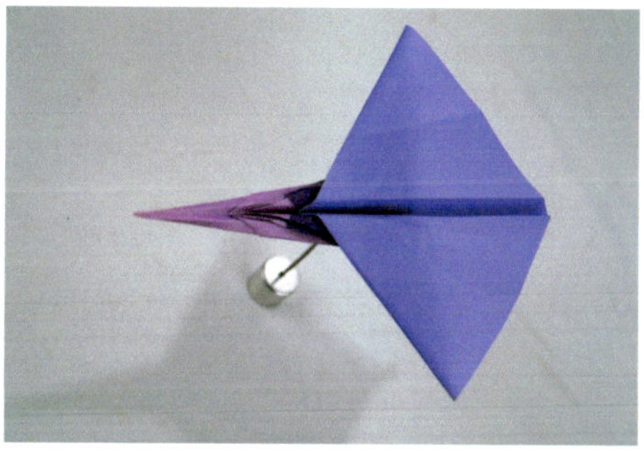

Iota Eight

Omicron Eight

Back Piece Eight

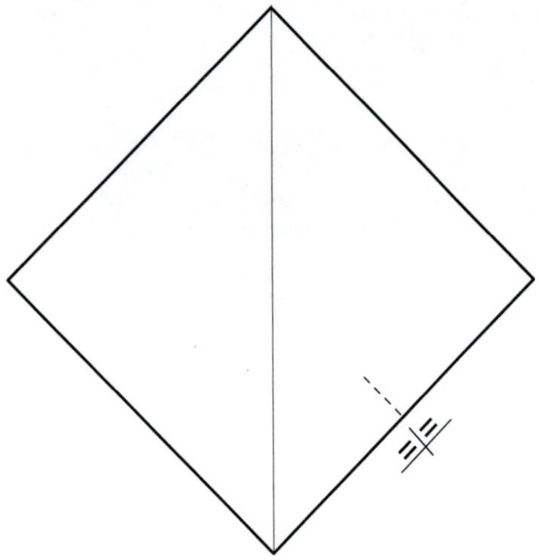

1. Mark the middle of one side with a short crease.

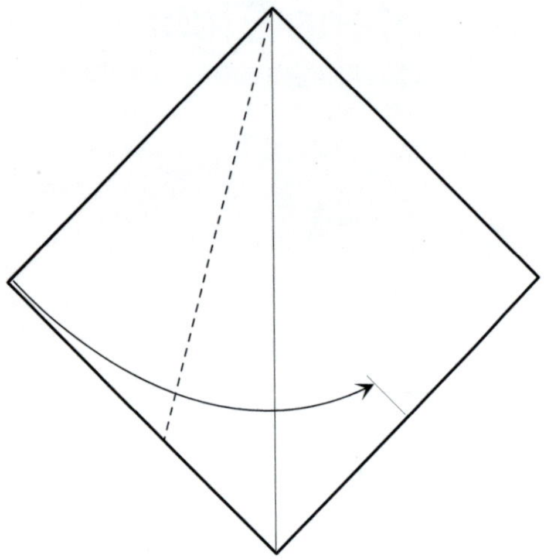

2. Valley fold the corner so that it touches the crease.

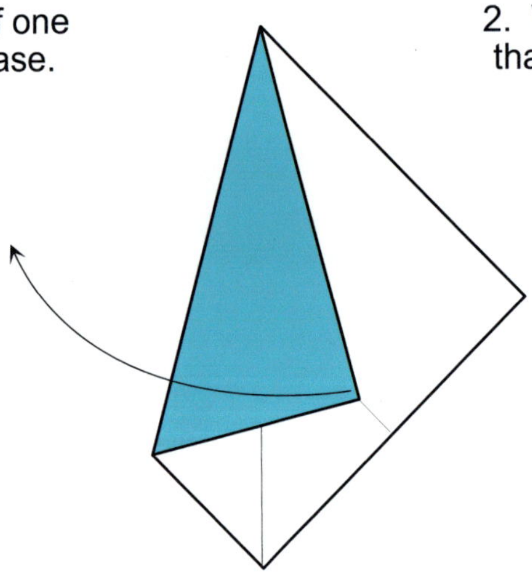

3. Unfold.

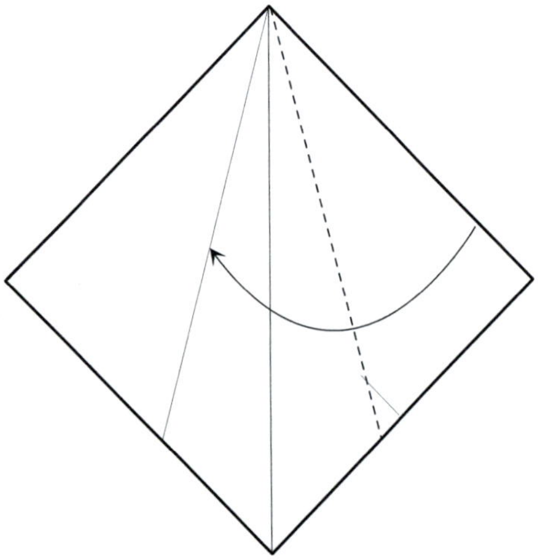

4. Valley fold so the adjacent side lies on the crease and unfold.

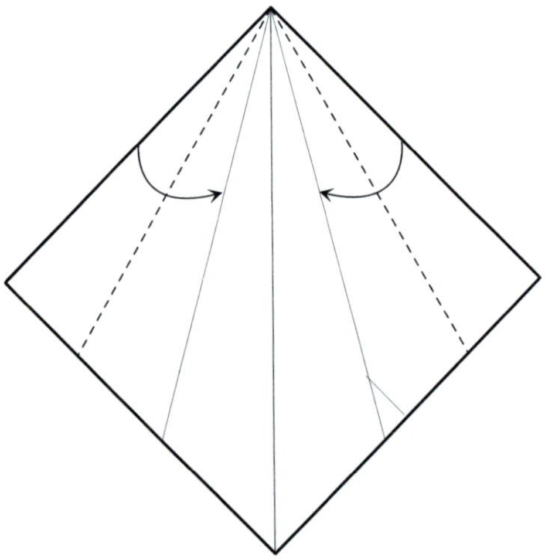

5. Valley fold the sides so their edges lie on the creases.

Back Piece Eight Page 2

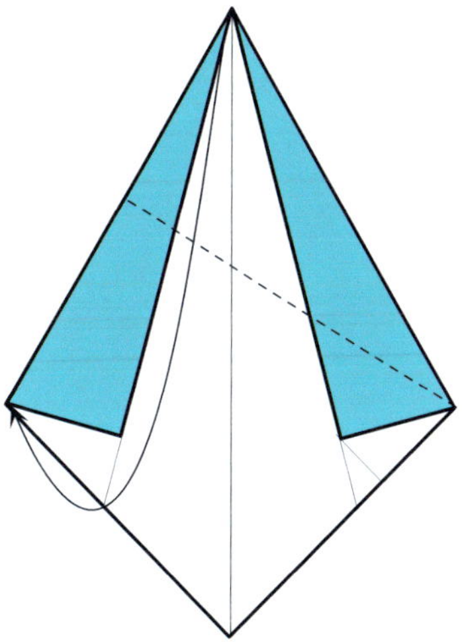

6. Valley fold so that the top point lies on the side corner.

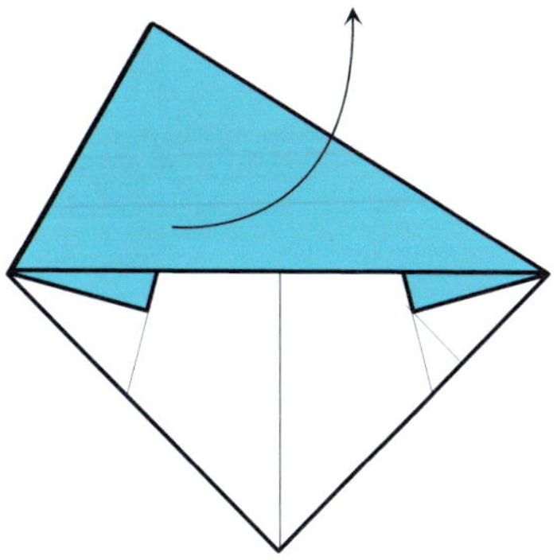

7. Unfold.

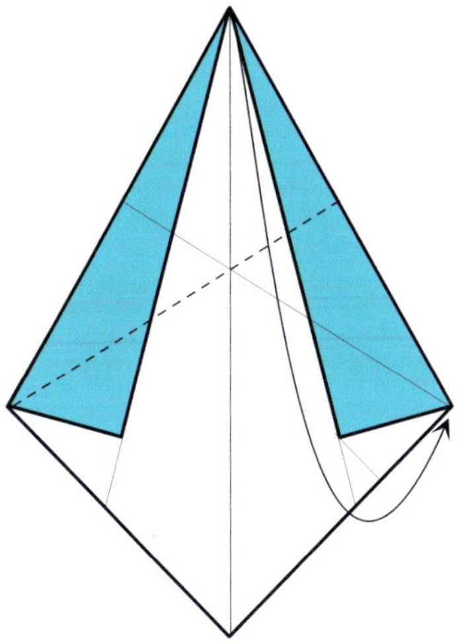

8. Repeat on the other side.

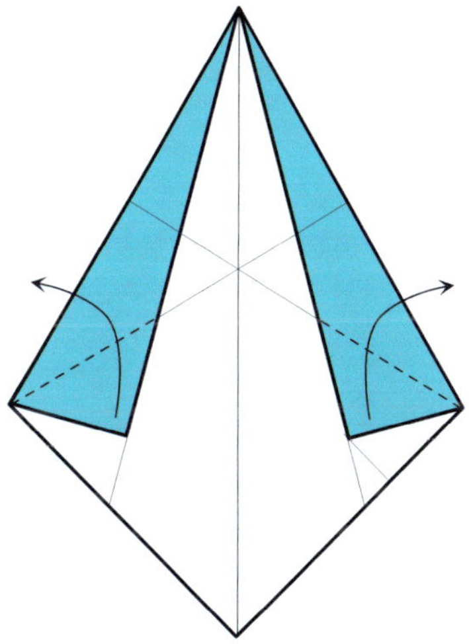

9. Valley fold the corners along preexisting creases.

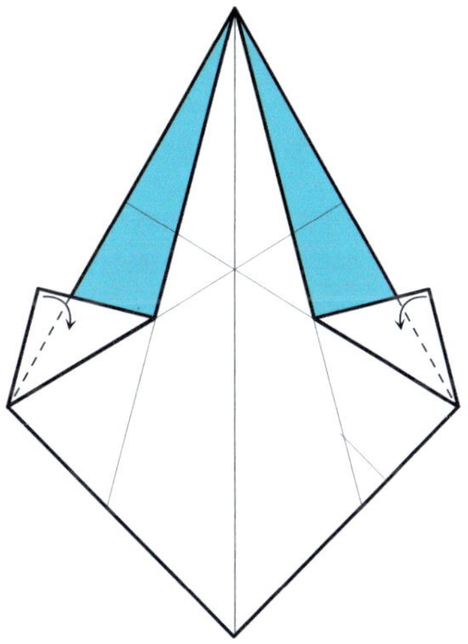

10. Valley fold the corners back where they meet the folded edge beneath.

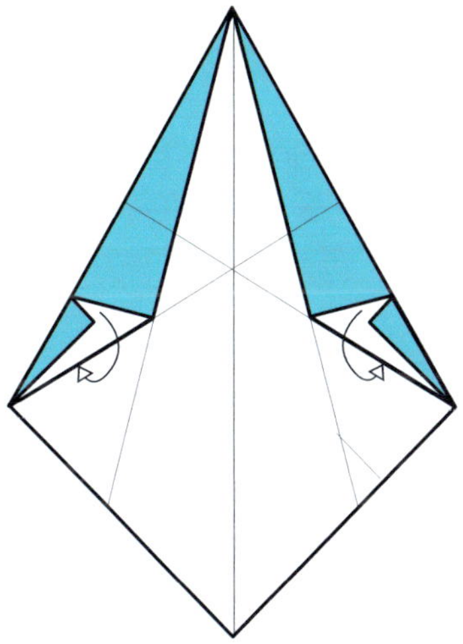

11. Mountain fold the flaps beneath.

Back Piece Eight

Page 3

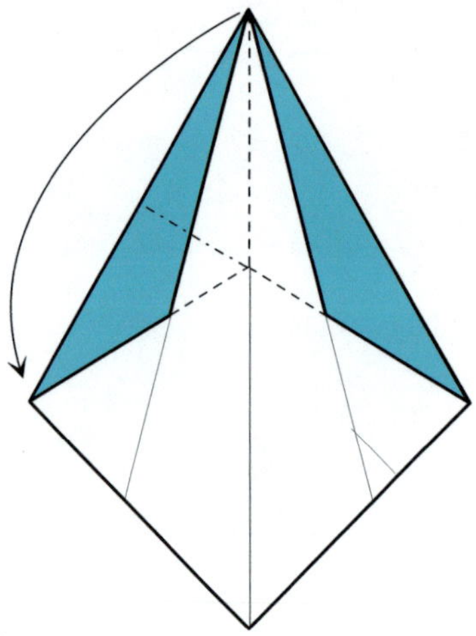

12. Rabbit ear the flap following creases made in steps 8 and 9.

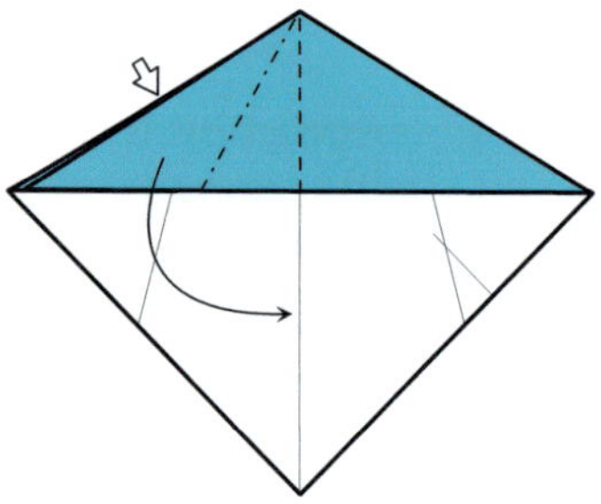

13. Squash fold the flap downwards.

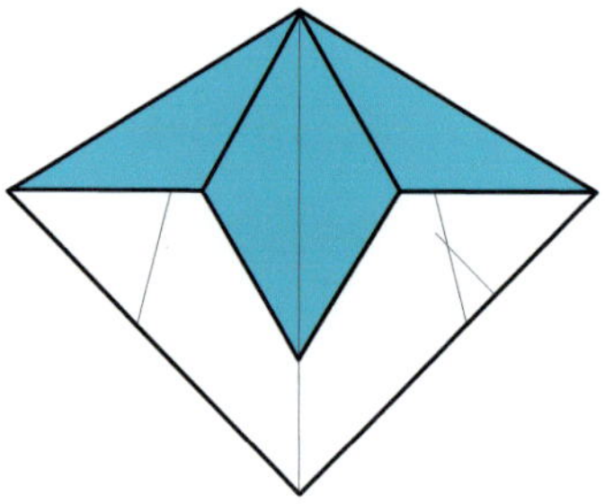

14. Unfold.

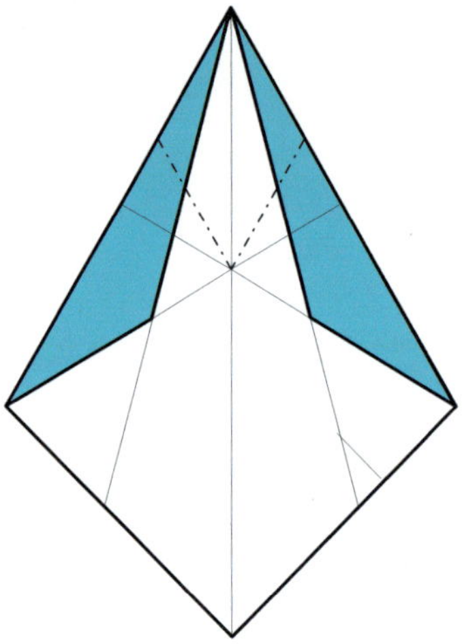

15. Change the inner creases to mountain folds.

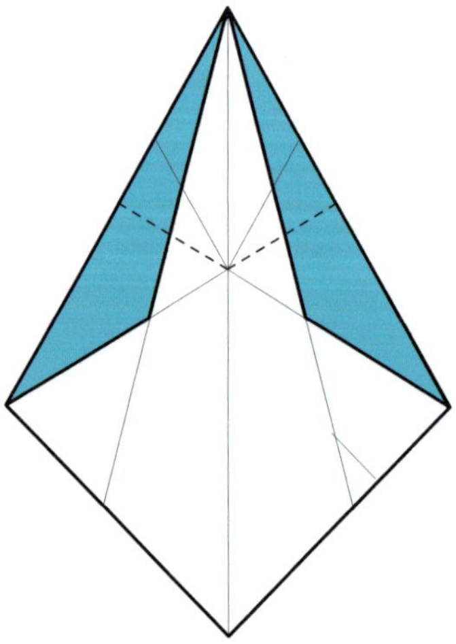

16. Change the outer creases to valley folds.

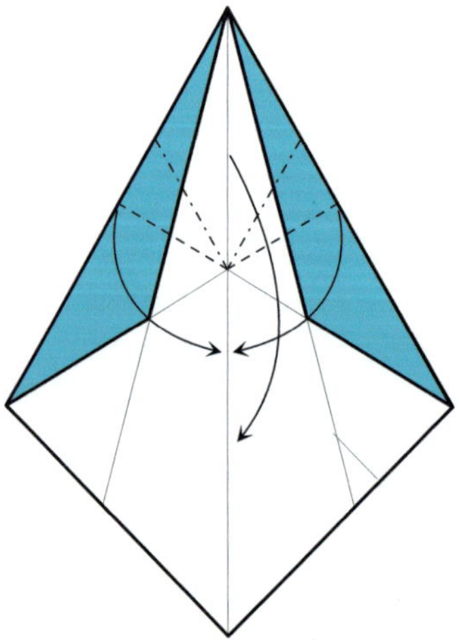

17. Reassemble following the new creases.

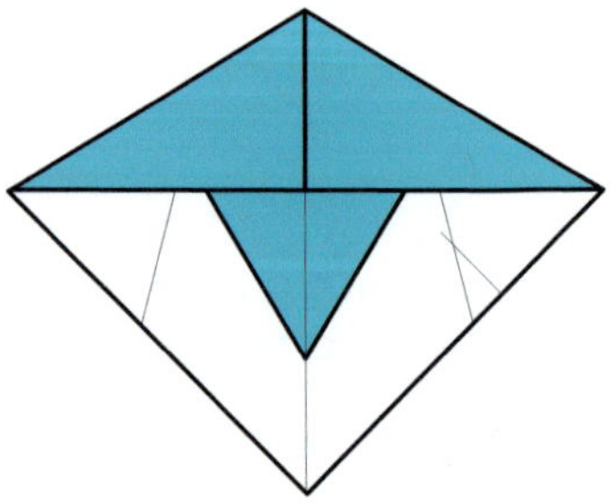

18. Ready for assembly.

Back Piece Eight Page 4

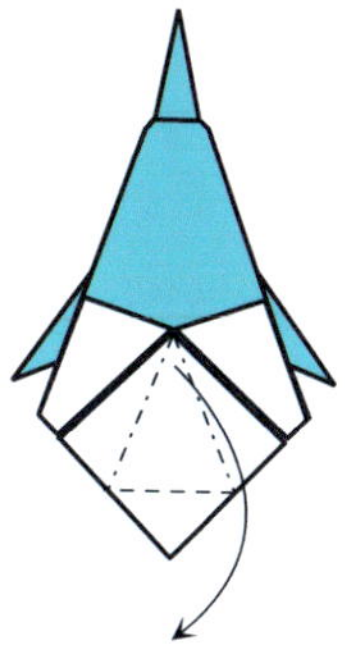

19. Petal fold the join on Front Piece Omicron.

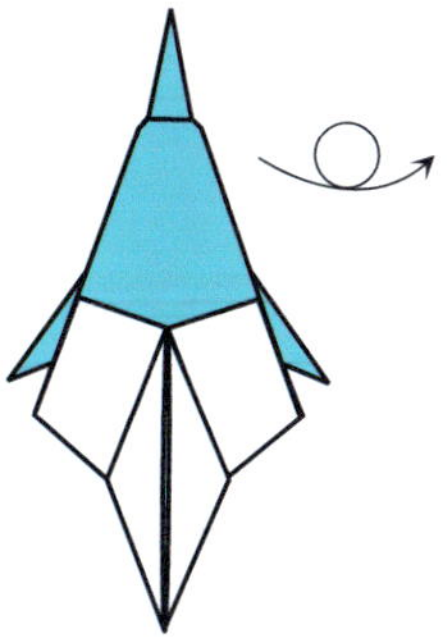

20. Turn over.

21. Slide the sides of the join inside the flaps of the back piece.

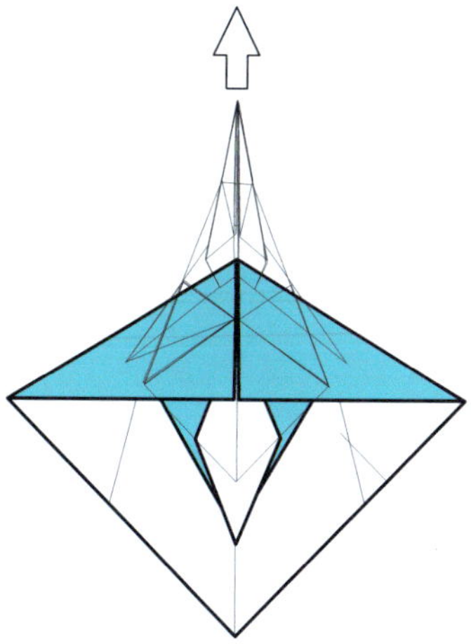

22. Slide the Front Piece forward as far as it will easily go.

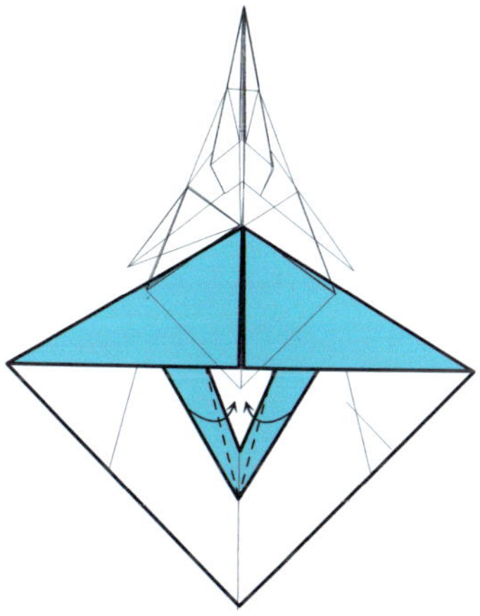

23. Valley fold the sides of the lower flap into the middle.

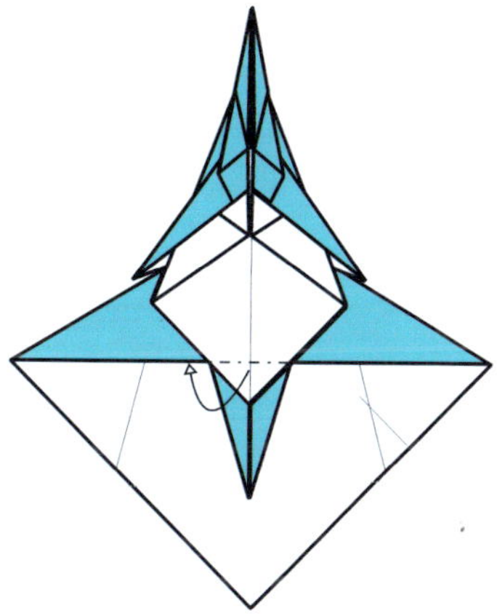

24. Mountain fold the lower flaps upwards inside the back piece.

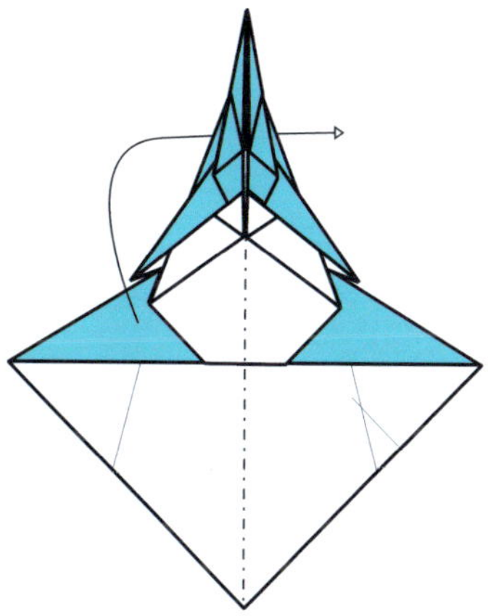

25. Mountain fold the aircraft in half.

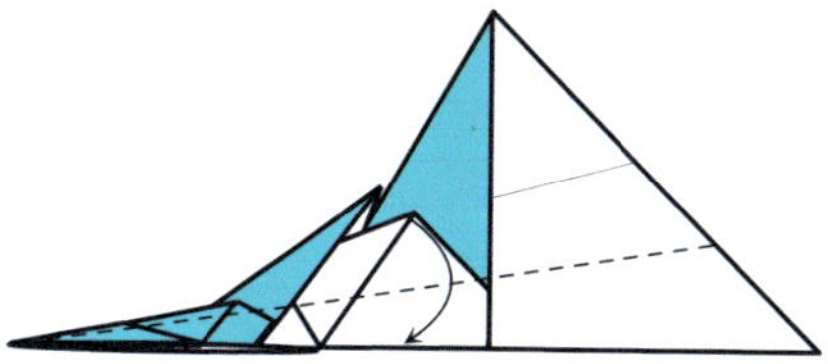

To fold the wings on Omicron, split the back part in half, and follow the fold through to the front. The angle of the rear wings should be the same as the canard wings.

Omicron Eight

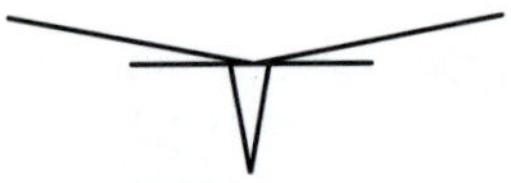

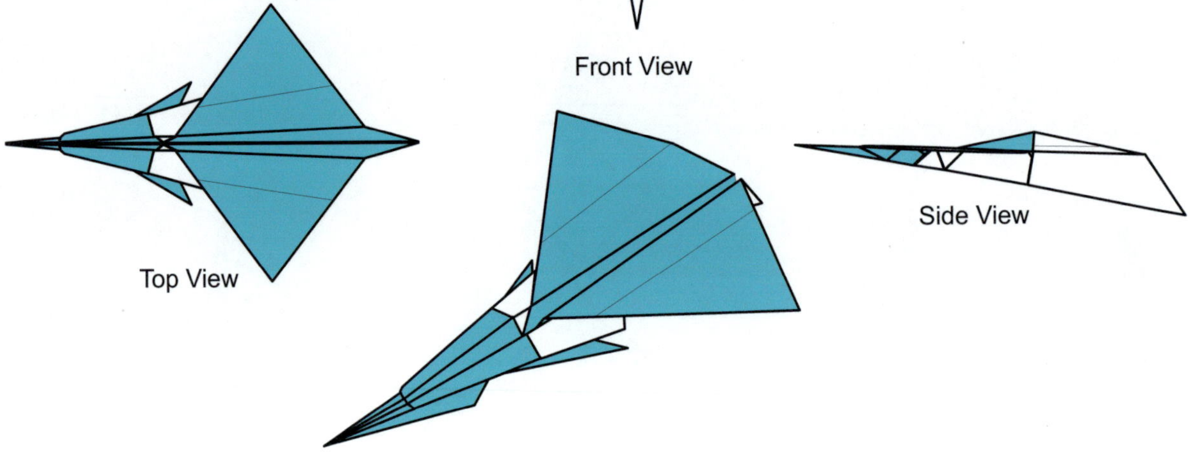

Alpha Eight

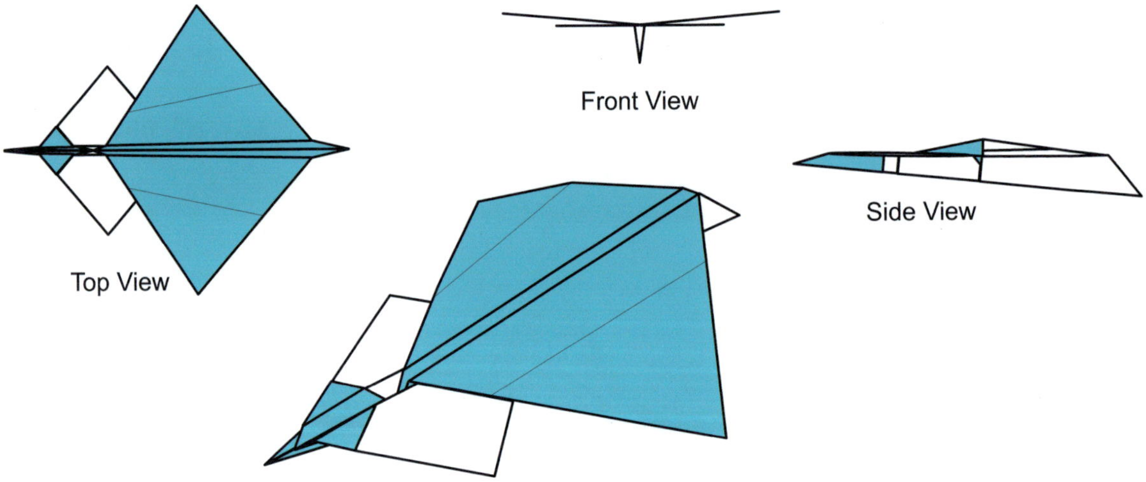

Beta Eight

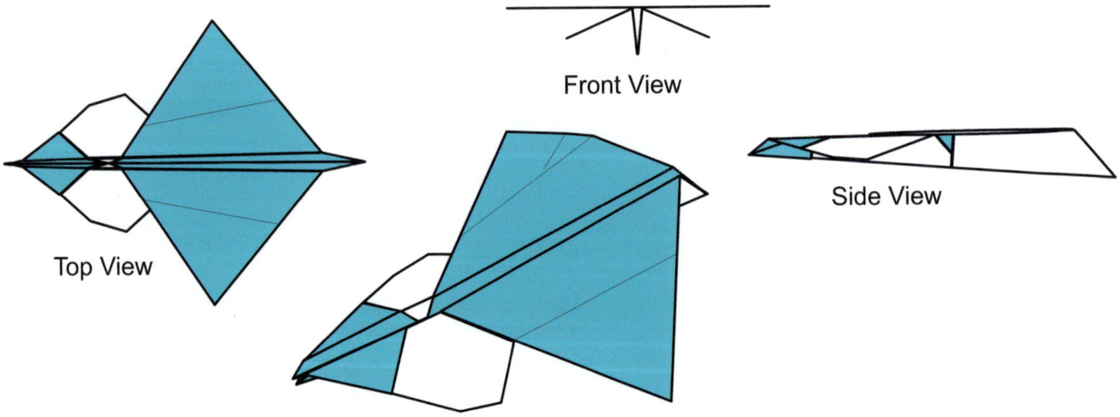

Delta Eight

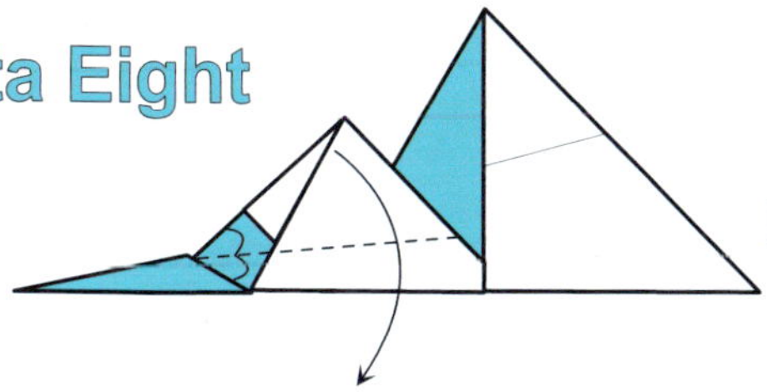

Front piece Delta is folded differently when combined with Back Piece Eight. Fold the bisector between the leading edge of the wing and the adjacent layer.

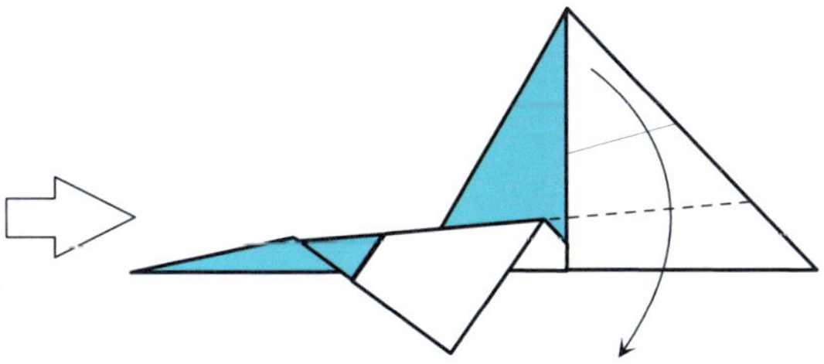

Once the Canard (Forward) wings are folded, match the angle of the front wings when folding the back wings.

Front View

Side View

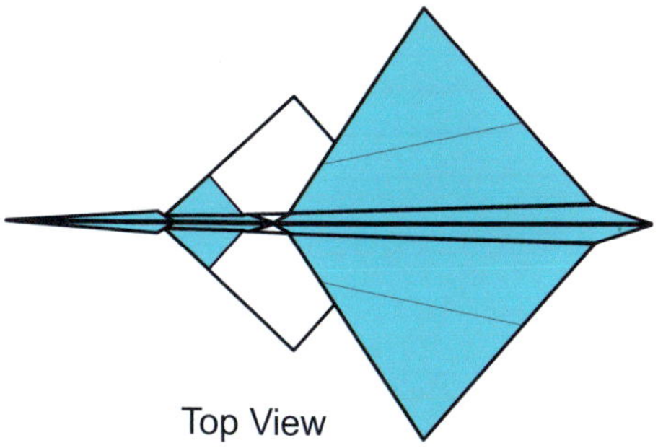

Top View

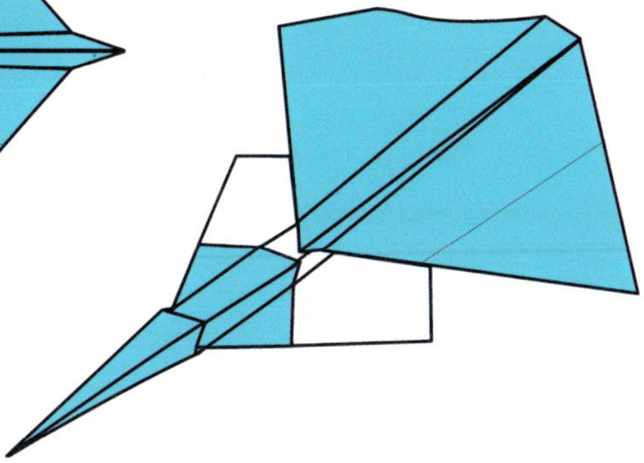

Epsilon Eight

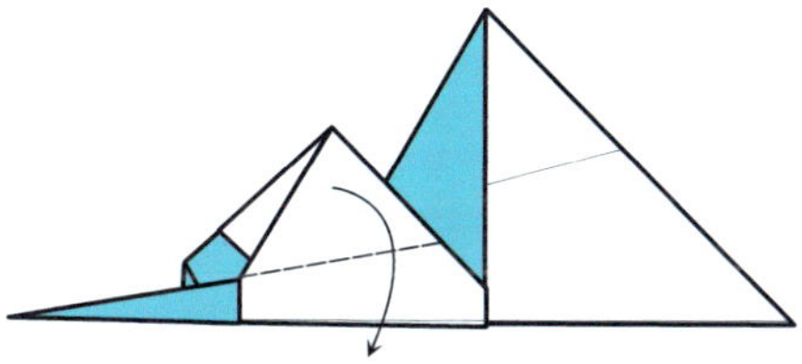

Fold the front wings by following the angle of the nose.

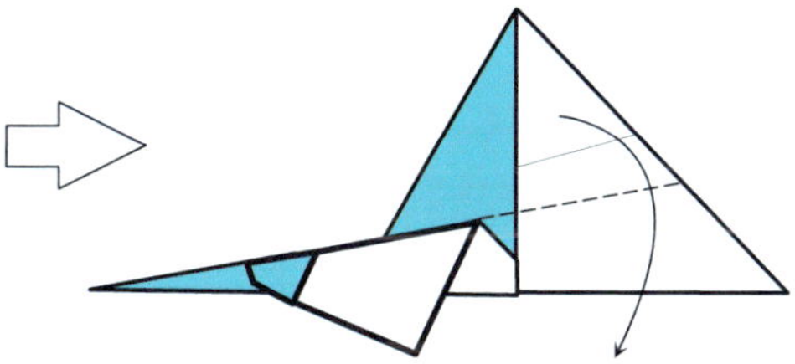

Fold the rear wings by following the angle of the front wings.

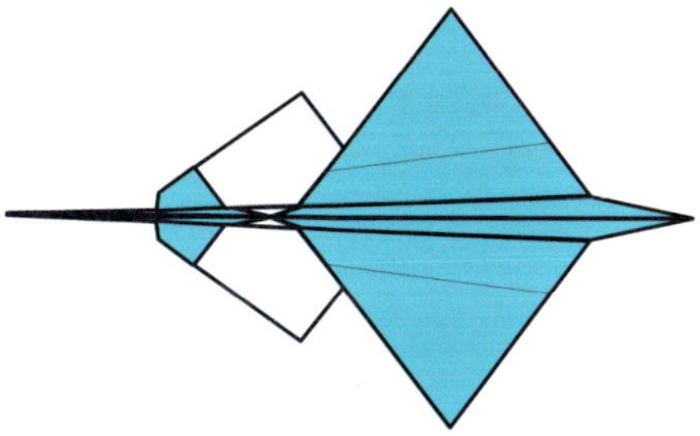

Top View

Front View

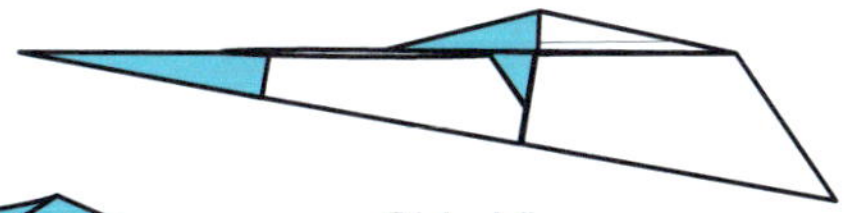

Side View

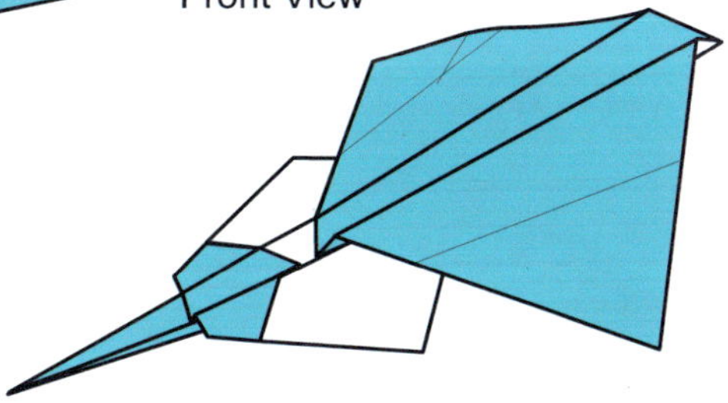

Gamma Eight

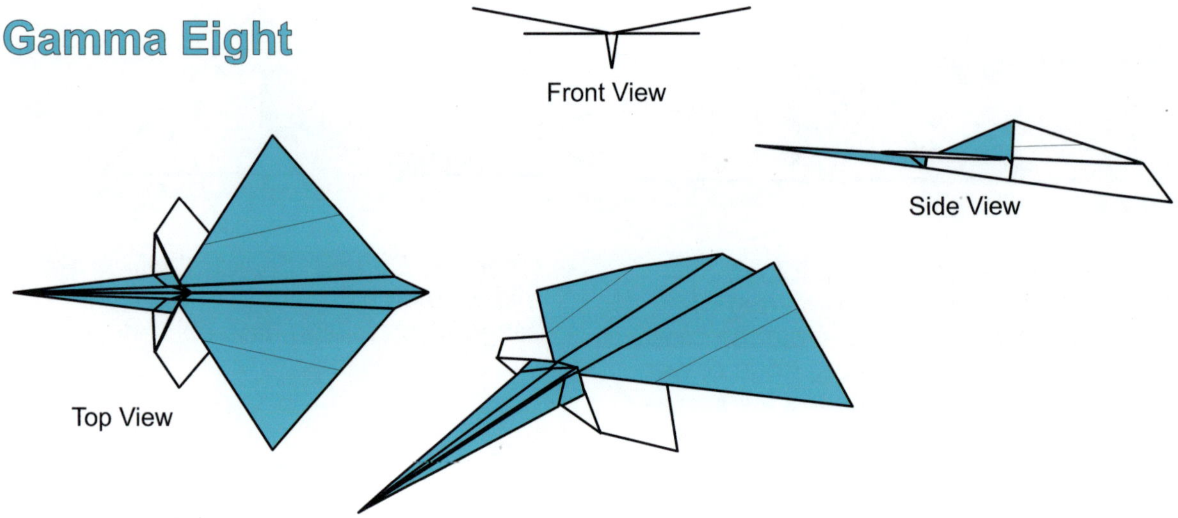

Zeta Eight

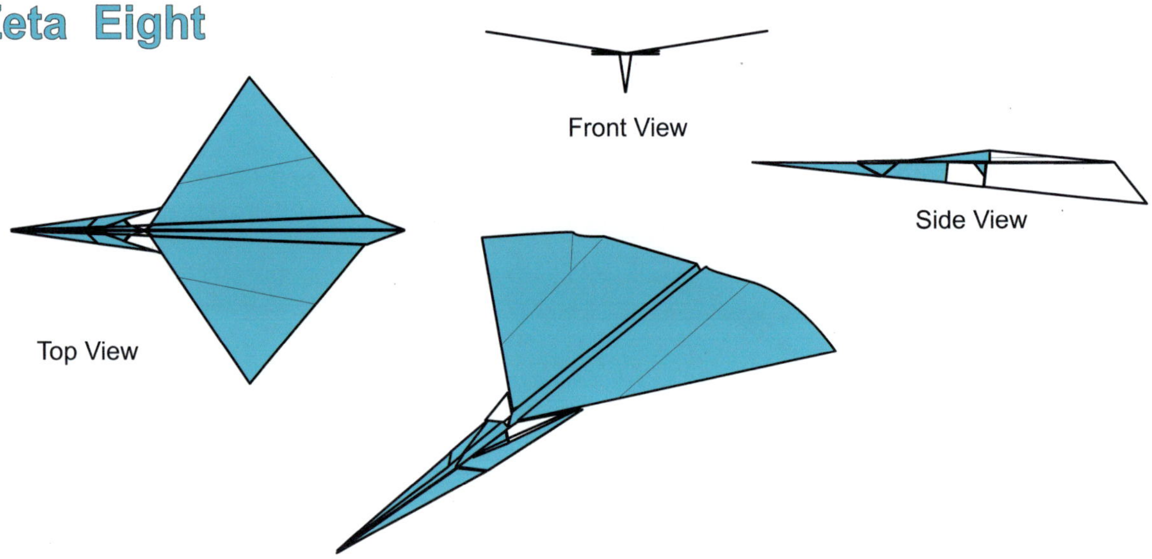

Iota Eight

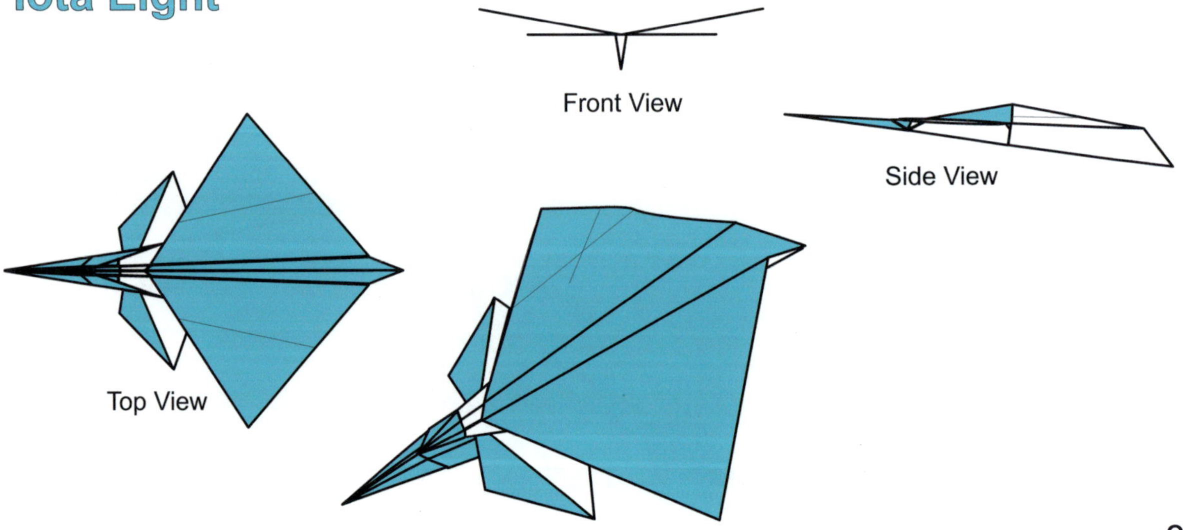

Back Piece Nine Airplanes

Alpha Nine

Beta Nine

Gamma Nine

Delta Nine

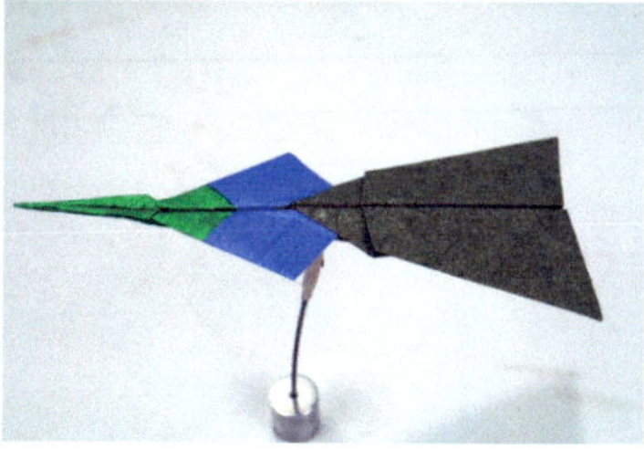

Epsilon Nine

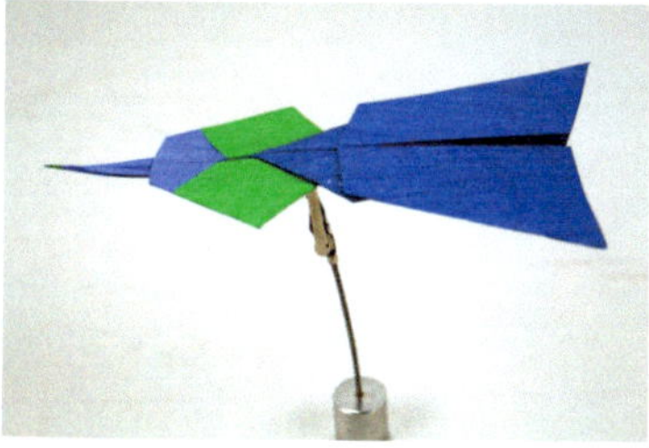

Iota Nine

Kappa Nine

Omicron Nine

Lambda Nine

Back Piece Nine

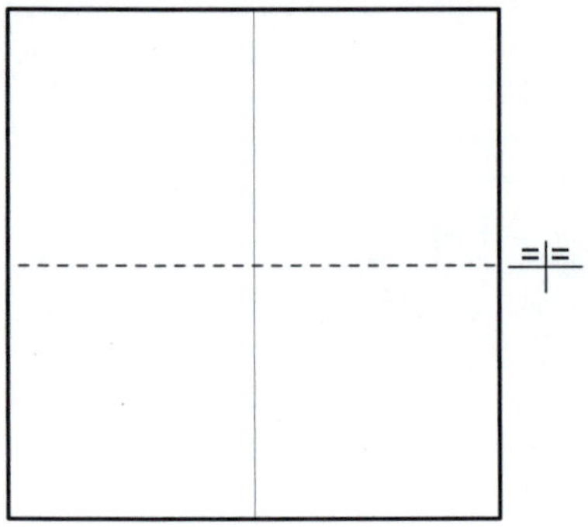

1. Valley fold in half and unfold.

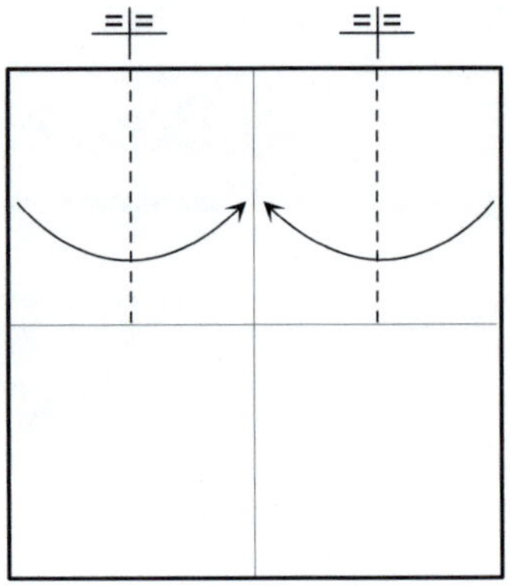

2. Valley fold each edge into the middle and unfold. The crease should only extend halfway.

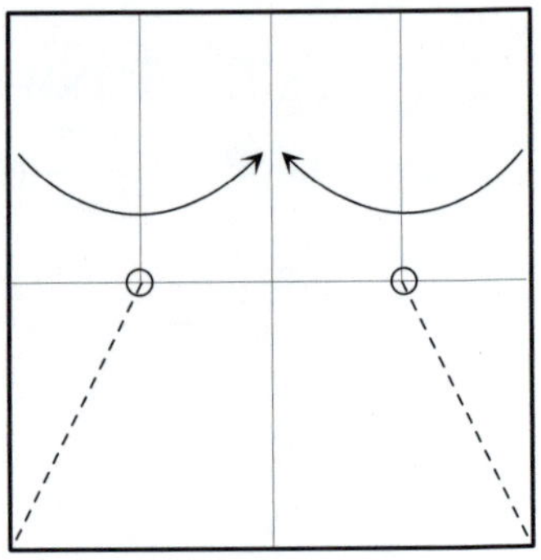

3. Valley fold from the intersection of the two creases and the adjacent corner. Unfold.

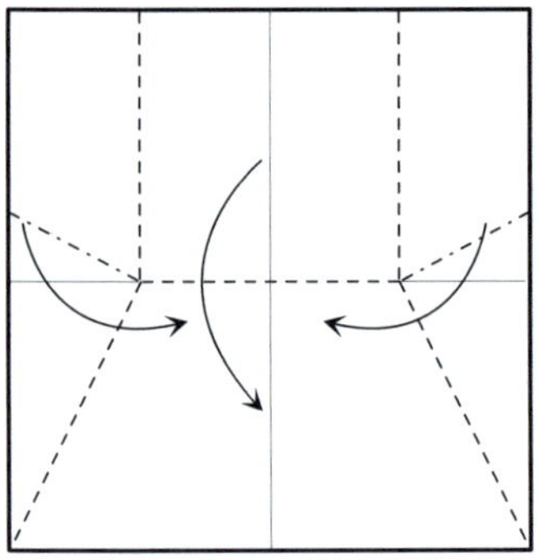

4. Use the creases made in the previous steps to collapse. Extra mountain folds will be made between layers.

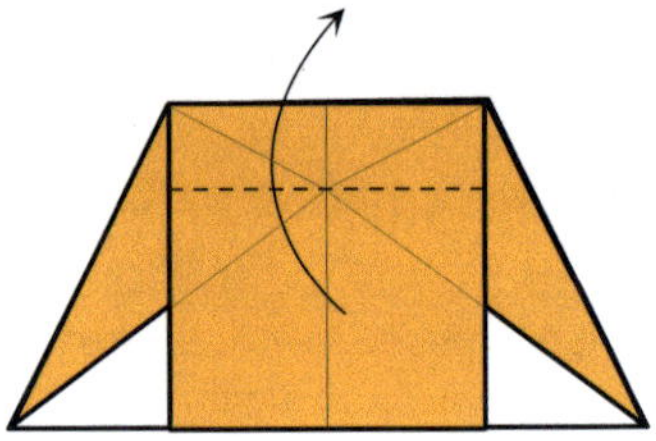

5. Valley fold the top layer up as far as it will go.

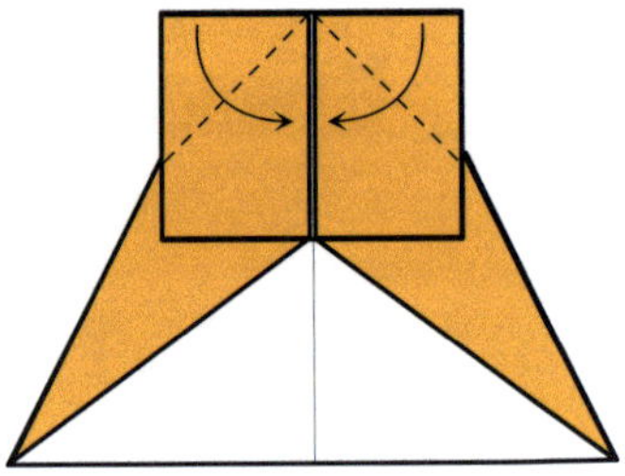

6. Valley fold the top corners into the middle and unfold.

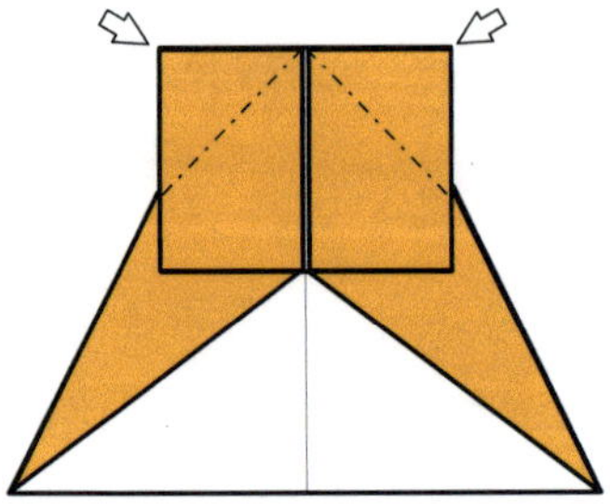

7. Reverse fold the corners inside.

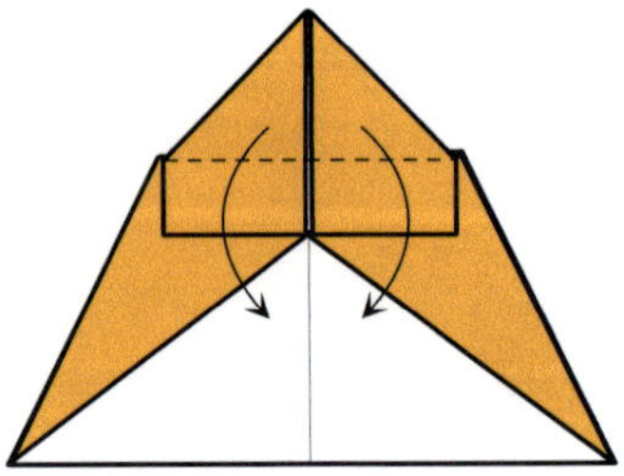

8. Valley fold the corner flaps downwards.

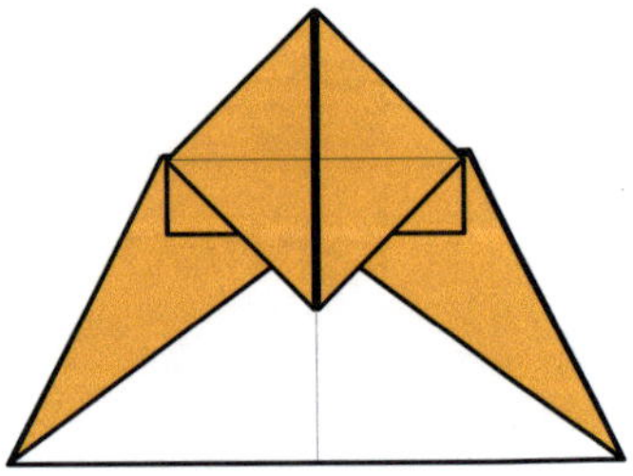

9. Time for assembly.

Back Piece Nine Page 2

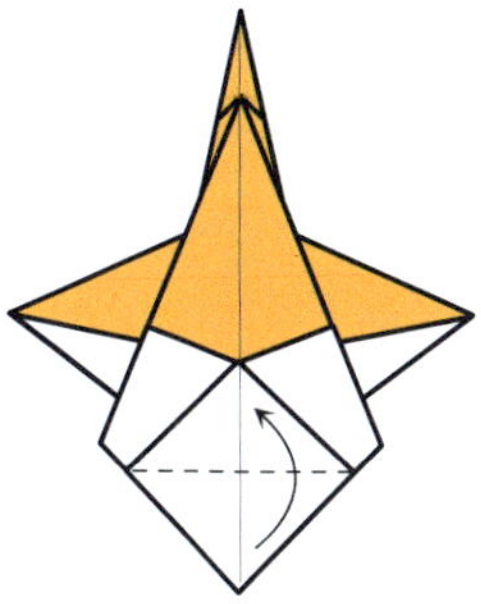

10. Valley fold the join diagonally on Front Piece Kappa in half and unfold.

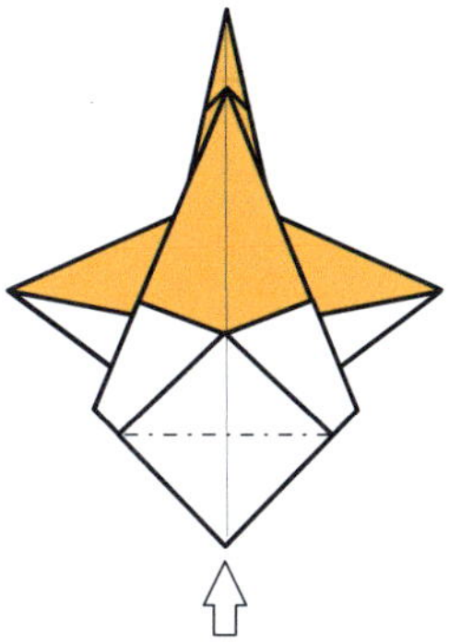

11. Open sink fold.

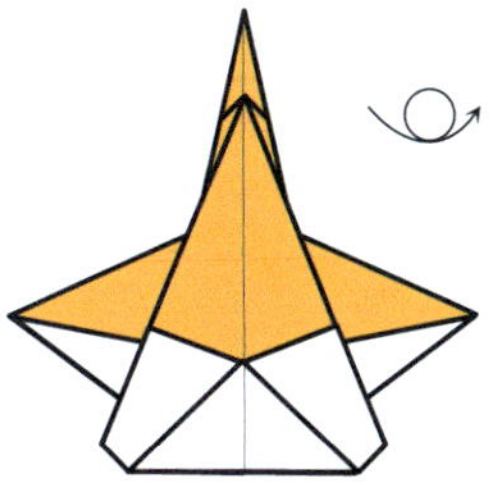

12. Turn over.

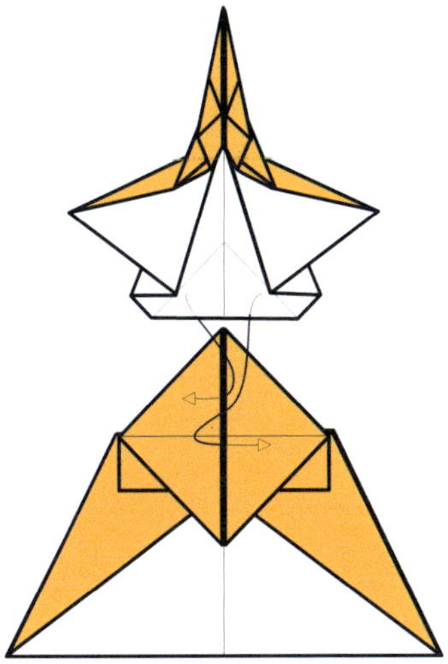

13. Insert the flaps of the join of the Front Piece into the pockets of the back piece.

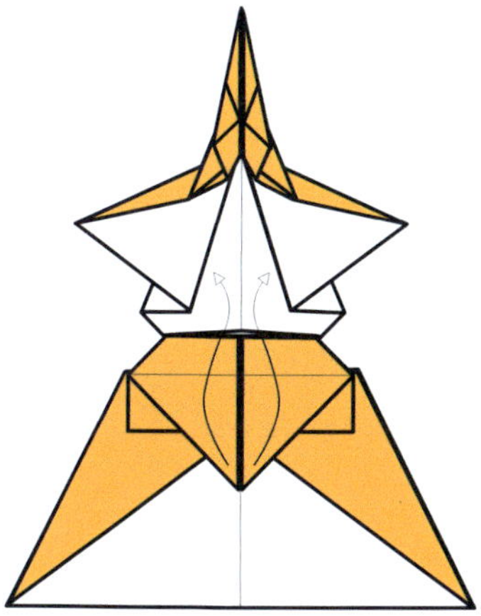

14. Insert the flaps of the back piece into the pocket in the front piece.

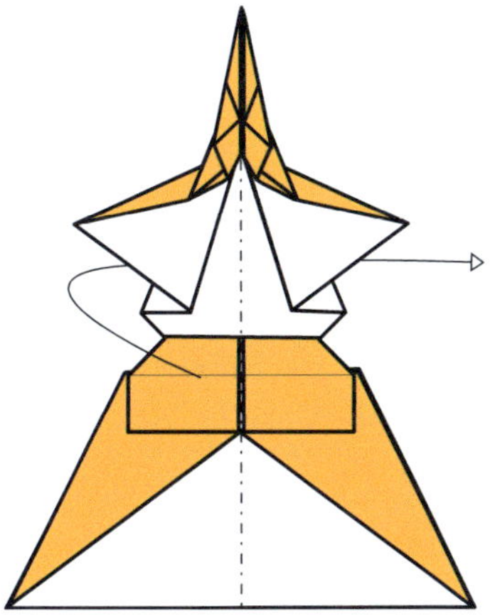

15. Mountain fold in half.

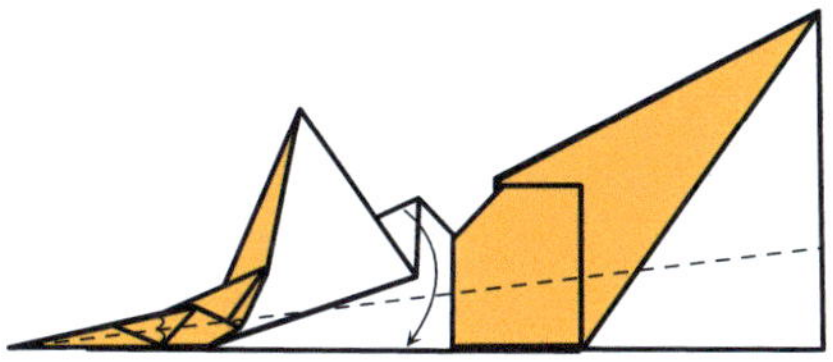

16. To fold the wings, bisect the angle of the nose. You should wind up splitting Front Piece Kappa right down the middle. Continue the fold on the back piece.

Kappa Nine

Front View

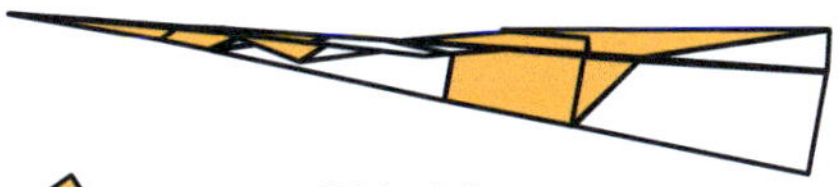

Side View

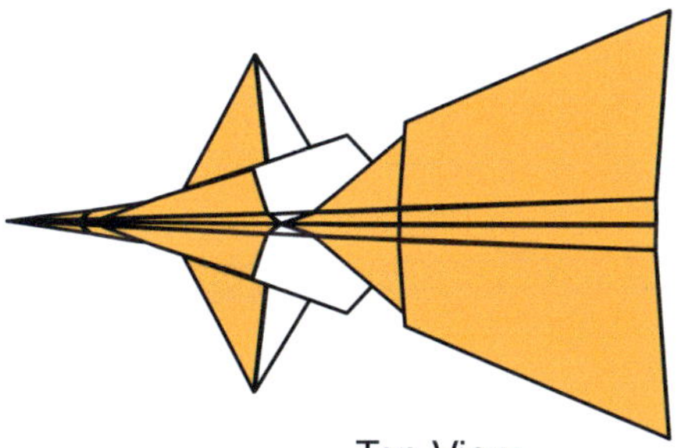

Top View

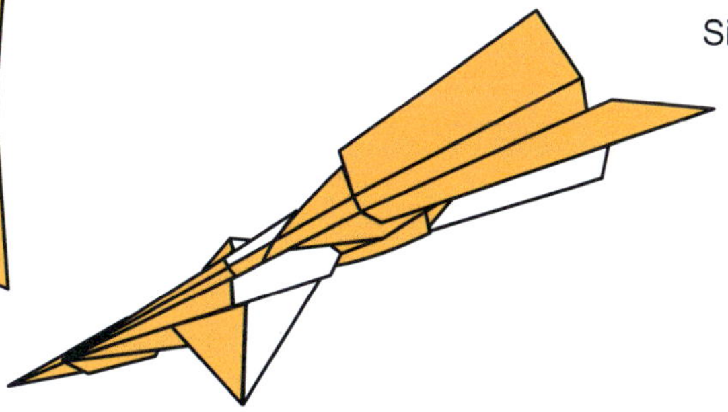

Alpha Nine

Front View

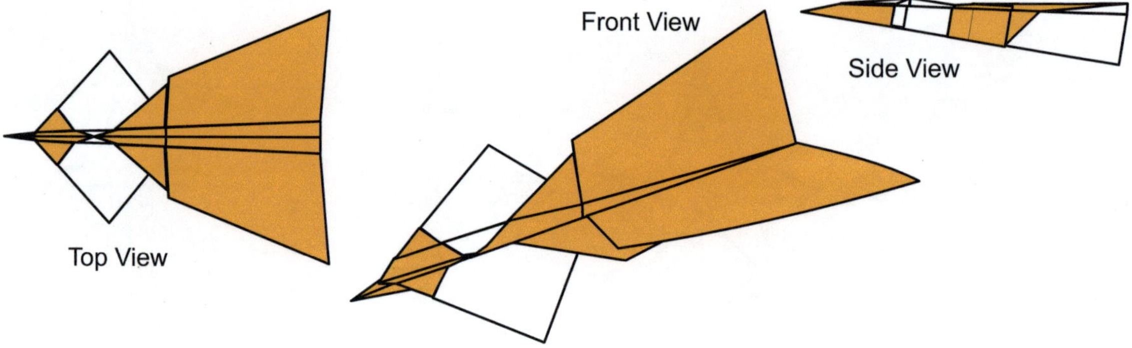

Beta Nine

Front View

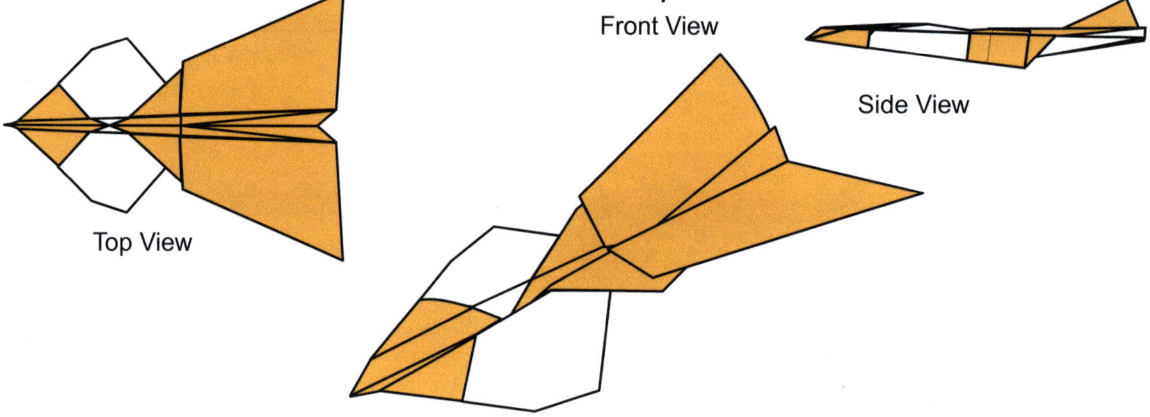

Gamma Nine

Front View

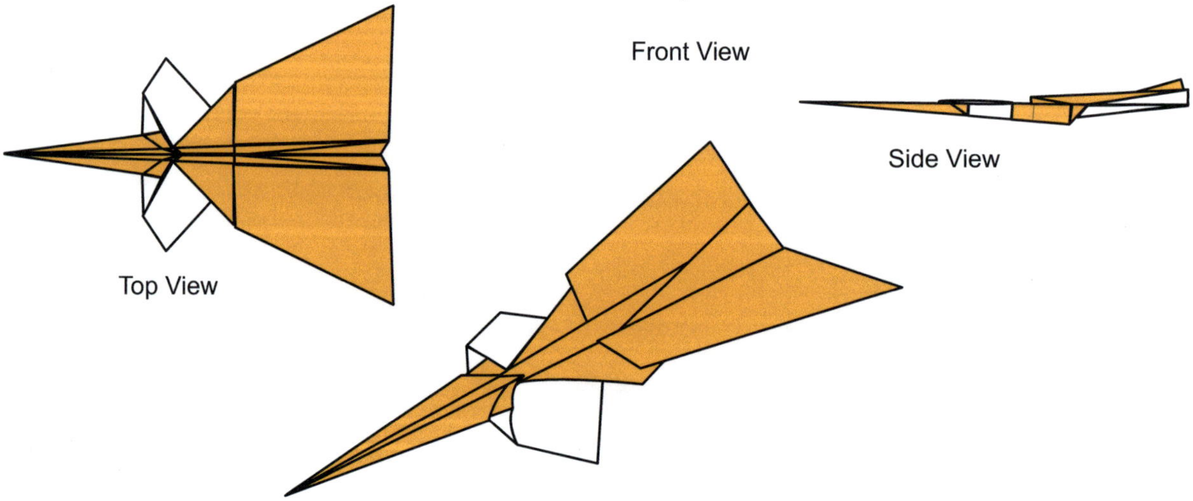

Delta Nine

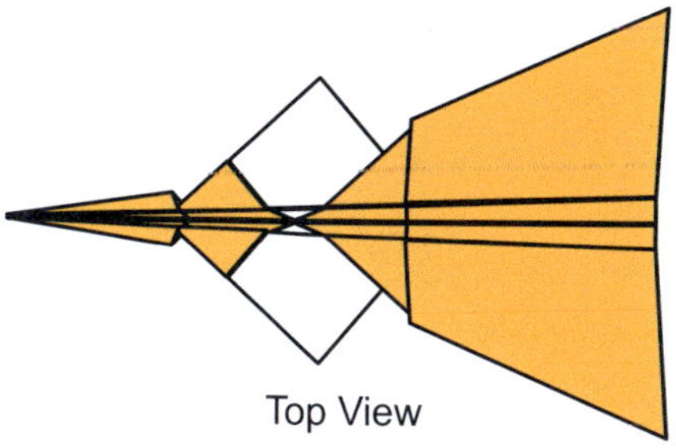

Top View

Front View

Side View

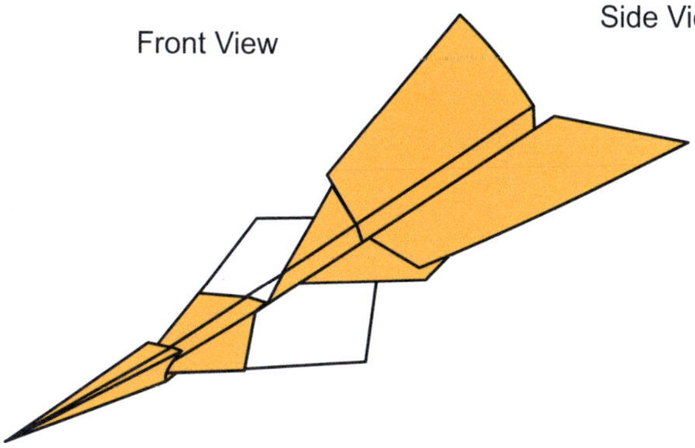

Epsilon Nine

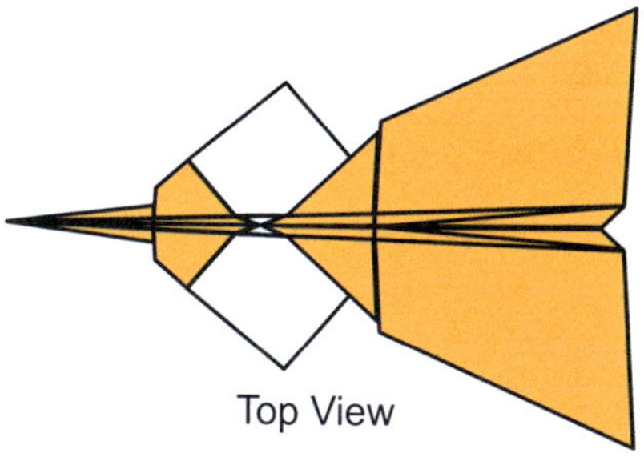

Top View

Front View

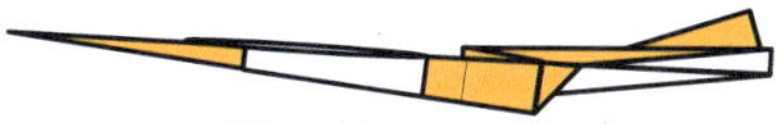

Side View

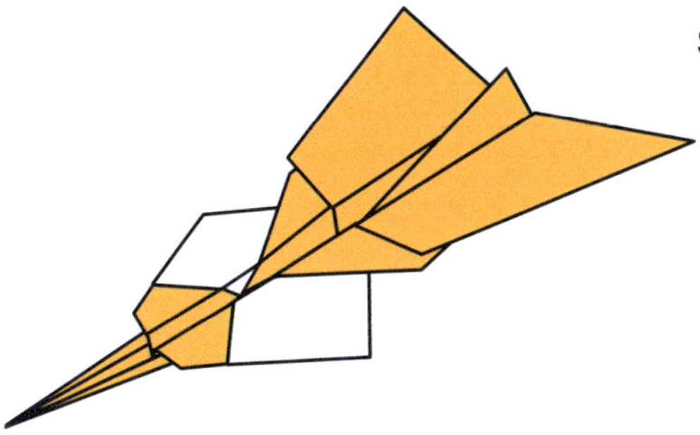

Iota Nine

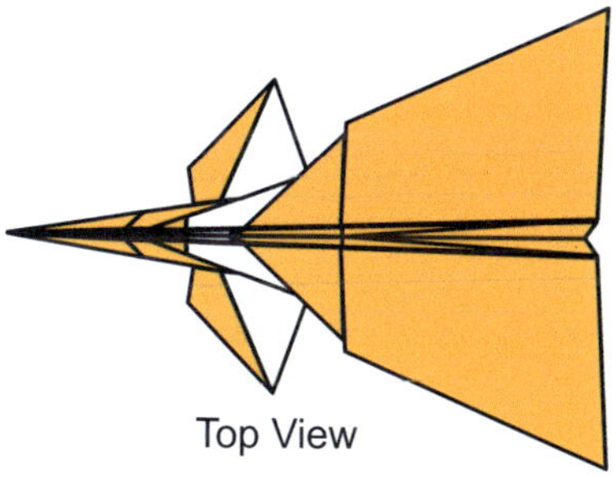

Top View

Front View

Side View

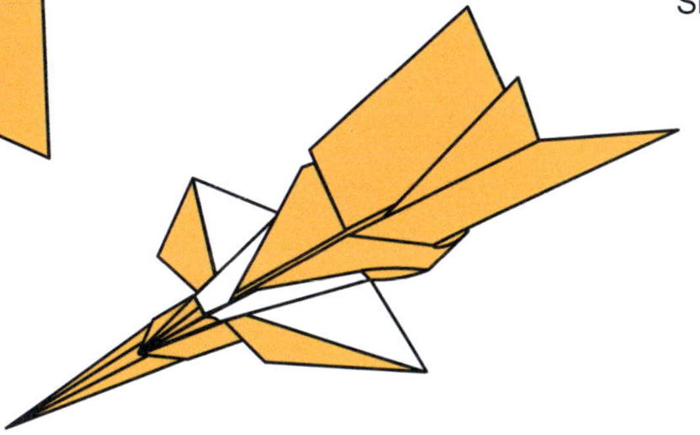

Omicron Nine

Front View

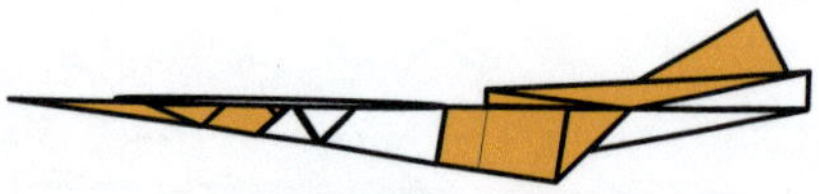

Side View

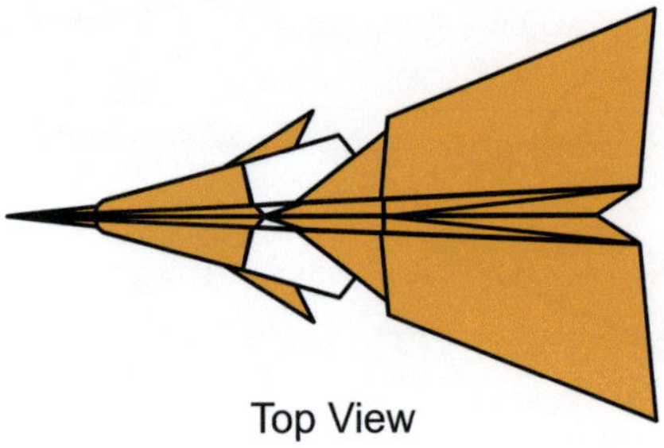

Top View

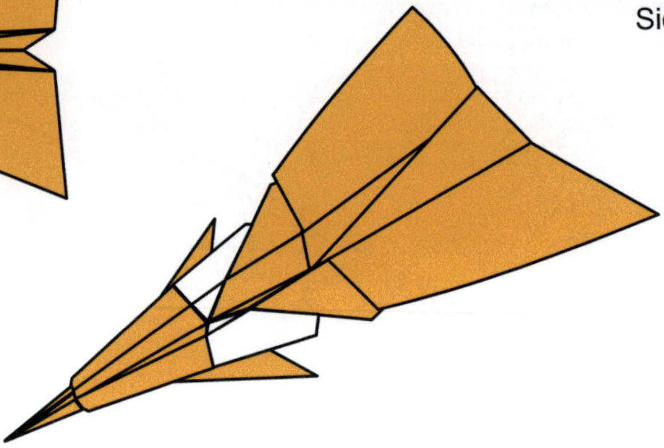

Lambda Nine

Front View

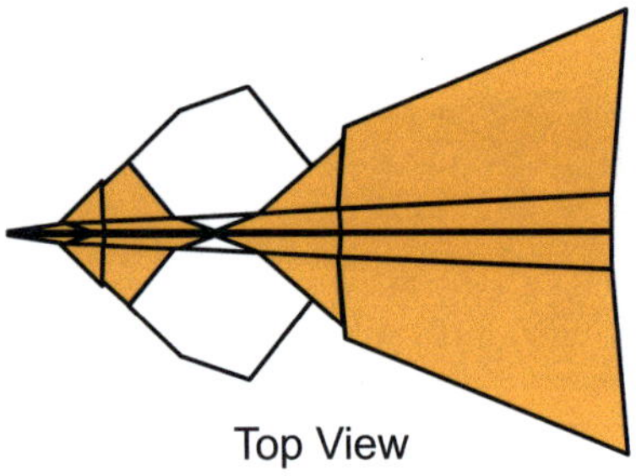

Top View

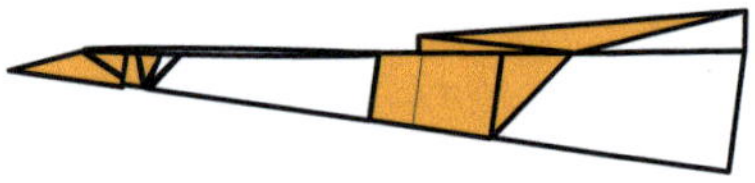

Side View

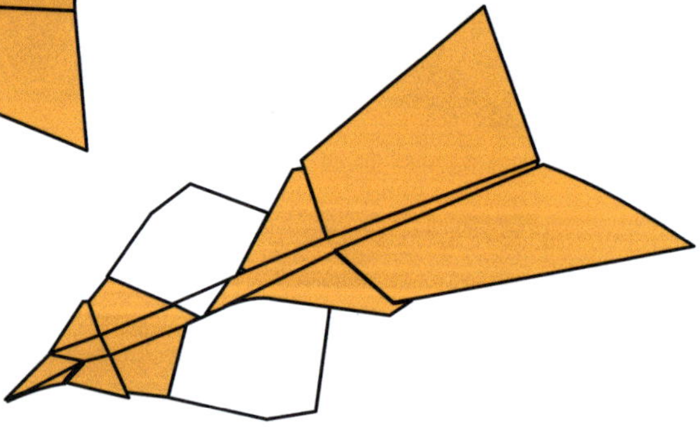

Back Piece Ten Airplanes

Alpha Ten

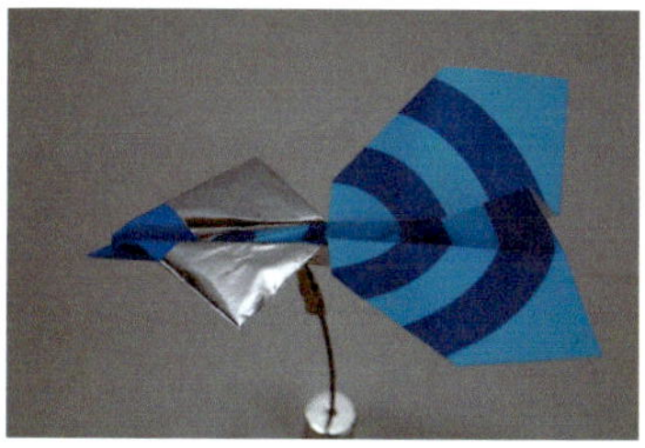

Beta Ten

Gamma Ten

Delta Ten

Epsilon Ten

Iota Ten

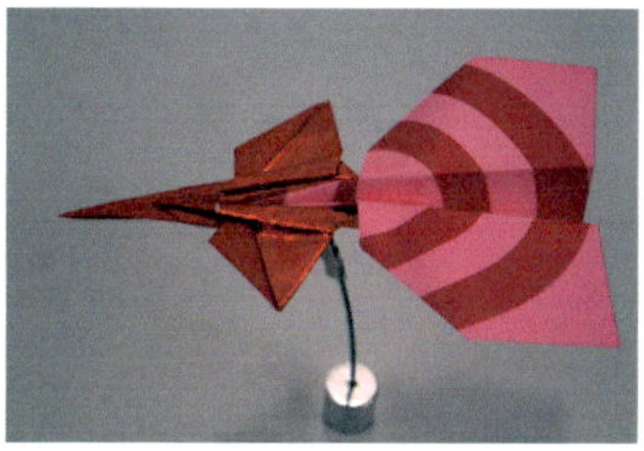

Kappa Ten

Omicron Ten

Lambda Ten

Back Piece Ten

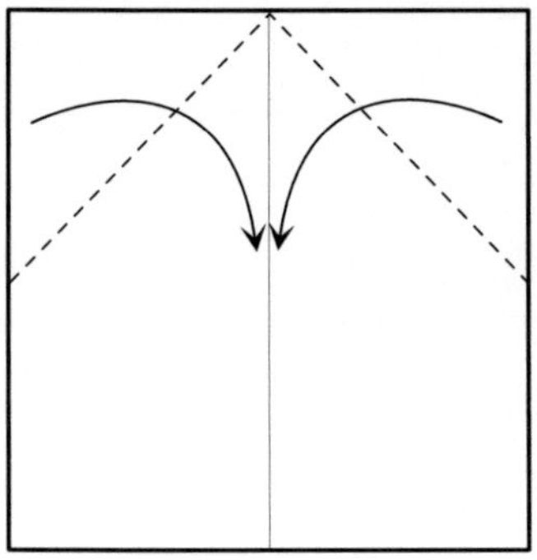

1. Valley fold the two corners into the middle.

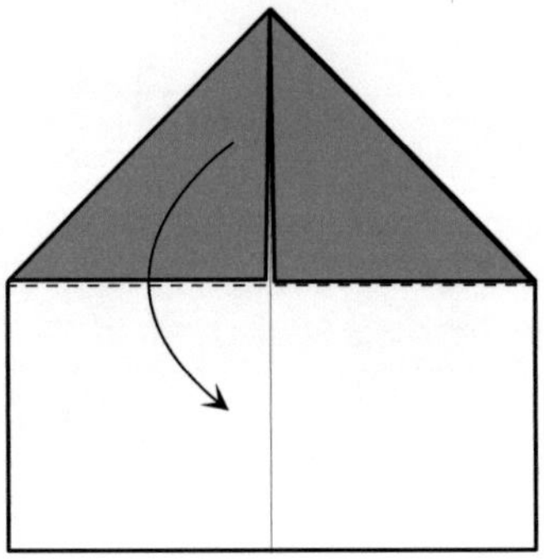

2. Valley fold the top down along the color change.

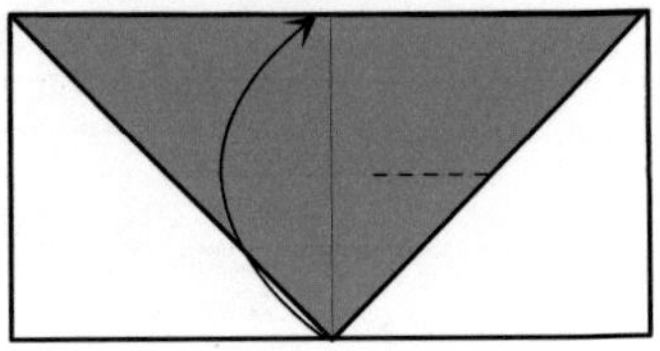

3. Valley fold the tip to the top middle. Make a short partial crease and unfold.

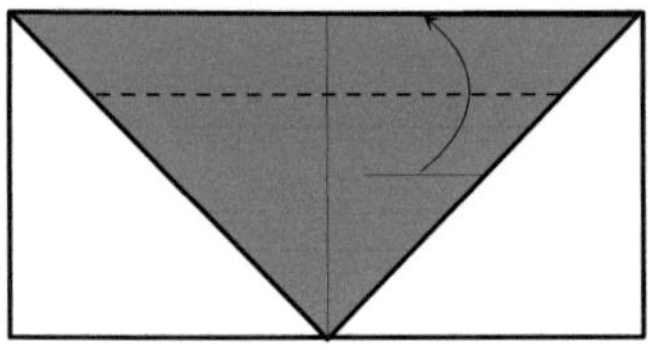

4. Valley fold so that the crease made in step 3 touches the top folded edge.

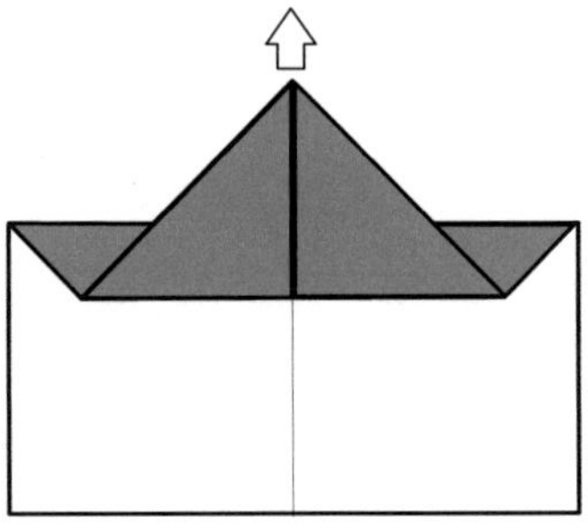

5. Unfold.

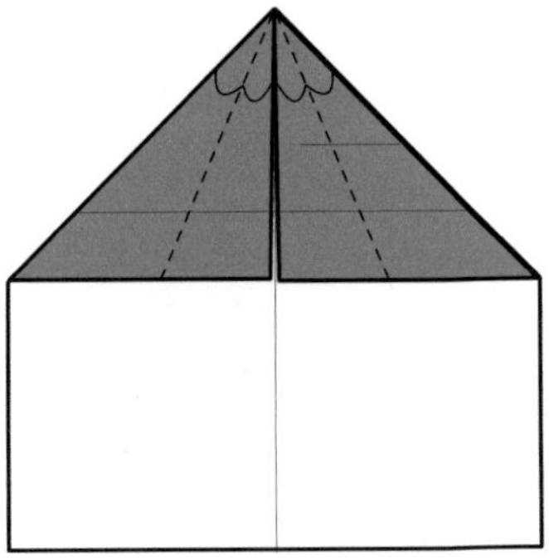

6. Crease angle bisectors along the top only.

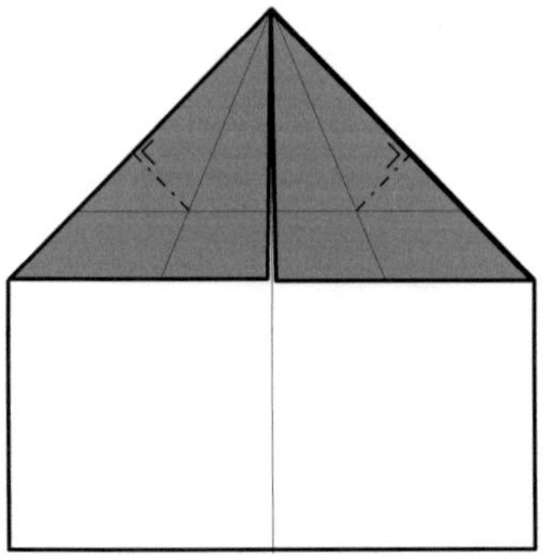

7. Precrease a pair of mountain folds. These will hit the intersection shown and be perpendicular to the edge.

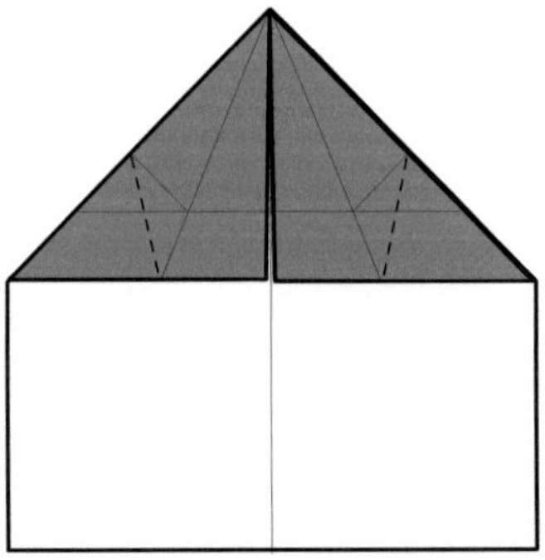

8. Precrease two valley folds between the intersections of creases and edges.

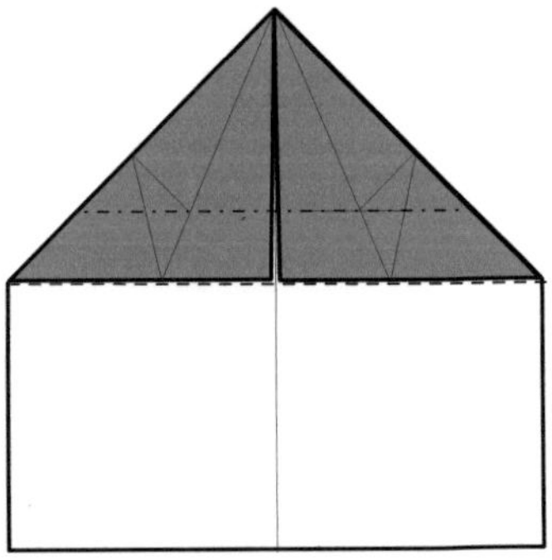

9. Refold Step 4.

Back Piece Ten Page 2

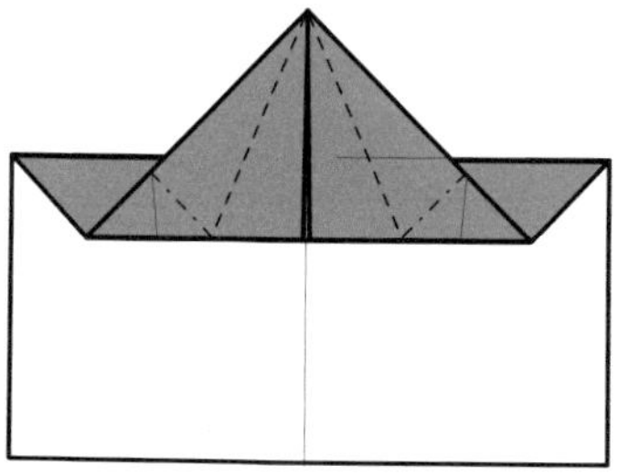

10. Time for the hardest fold in this book. Start by refolding the diagonals and the adjacent mountain folds. The mountain folds will lie at the base of the colored layer.

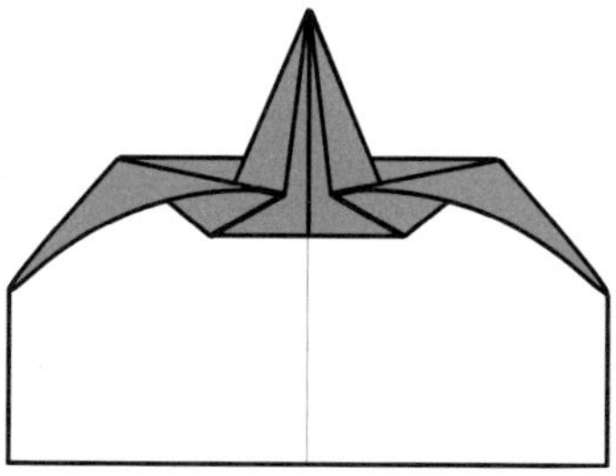

11. Flatten, making new diagonal creases to form the front edges of the wings.

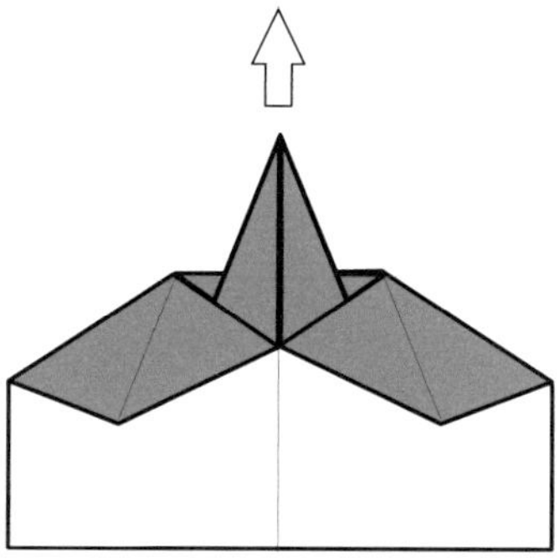

12. Back Piece Ten is almost done, and time for assembly. Unfold to step 2.

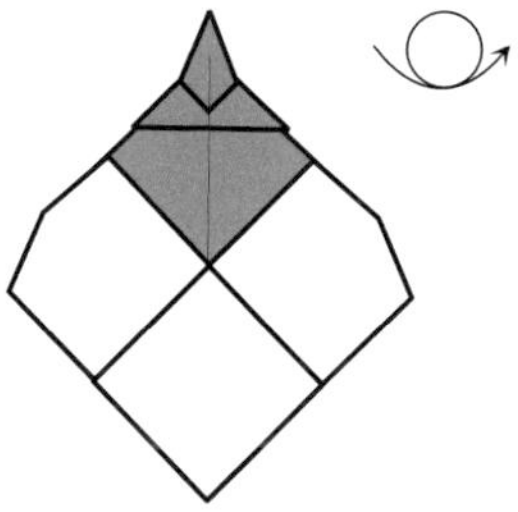

13. Turn Front Piece Lambda to the top side.

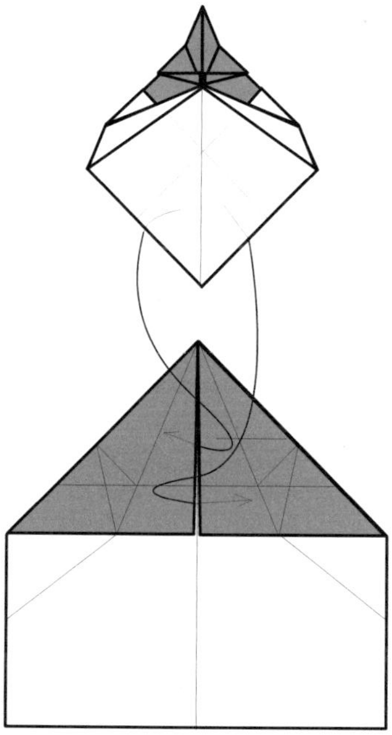

14. The flaps of the join go inside the flaps of the back piece.

Back Piece Ten Page 3

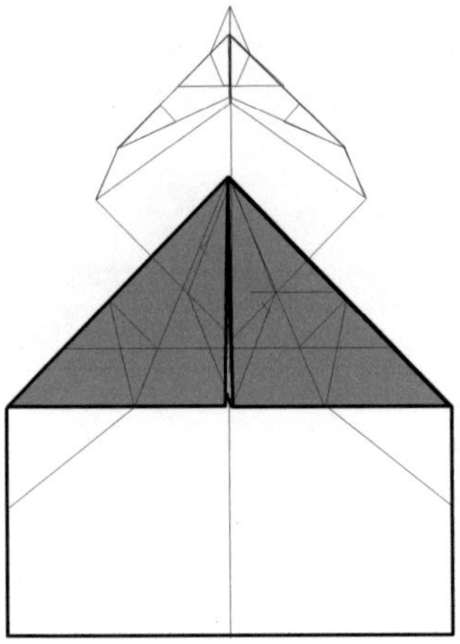

15. Slide the Front Piece so the join is at the front of the Back Piece.

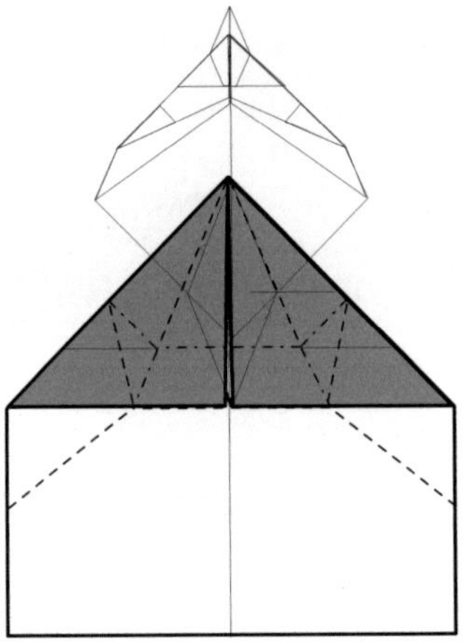

16. Refold the back piece around the join of the front piece.

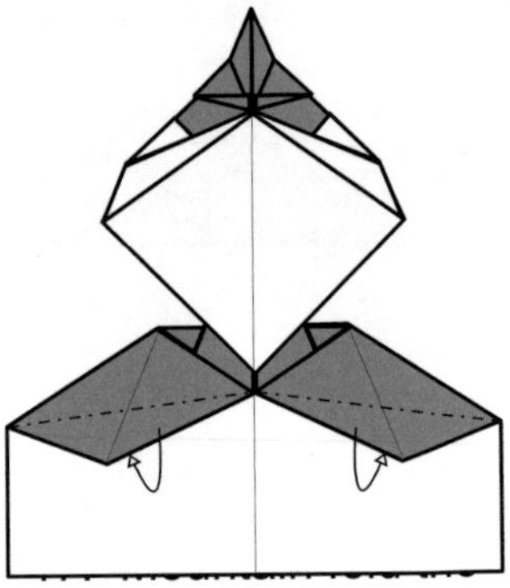

edges behind to move weight forward.

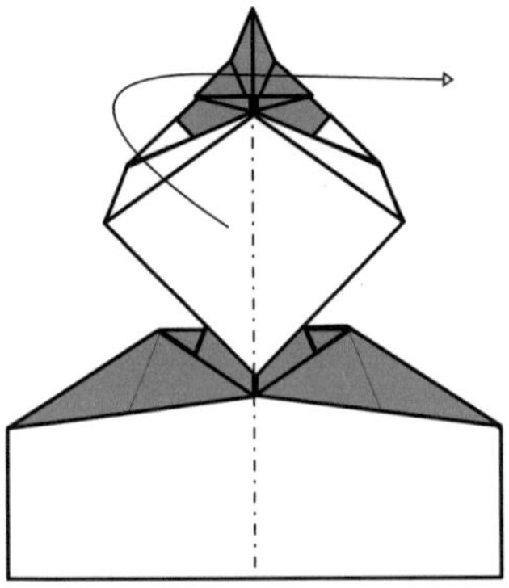

18. Mountain fold in half.

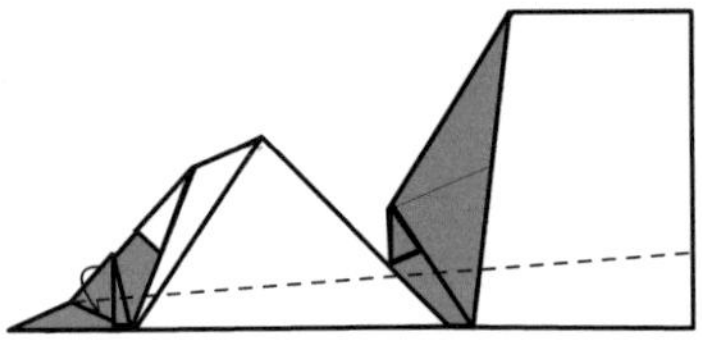

19. For Front Piece Lambda the front edge is lined up with a folded edge at the front. Fold the back piece along the same angle.

Lambda Ten

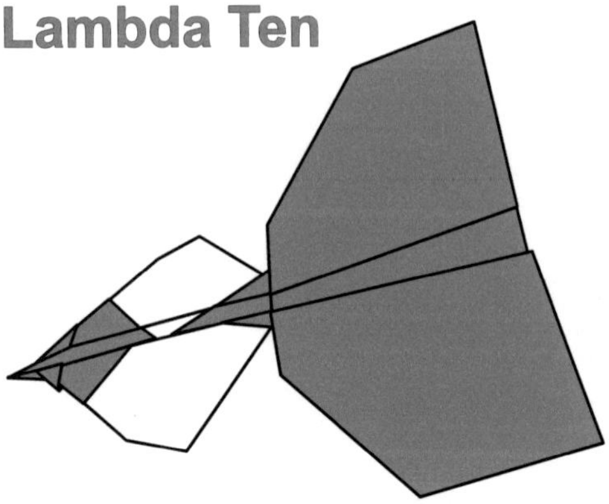

Front View

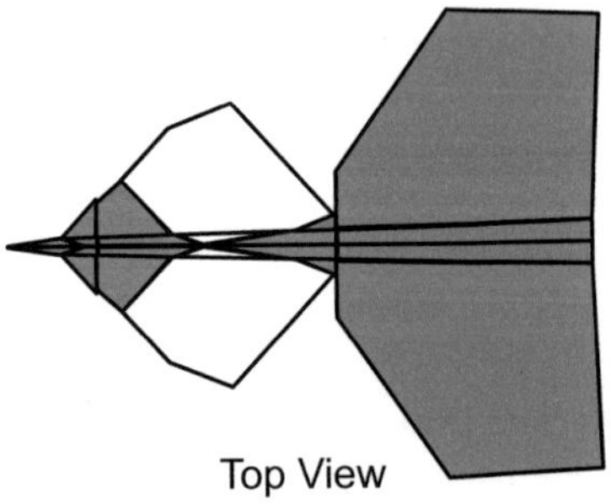

Top View

Side View

Alpha Ten

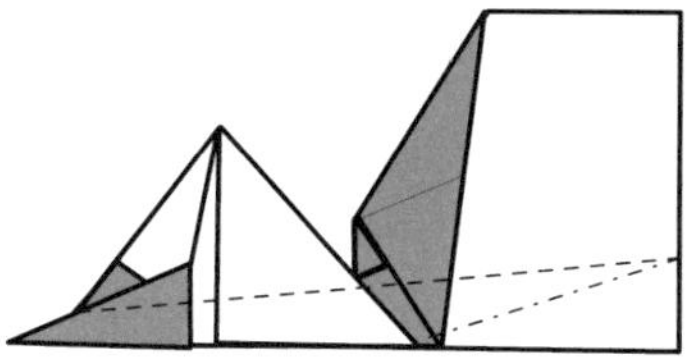

Front View

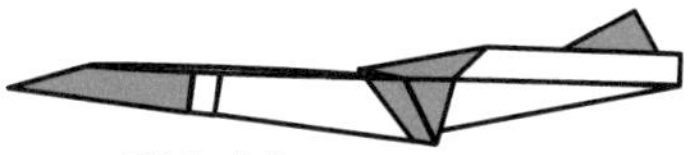

Side View

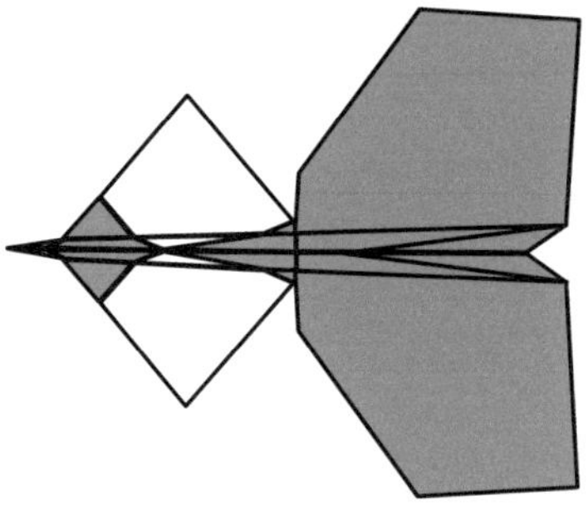

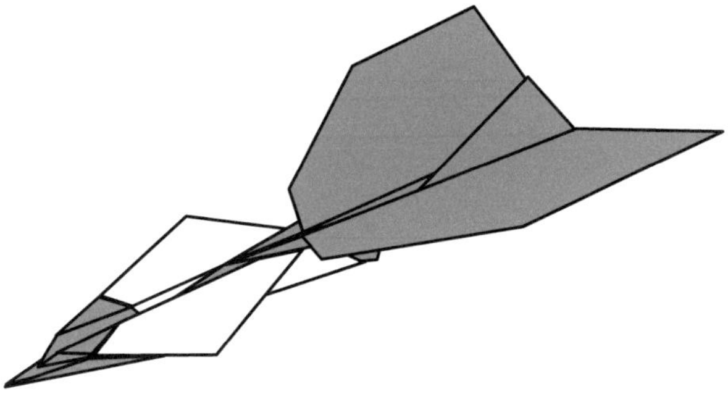

Top View

Beta Ten

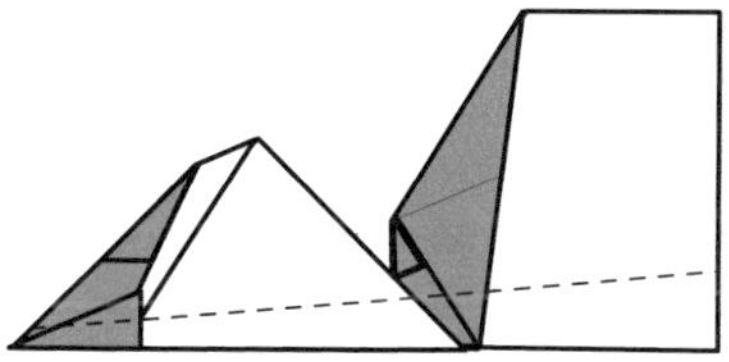

Front View

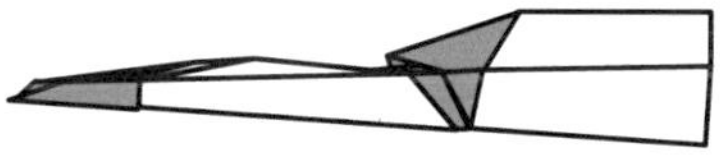

Side View

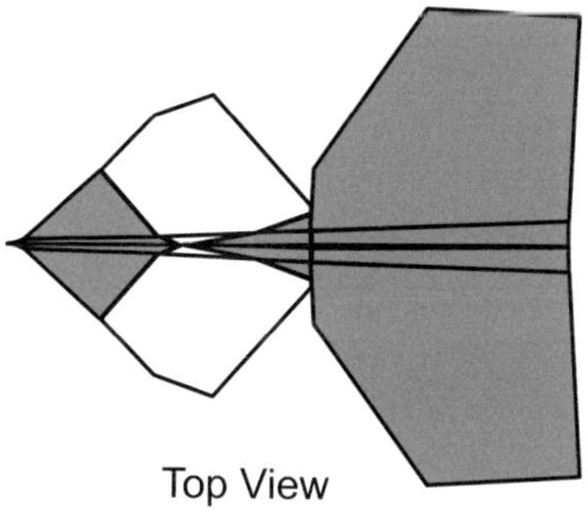

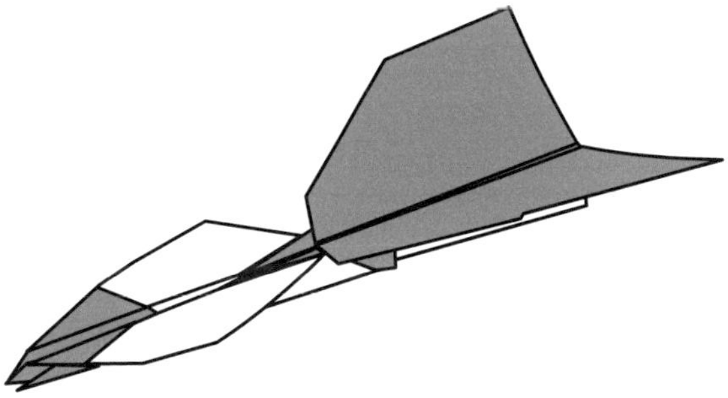

Top View

Gamma Ten

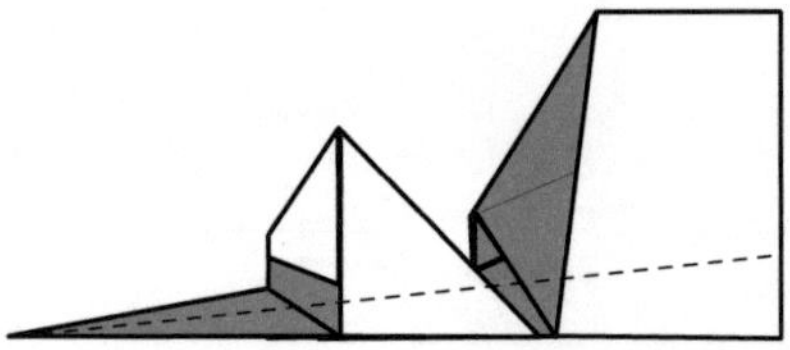

Front View

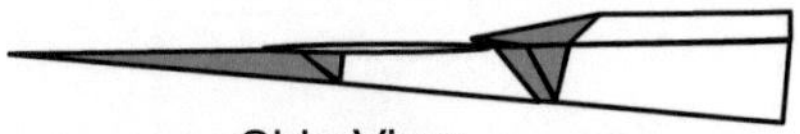

Side View

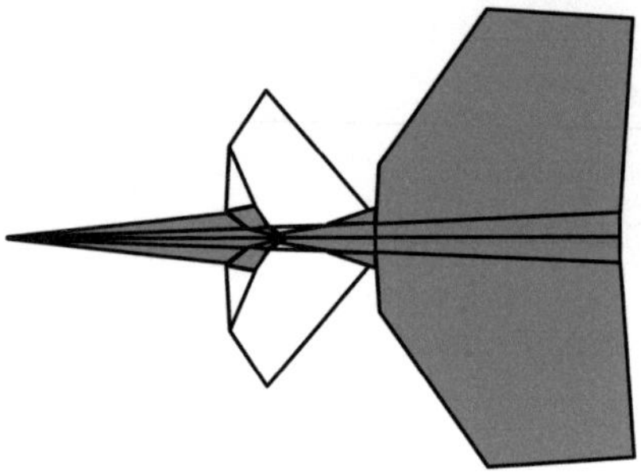

Top View

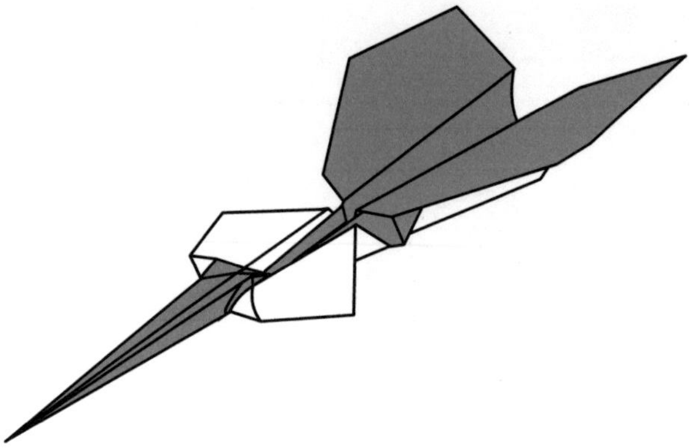

Delta Ten

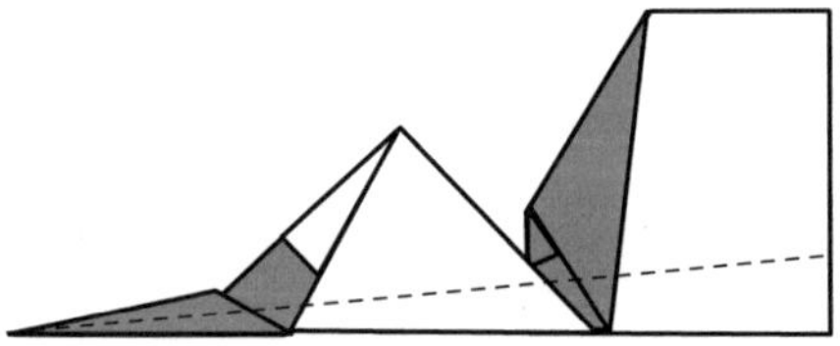

Front View

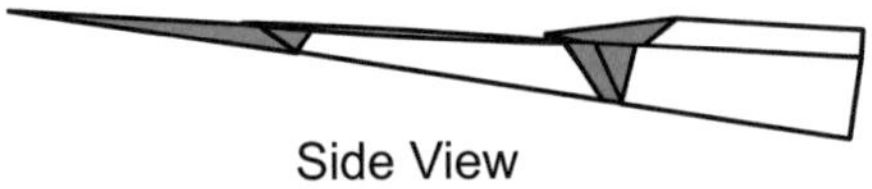

Side View

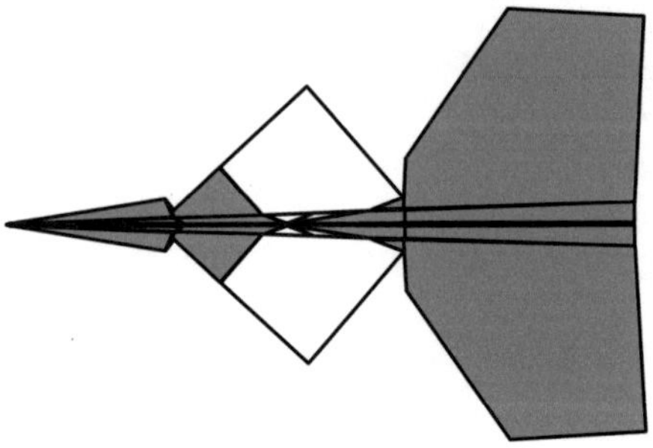

Top View

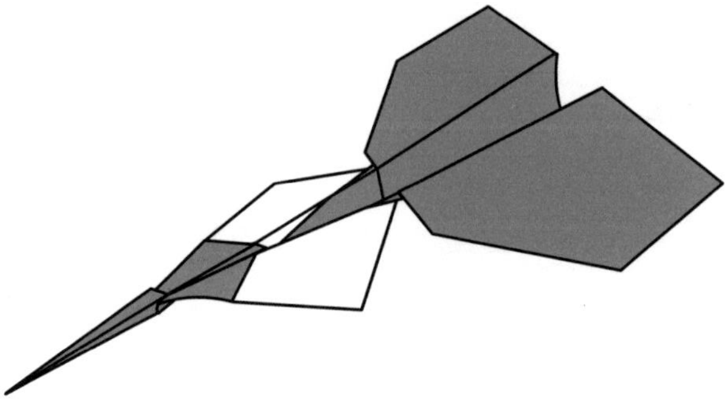

Epsilon Ten

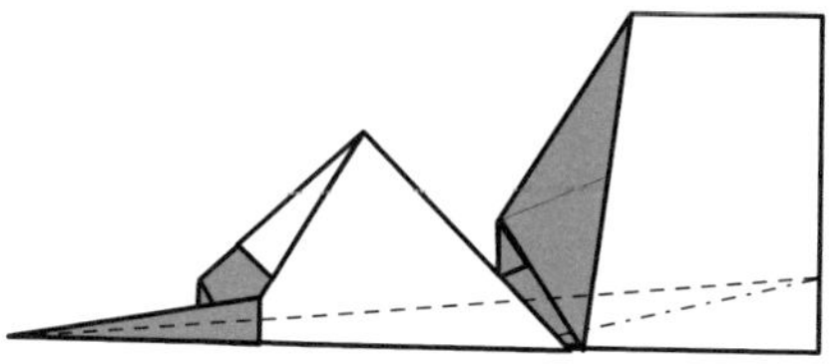

Front View

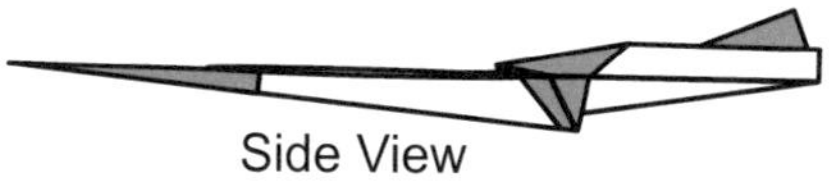

Side View

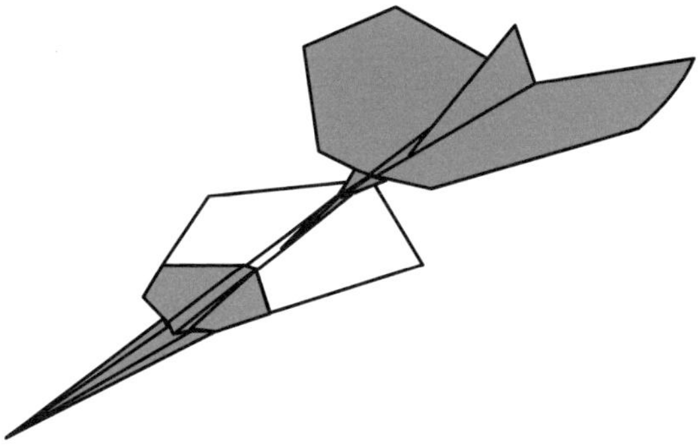

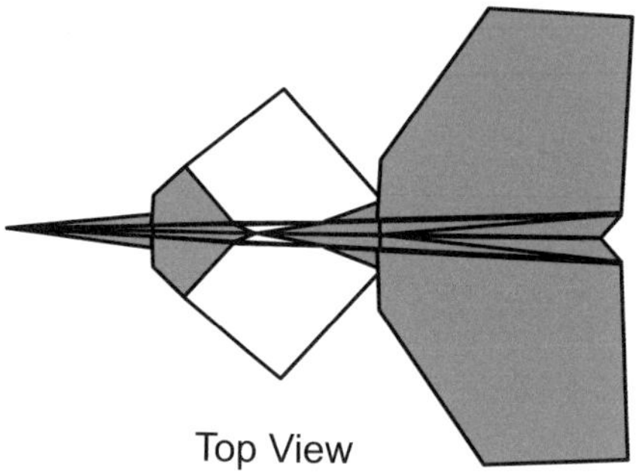

Top View

IotaTen

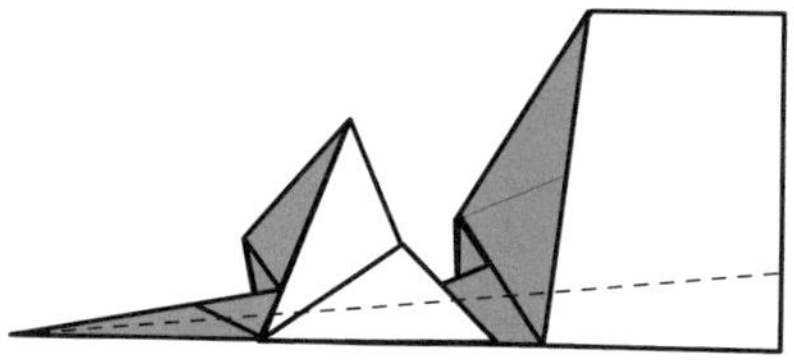

Front View

Side View

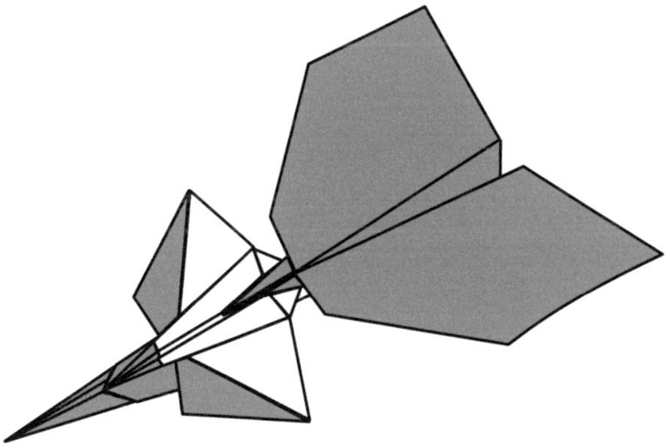

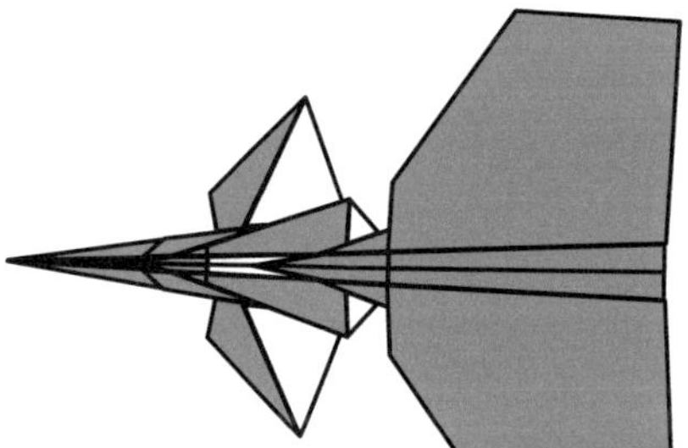

Top View

Kappa Ten

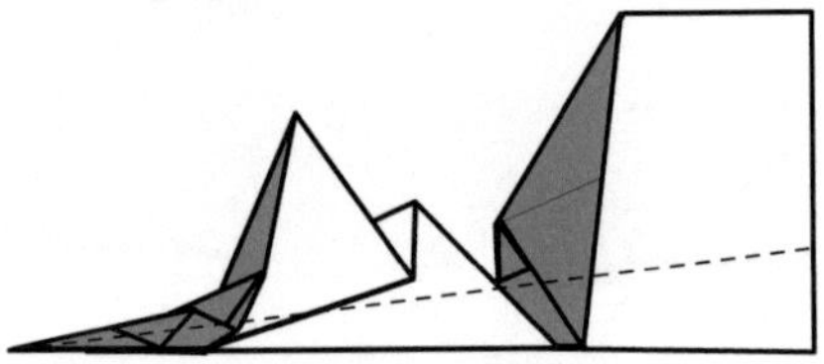

Front View

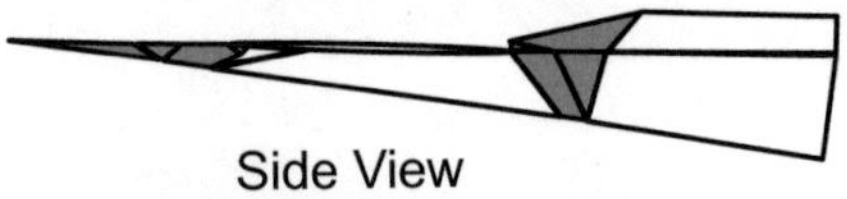

Side View

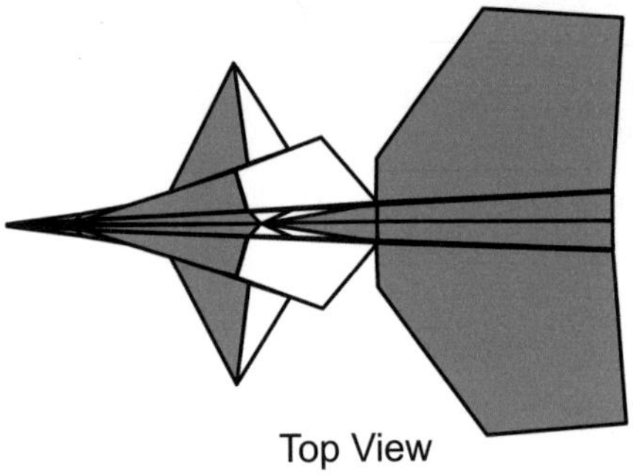

Top View

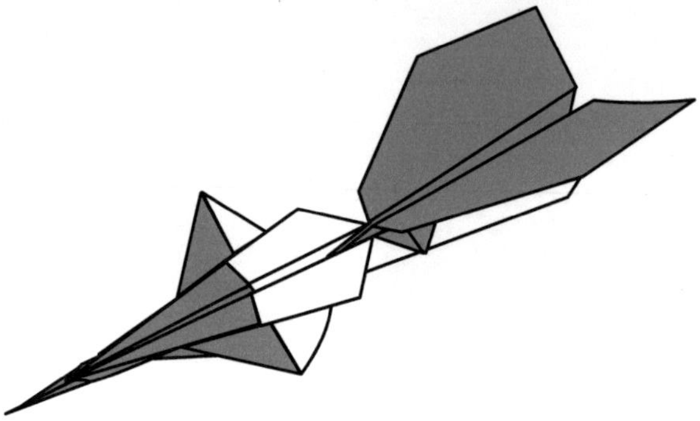

Omicron Ten

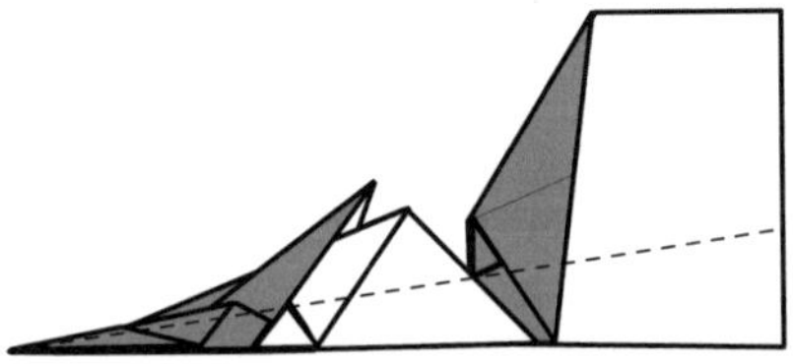

Front View

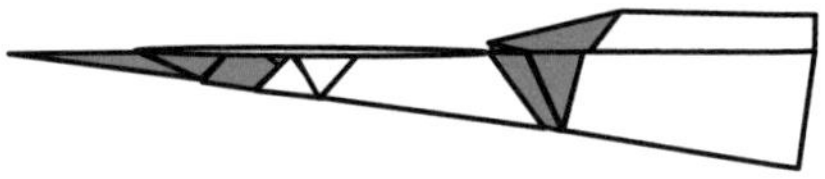

Side View

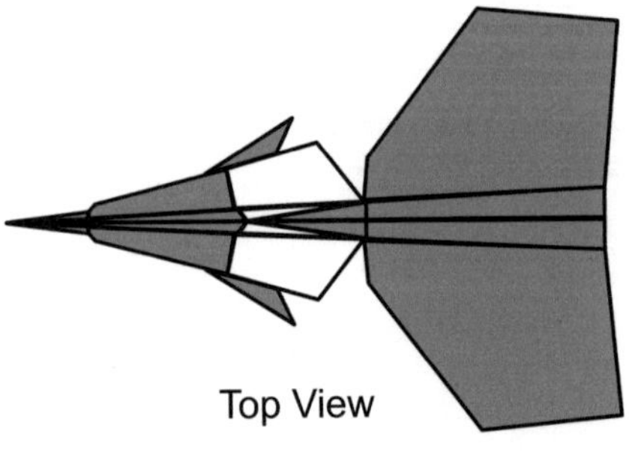

Top View

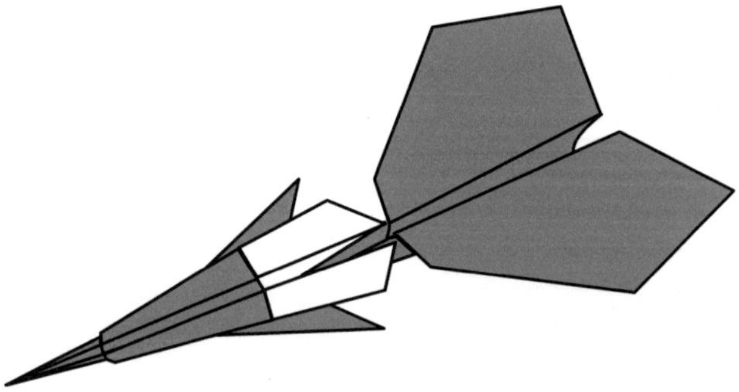

Three Piece Airplanes

The following aircraft are all constructed from three pieces, and all fly. There are 10x10x10 possible combinations of the different front and back pieces (and remember, most Front Pieces make good Back Pieces. These are the combinations I've set up for you.

Zeta Kappa One

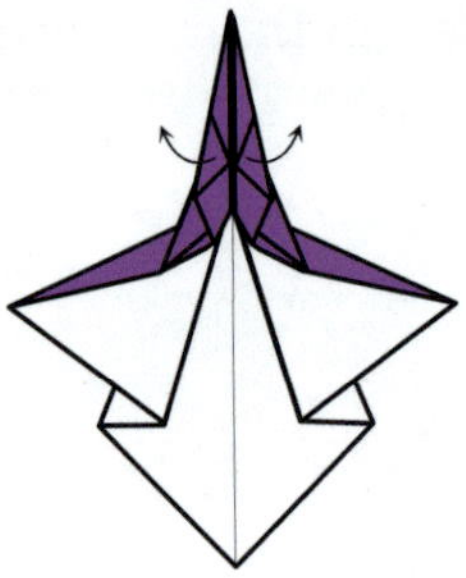

1. To insert into Front Piece Kappa, open up the front.

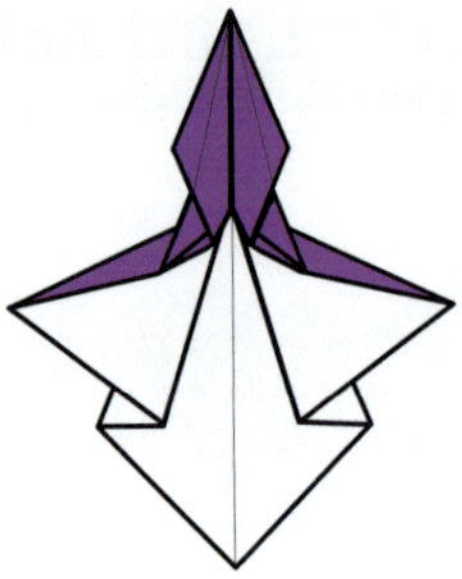

2. Like so.

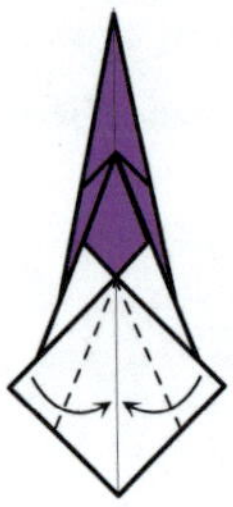

3. Valley fold the sides of the join of Front Piece Zeta into the middle.

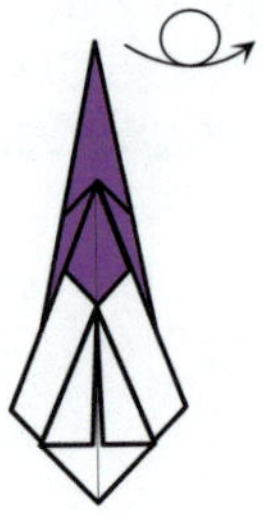

4. Turn over.

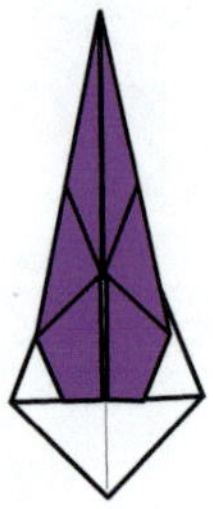

5. Like so.

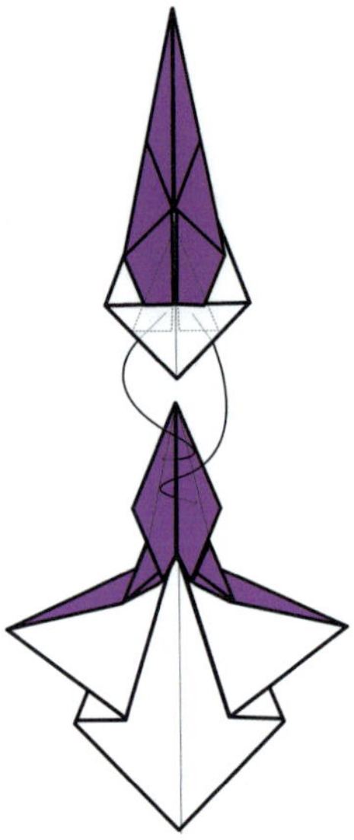

6. Insert the join of Front Piece Zeta inside the front flaps of Front Piece Kappa.

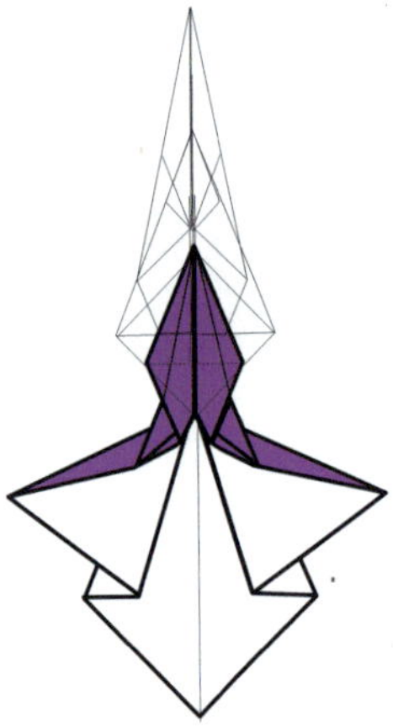

7. Once again narrow the front of Front Piece Kappa with valley folds.

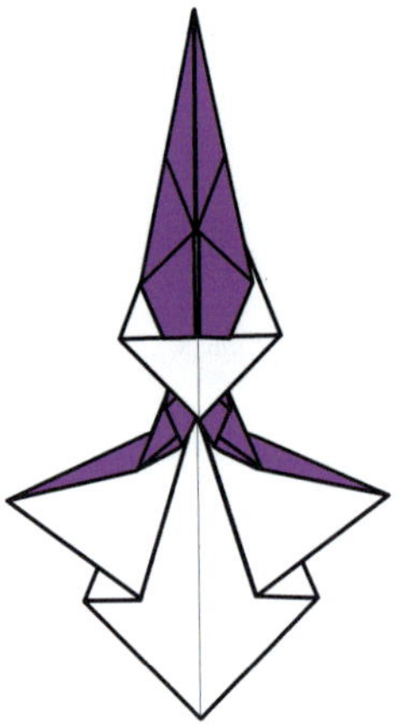

8. Insert Front Piece Kappa into Back Piece one.

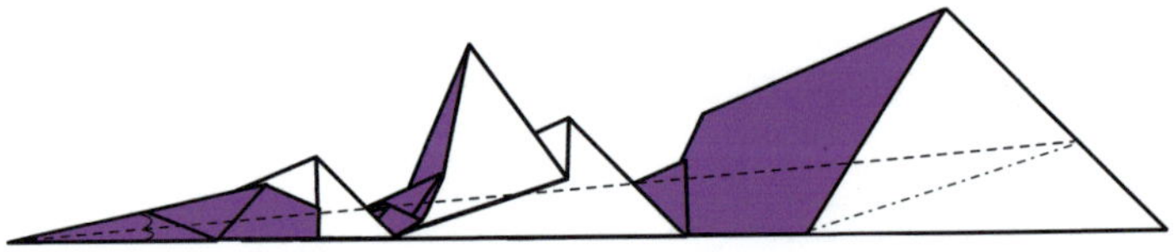

To fold the wings on this aircraft, bisect the angle in the front. Continue the fold on the middle piece. You'll need some sort of straight edge to keep the pitch angle on the back piece, since it isn't contiguous with the other parts of the aircraft.

Zeta Kappa One Page 2

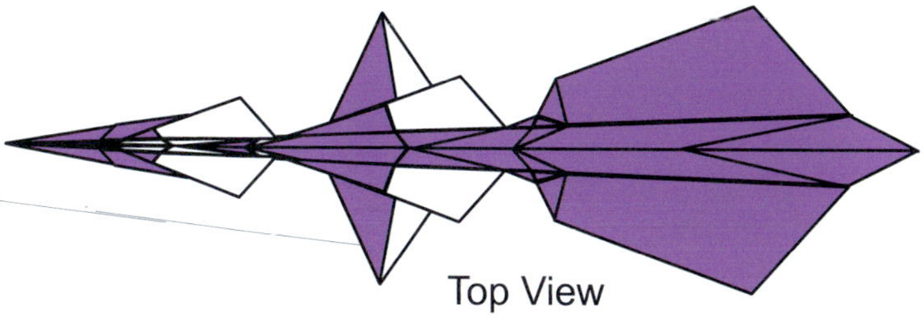

Top View

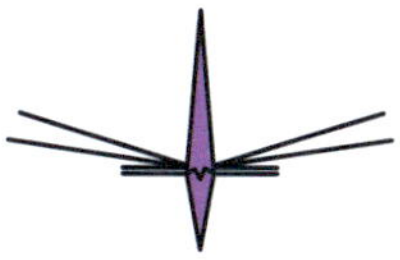

Front View

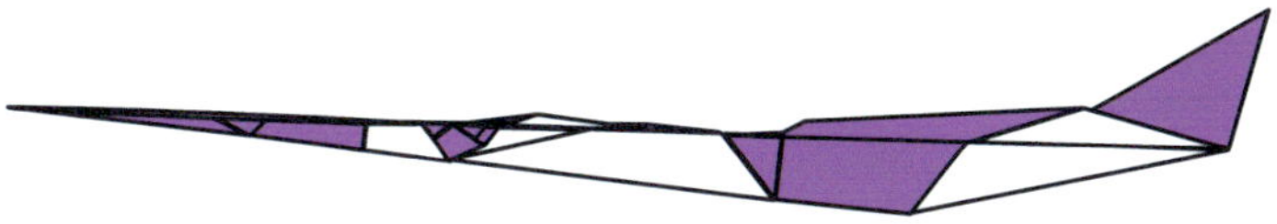

Side View

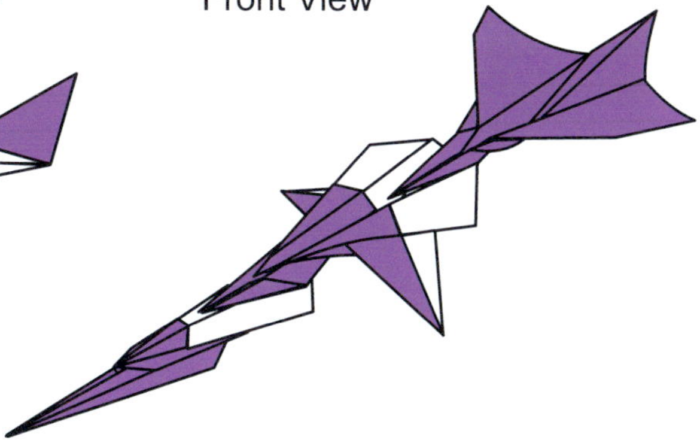

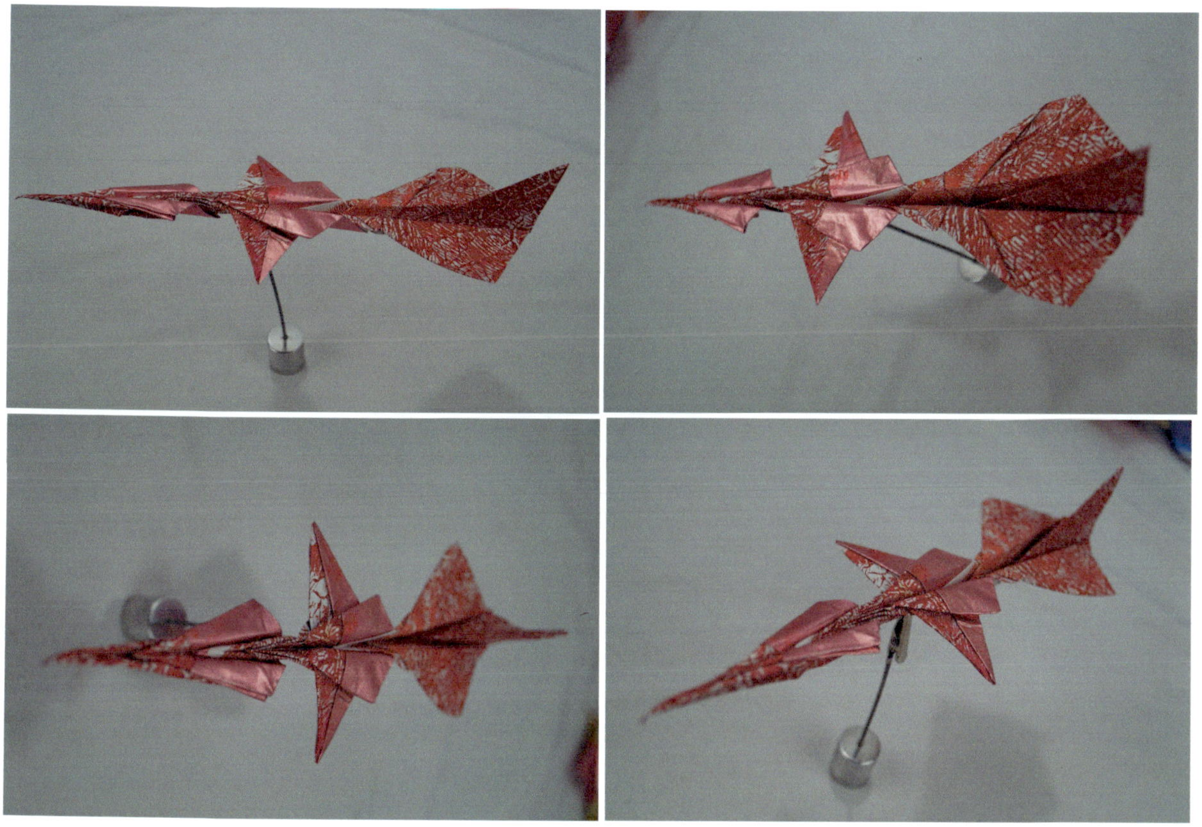

Delta Beta Two

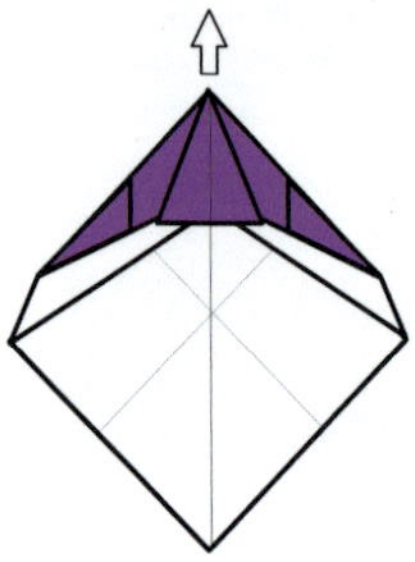

1. Unfold the front of Front Piece Beta.

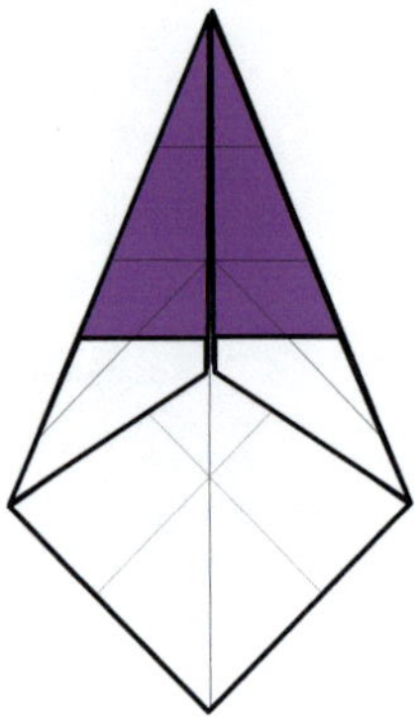

2. Like so.

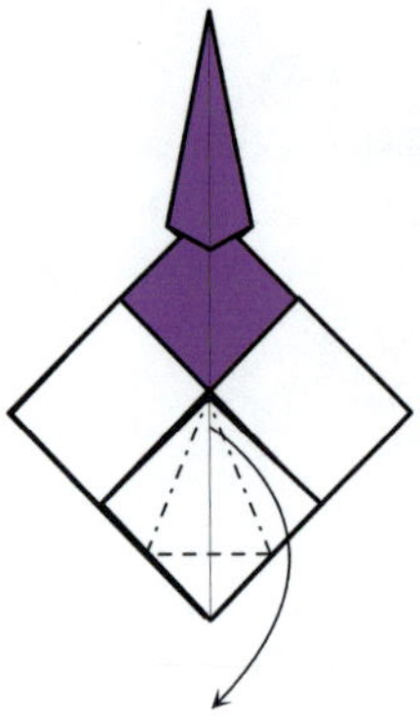

3. Petal fold the join on Front Piece Delta

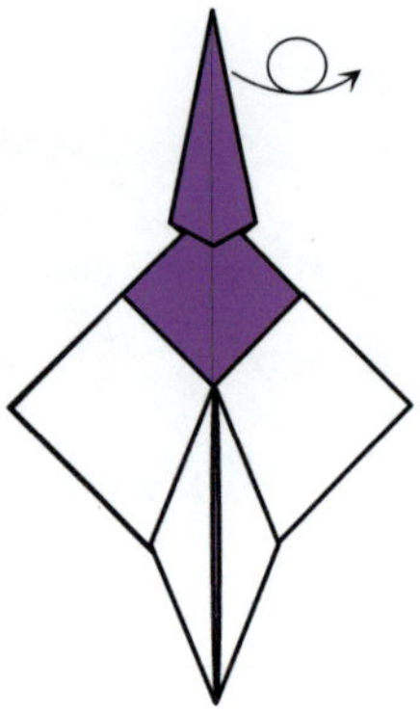

4. Turn over.

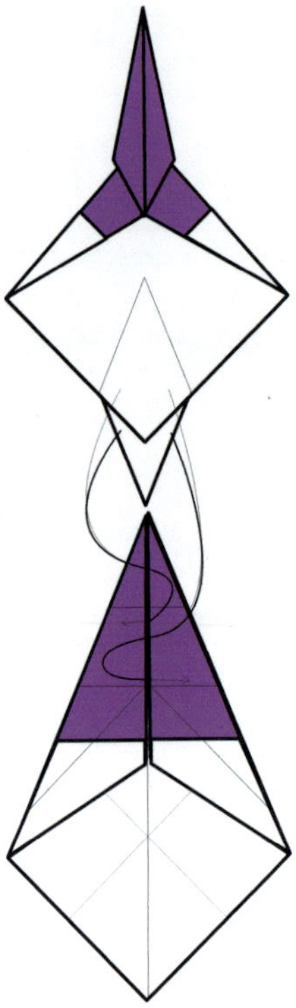

5. Insert the join of Front Piece Delta into the flaps of Front Piece Beta.

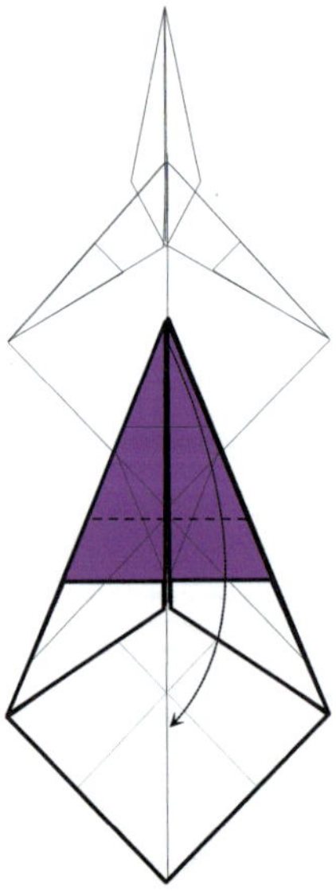

6. Begin refolding the front of Front Piece Beta with a valley fold along a preexisting crease.

7. Refold the diagonal flaps right over Front Piece Delta.

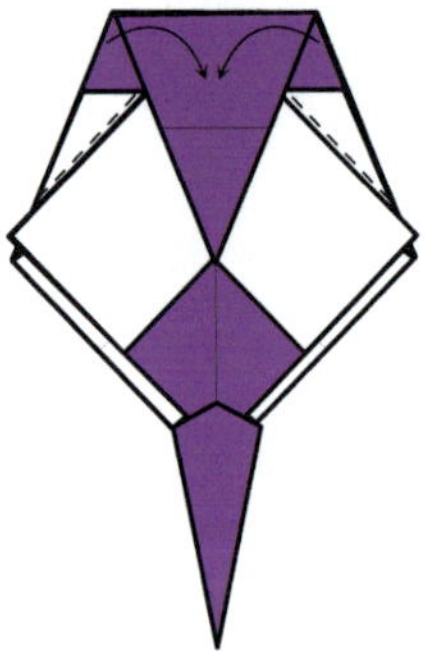

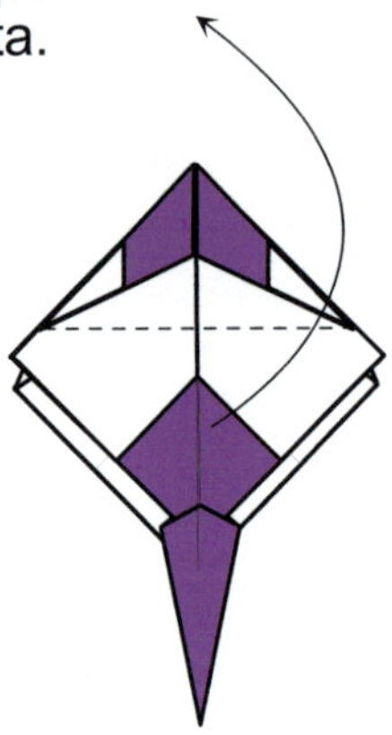

8. Valley fold Front Piece Beta upwards as far as it will go.

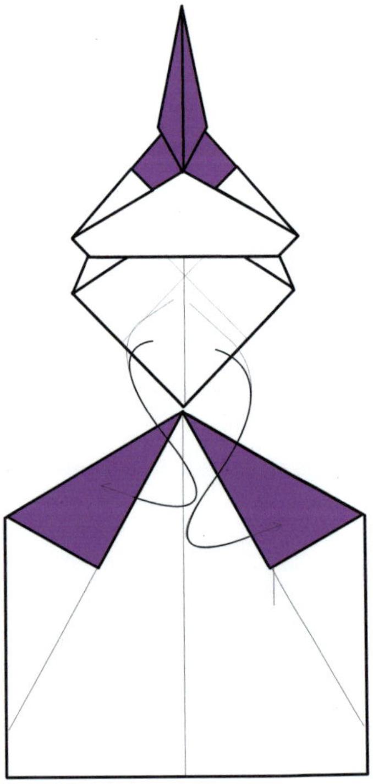

9. Insert the Delta Beta couple into Back Piece Two.

Delta Beta Two
Page 2

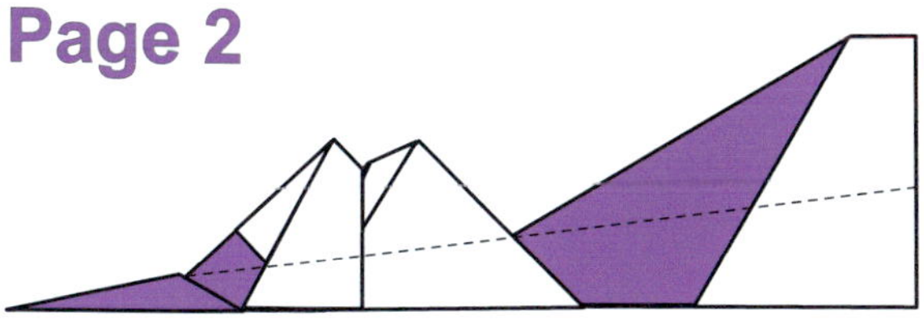

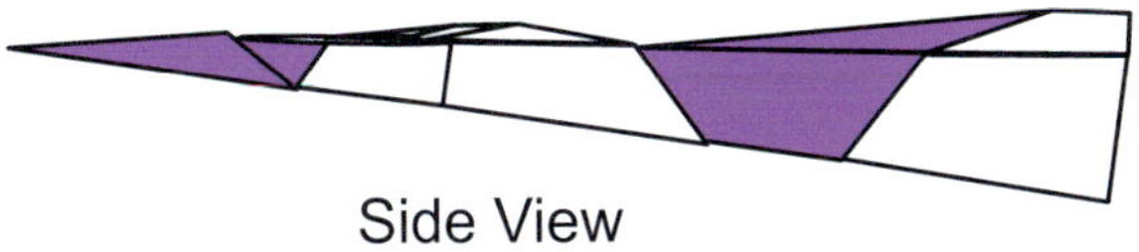

Side View

The fold for the wings starts behind the front diamond. It slants back, catching the intersection of Front Piece Beta and Back Piece Two.

Front View

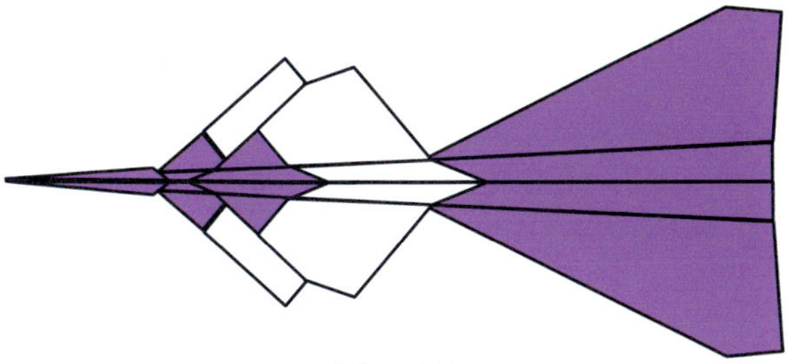

Top View

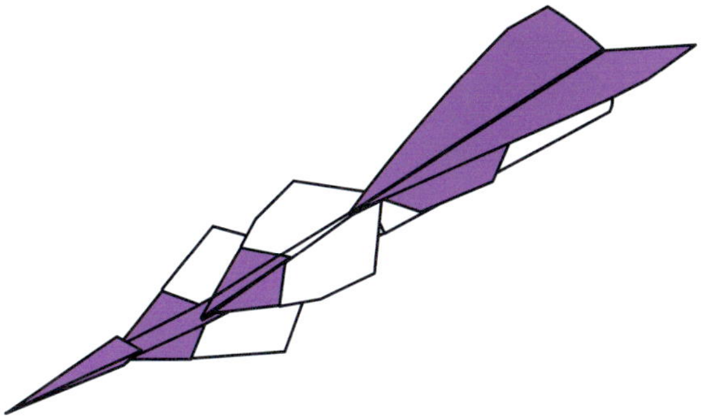

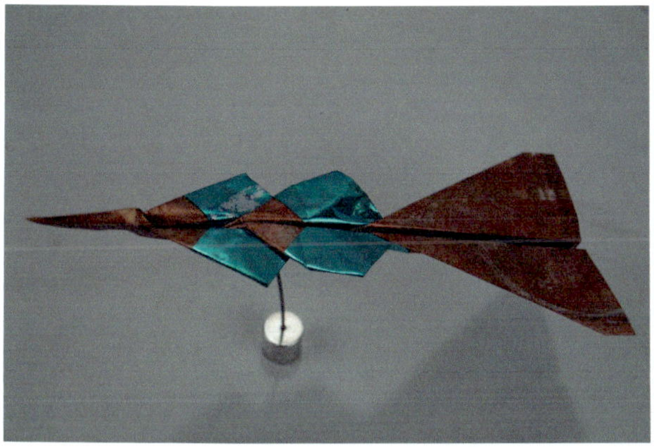

Gamma Alpha 3

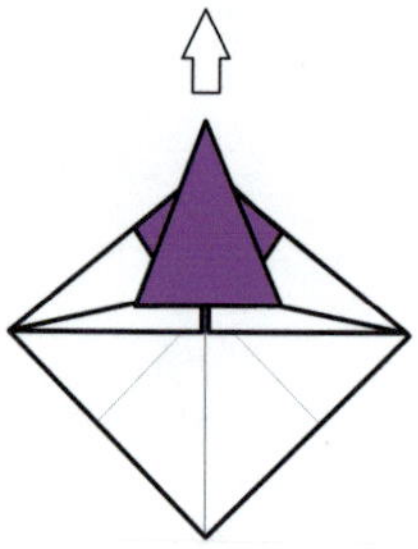

1. Undo the front of Front Piece Alpha.

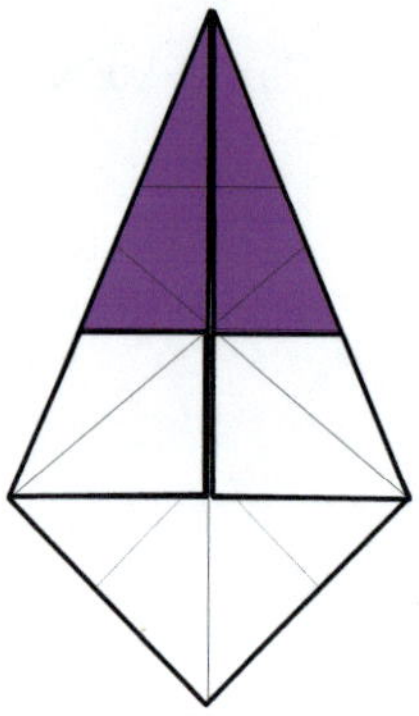

2. Like so.

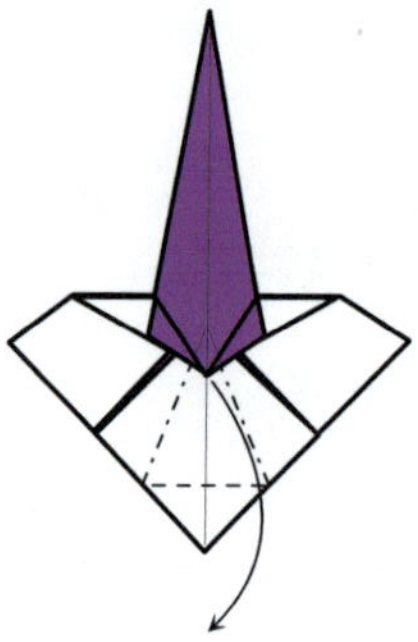

3. Petal fold the join of Front Piece Gamma. You'll have to fish a bit of paper from underneath a layer.

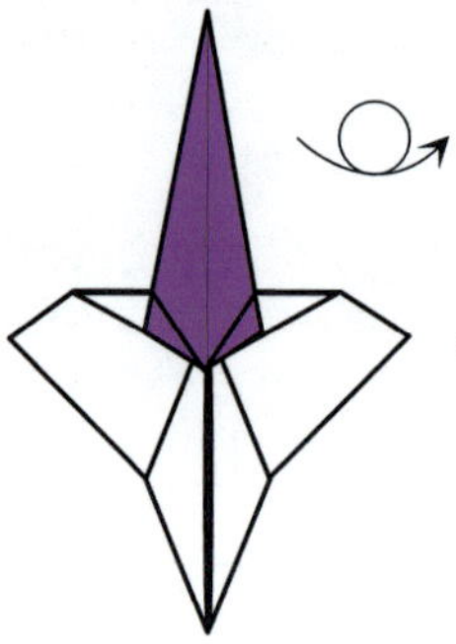

4. Turn over.

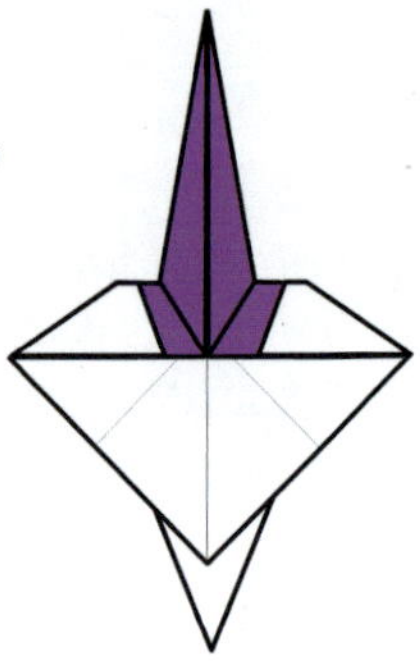

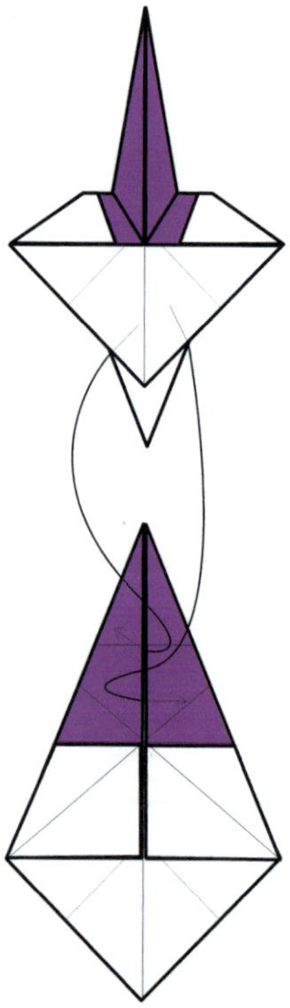

5. Insert the flaps of the join inside the flaps of Front Piece Alpha.

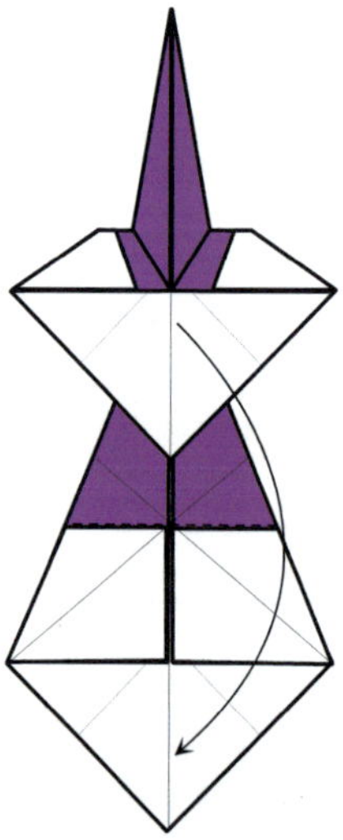

6. Valley fold down along a preexisting crease.

7. Refold the diagonals for Front Piece Alpha.

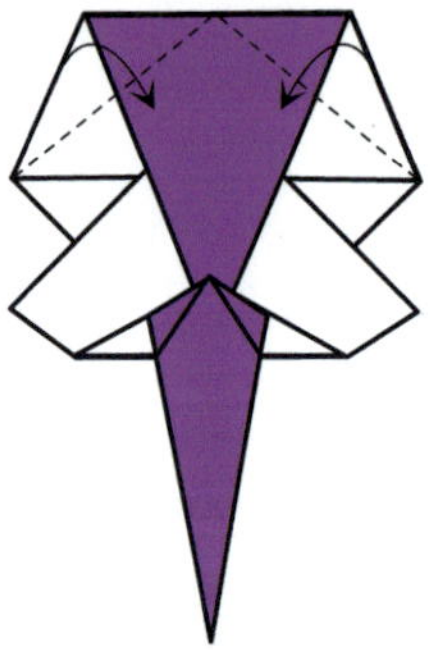

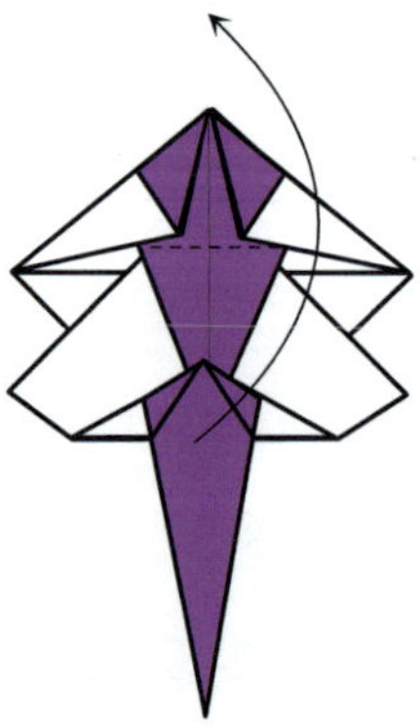

8. Valley fold upwards to complete Front Piece Alpha.

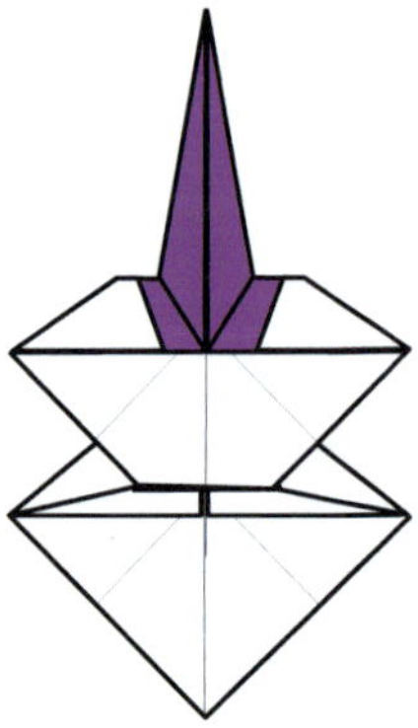

9. Insert into Back Piece 3.

Gamma Alpha 3 Page 2

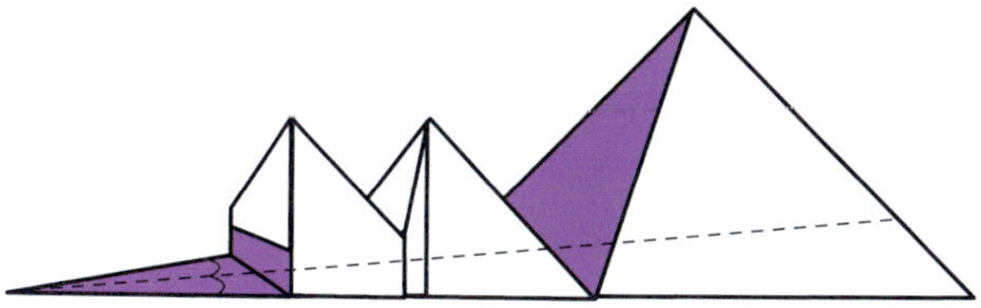

Bisect the angle of the nose, and keep the fold going through the other two pieces.

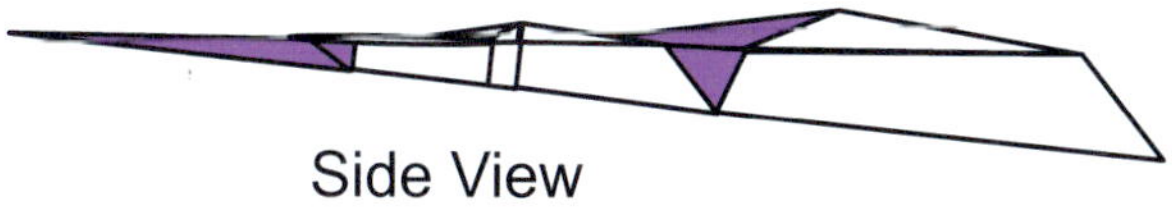

Side View

Front View

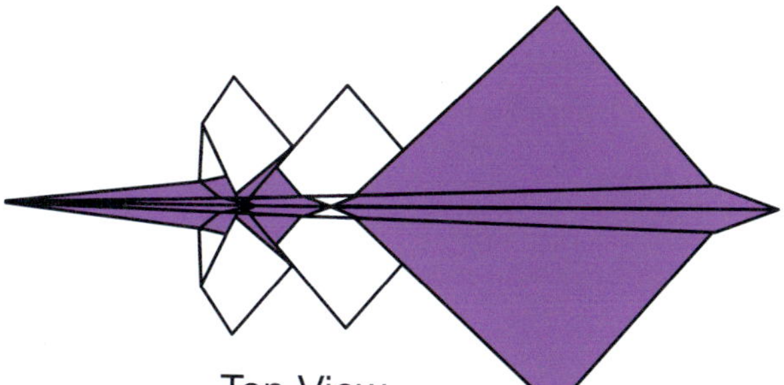

Top View

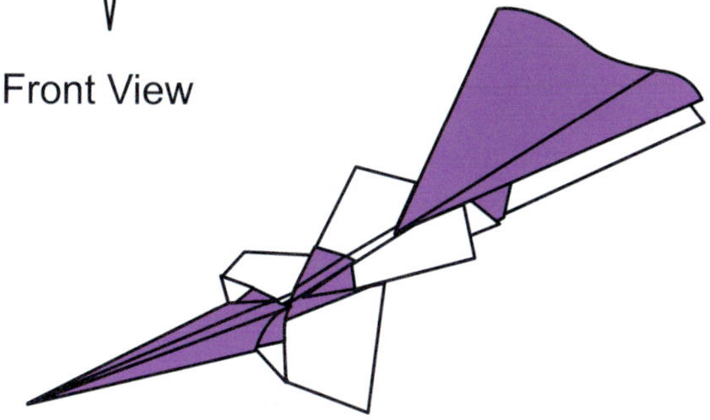

Omicron Epsilon Four

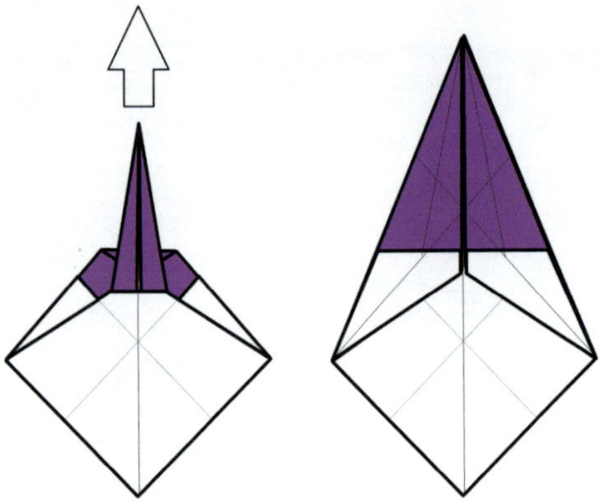

1. Unfold Front Piece Epsilon to step 1.

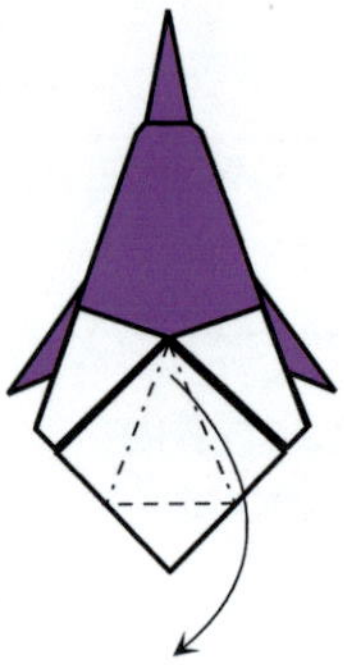

2. Petal fold the join on Front Piece Omicron.

3. Turn over.

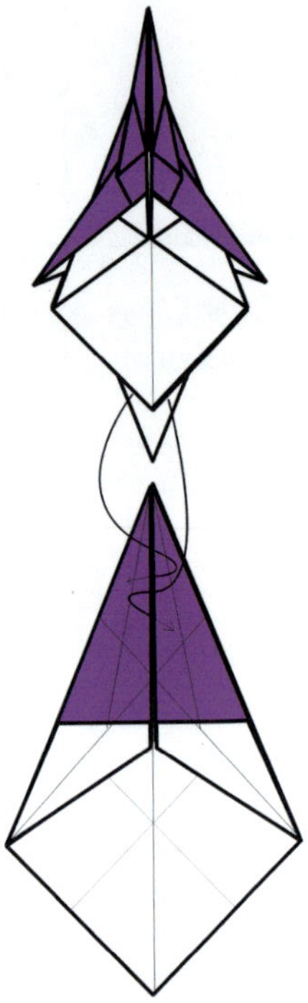

4. Insert the join of Front Piece Omicron into the flaps of Front Piece Epsilon.

5. Refold Front Piece Epsilon around the join of Front Piece Omicron.

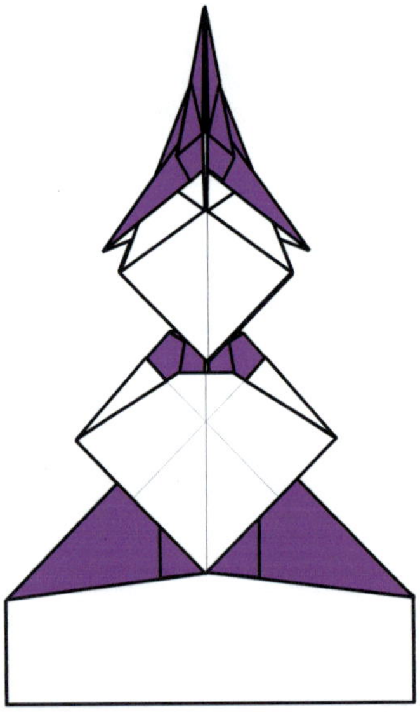

Omicron Epsilon Four Page 2

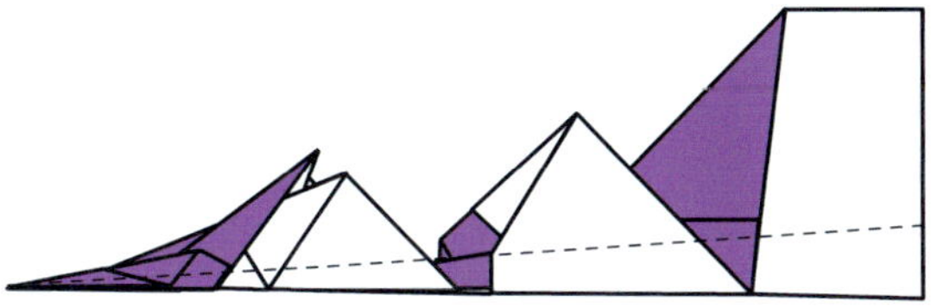

To fold the wings, fold Front Piece Omicron from the front to the base of Front Piece Epsilon. Continue the fold on the same angle all the way to the back.

Side View

Front View

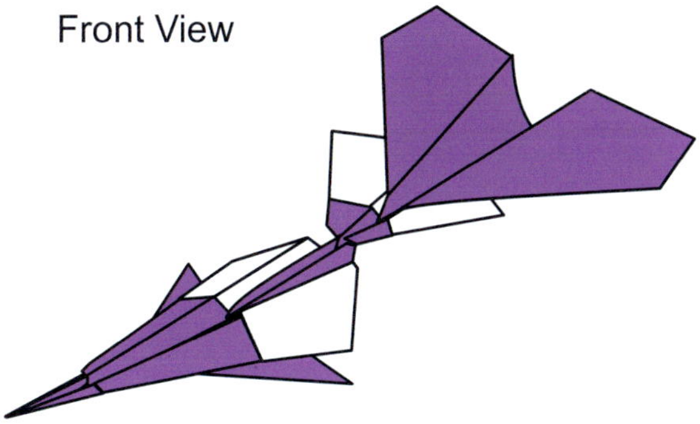

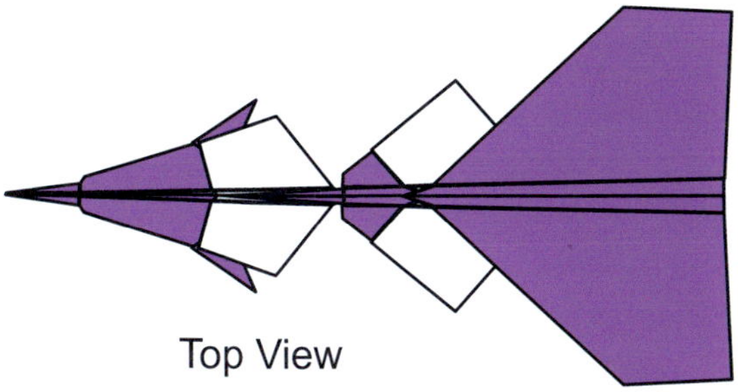

Top View

Epsilon 35

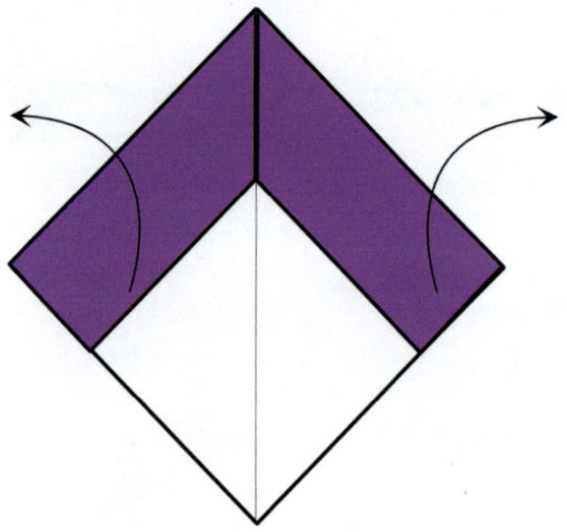

1. To use back piece 3 as a middle piece, start with step 9. Unfold the two flaps.

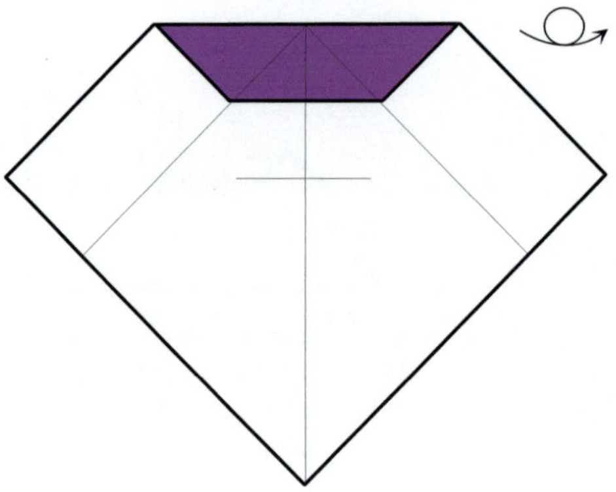

2. Turn over.

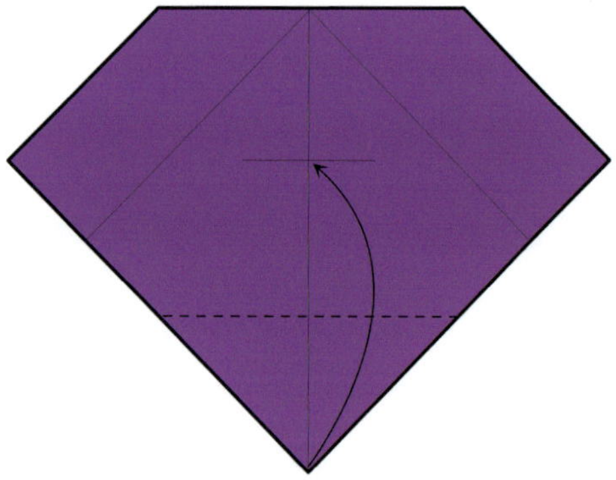

3. Fold the bottom tip up to the original halfway mark.

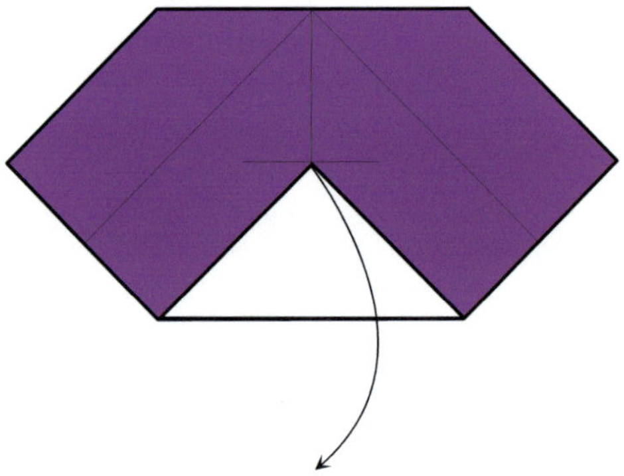

4. Unfold.

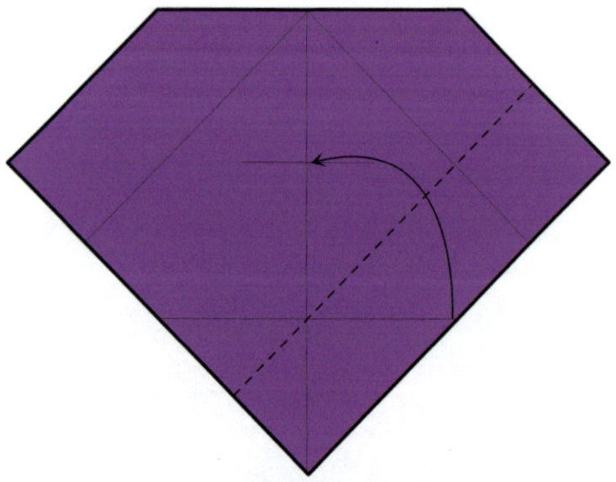

5. Valley fold a diagonal through the intersection shown

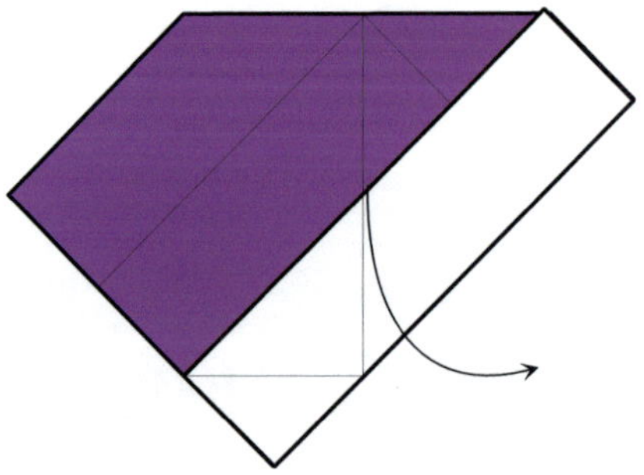

6. Unfold.

Epsilon 35 Page 2

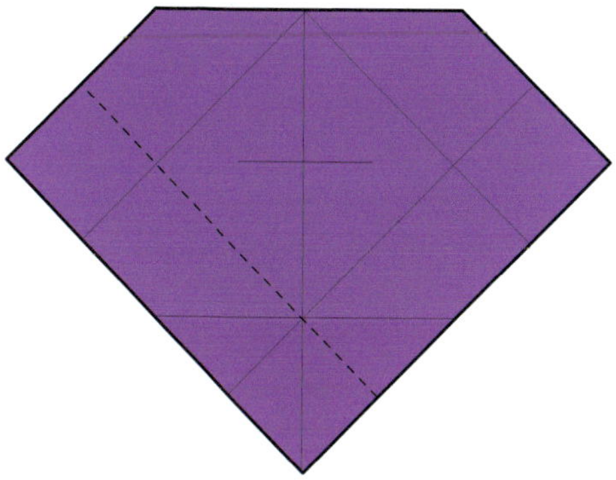

7. Repeat steps 5 and 6 on the other side.

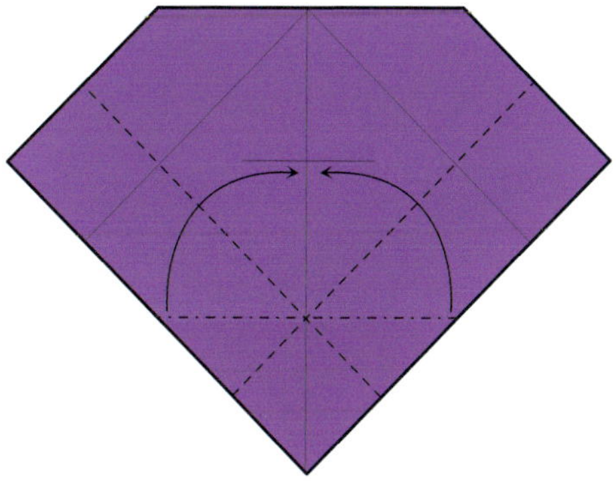

8. Assemble a joining unit on Back Piece 3.

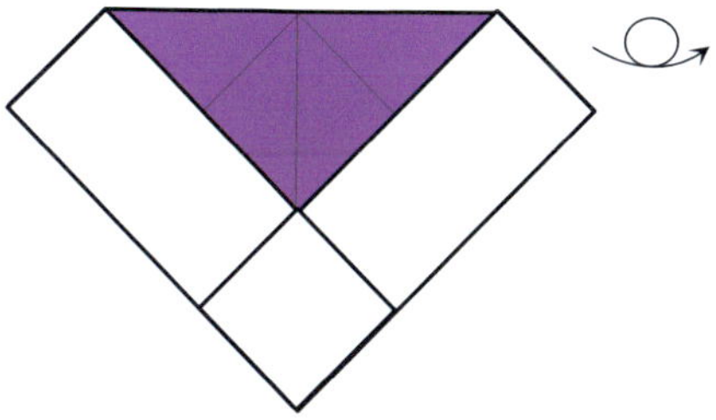

9. Turn over.

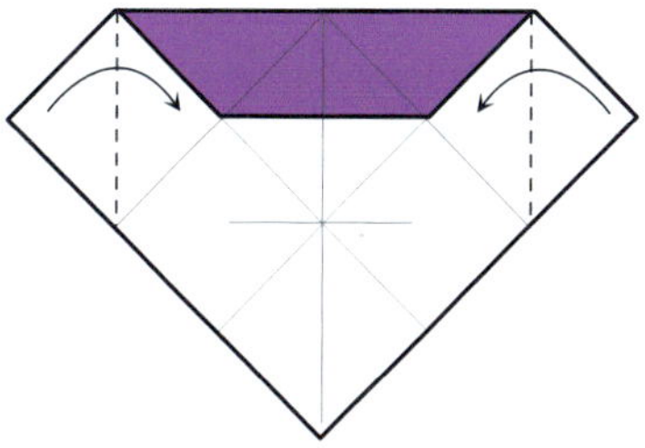

10. Valley fold in the corners.

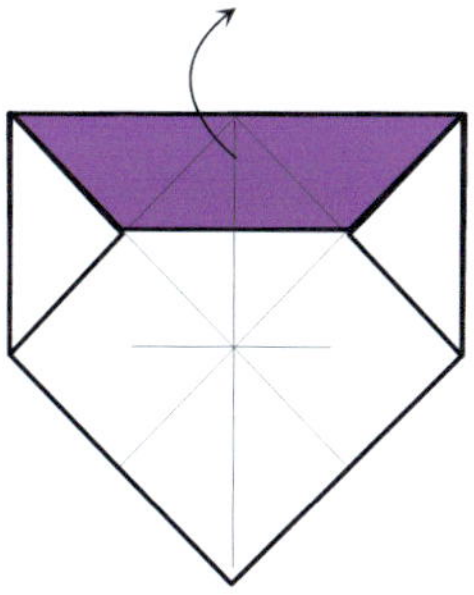

12. Unfold the top.

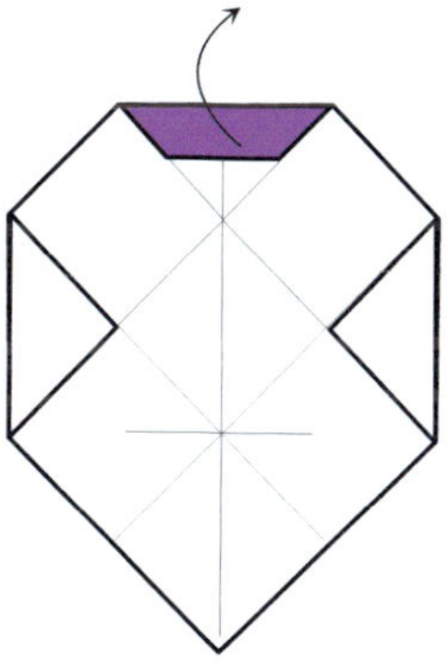

13. Unfold one more.

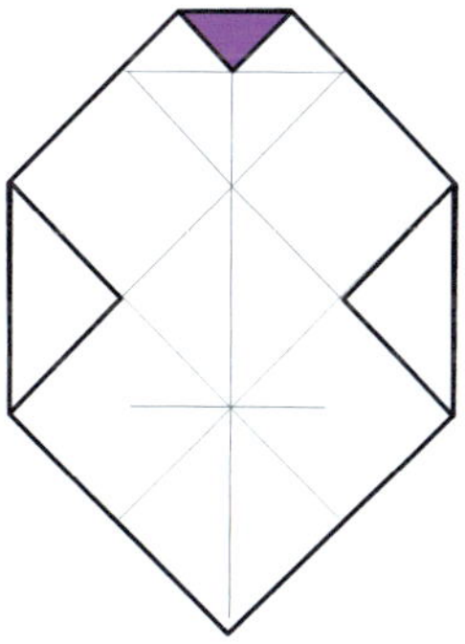

14. Assembly time.

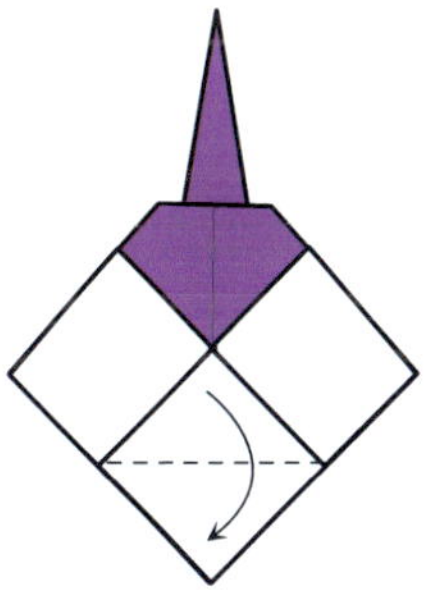

15. Valley fold the joining unit of Front Piece Epsilon.

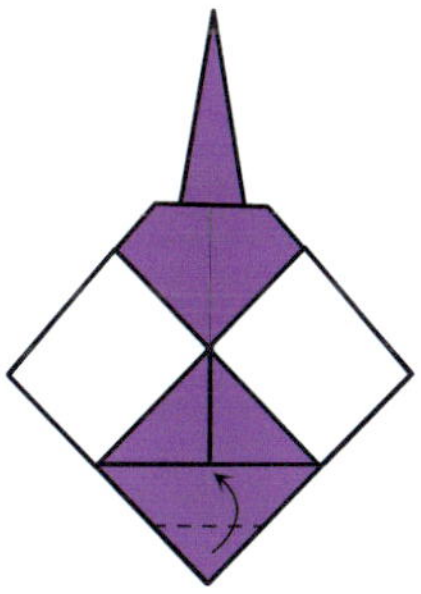

16. Valley fold the tip to the folded edge above.

Epsilon 35 Page 3

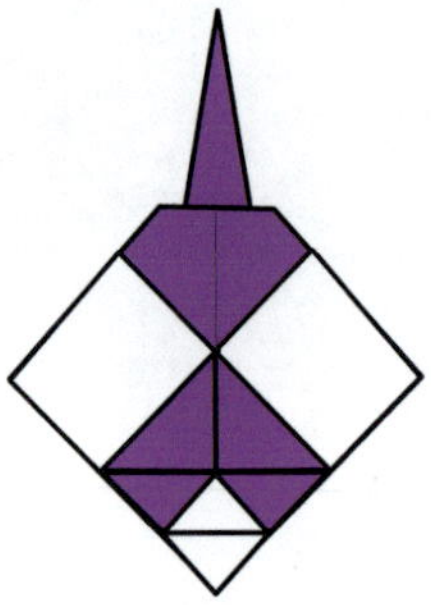

17. Rotate Front Piece Epsilon 180 degrees and place it on top of Back Piece 3.

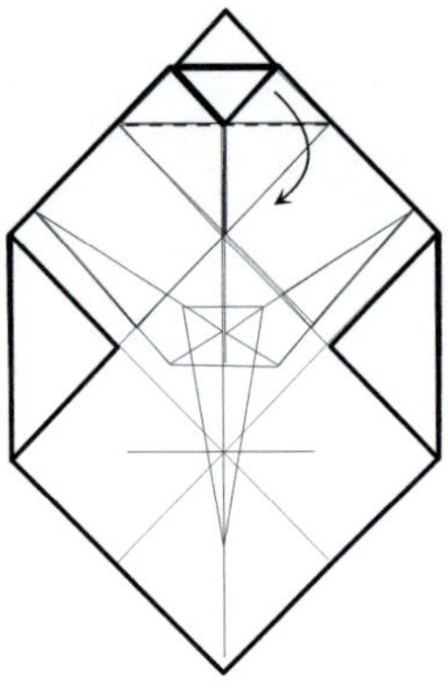

18. Overlap the flaps of Front Piece Epsilon and Back Piece 3. Valley fold them together along a preexisting crease.

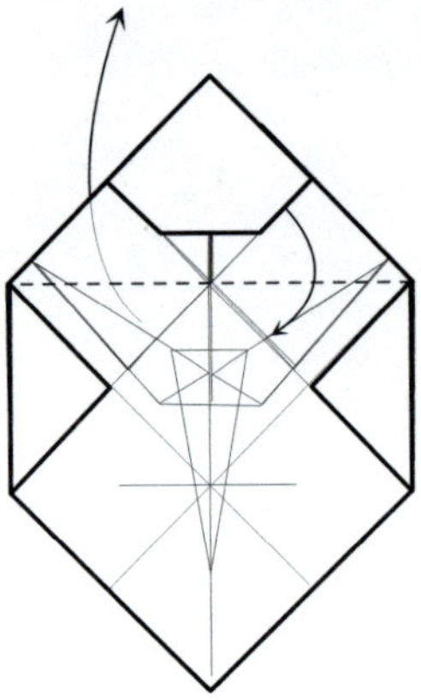

19. Valley fold again on a preexisting crease.

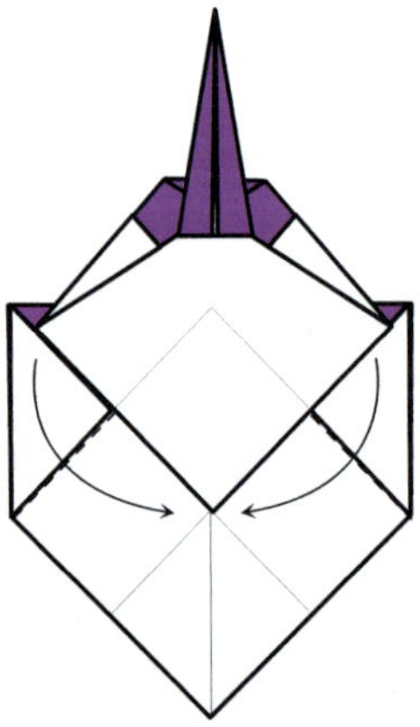

20. Valley fold the corners in on preexisting creases.

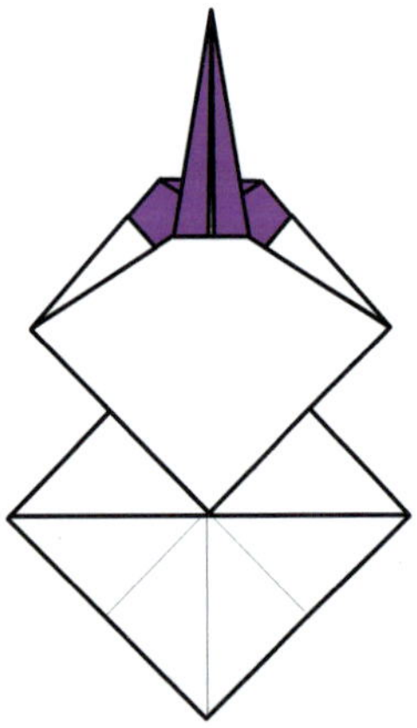

21. Insert into Back Piece 5.

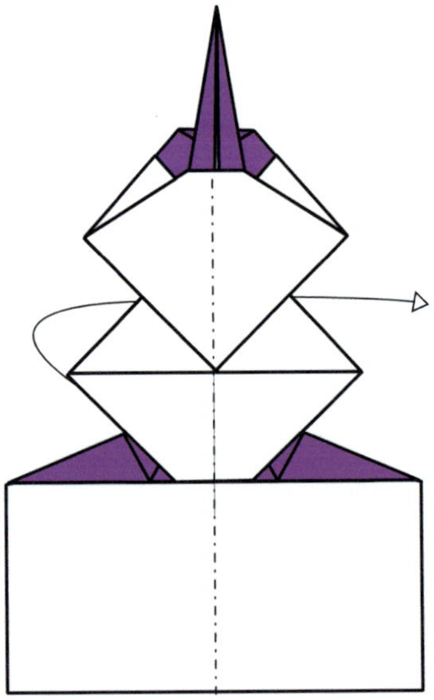

22. Mountain fold in half.

Epsilon 35 Page 4

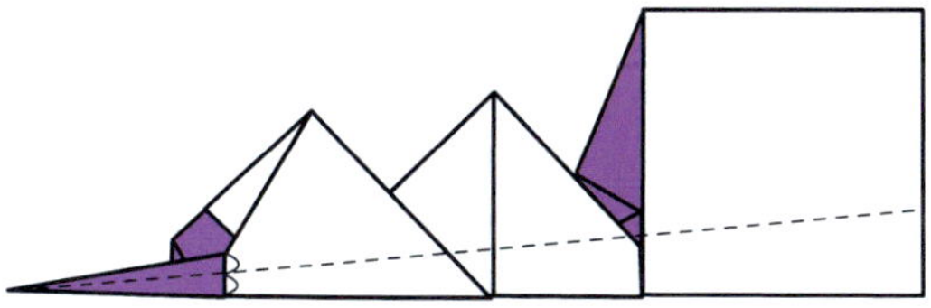

Front View

To fold the wings, split the angle in the nose in half. Continue the fold through the next two pieces.

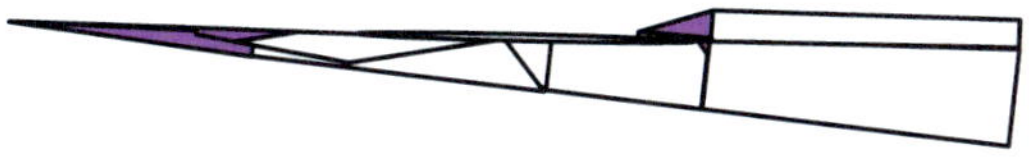

Side View

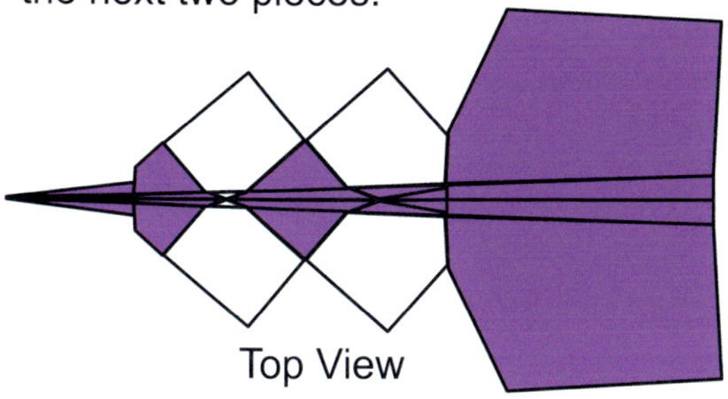

Top View

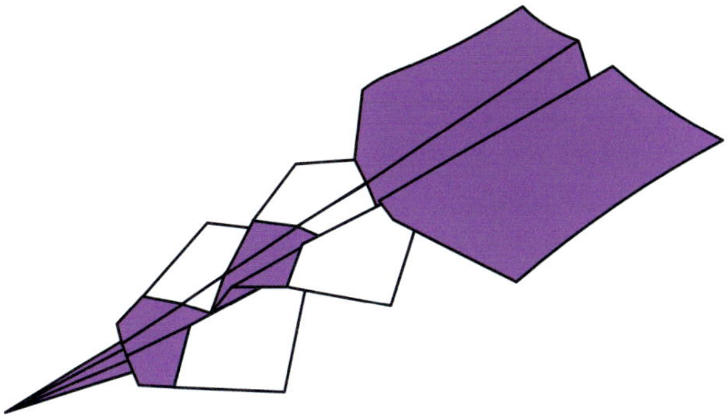

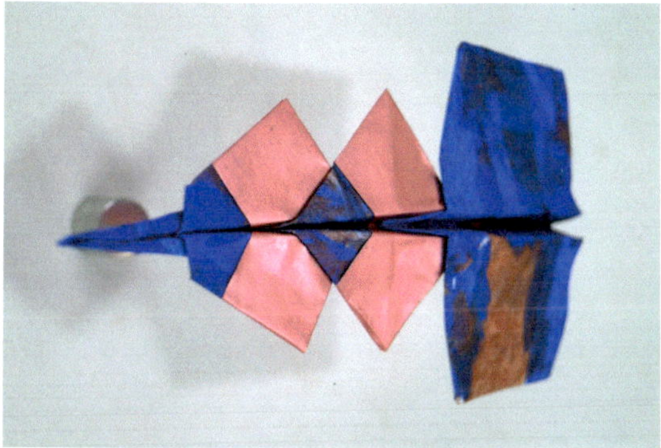

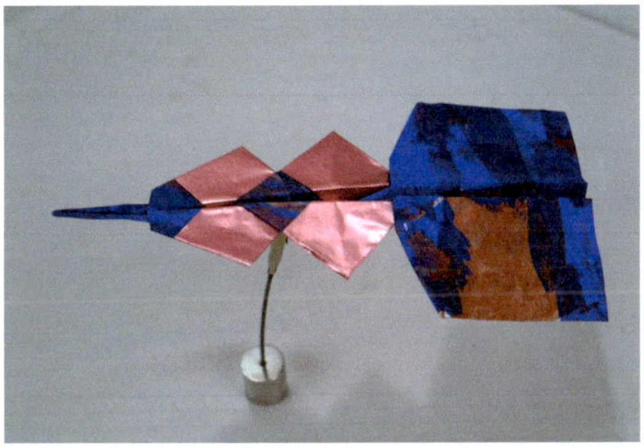

Iota Alpha Six

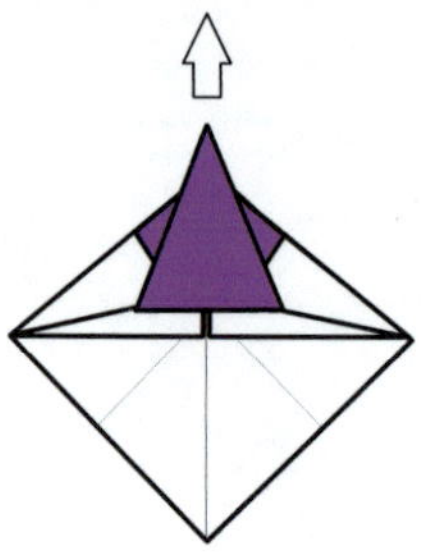

1. Undo the front of Front Piece Alpha.

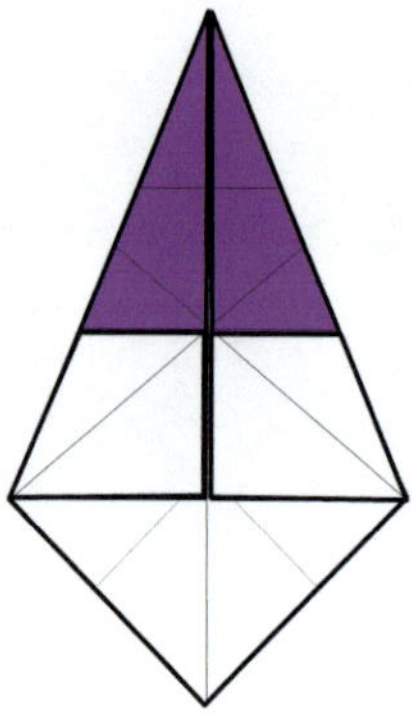

2. Like so.

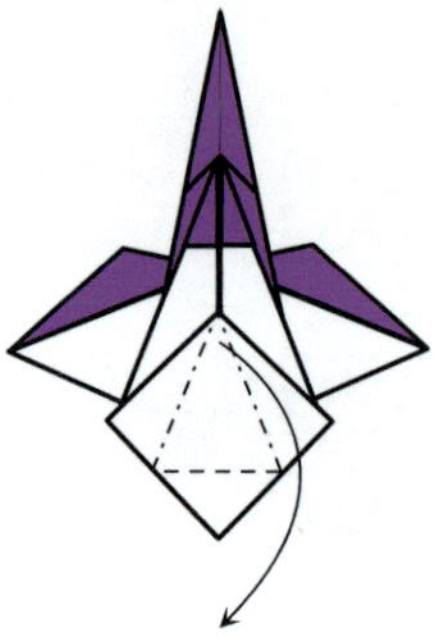

3. Petal fold the join of Front Piece Iota.

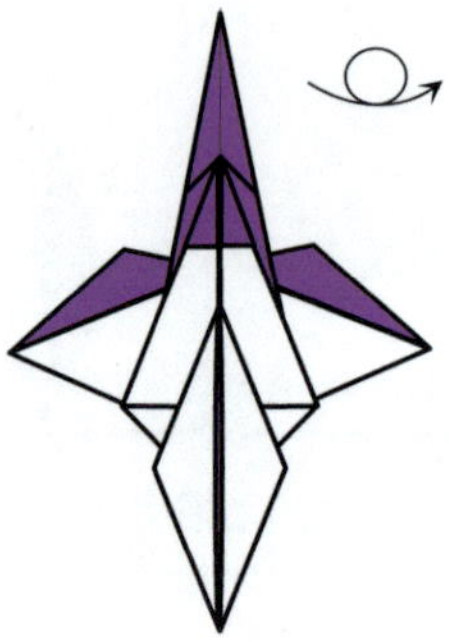

4. Turn over.

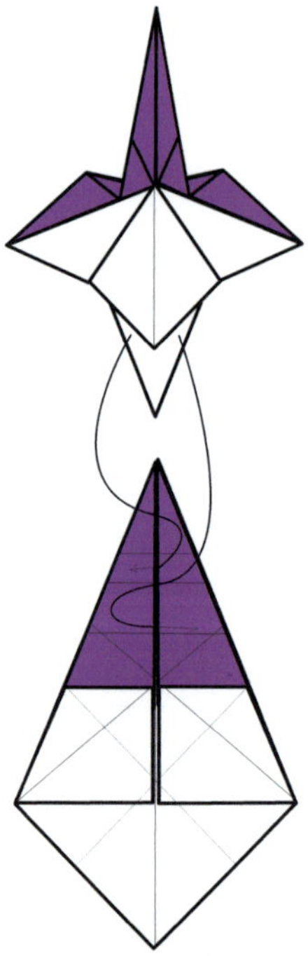

5. Insert the join of Front Piece Iota into the flaps of Front Piece Alpha.

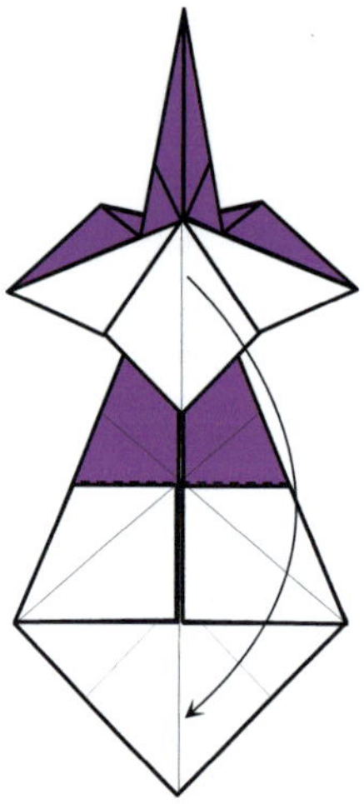

6. Valley fold down along a preexisting crease.

7. Refold the diagonals for Front Piece Alpha.

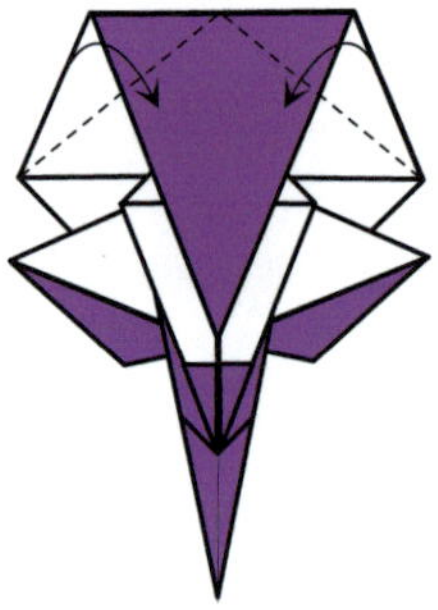

8. Valley fold upwards to complete Front Piece Alpha.

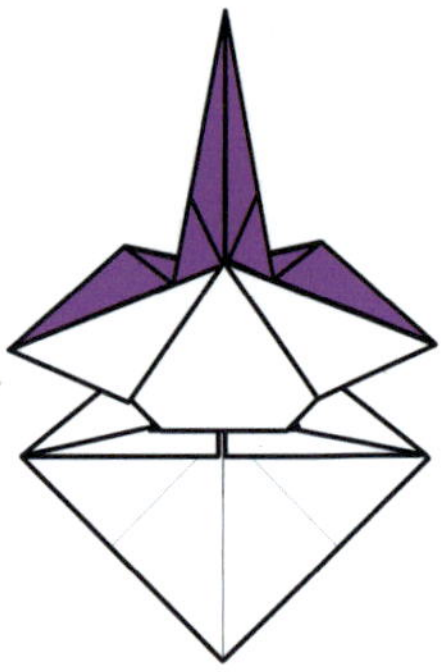

9. Insert into Back Piece 6.

Iota Alpha Six Page 2

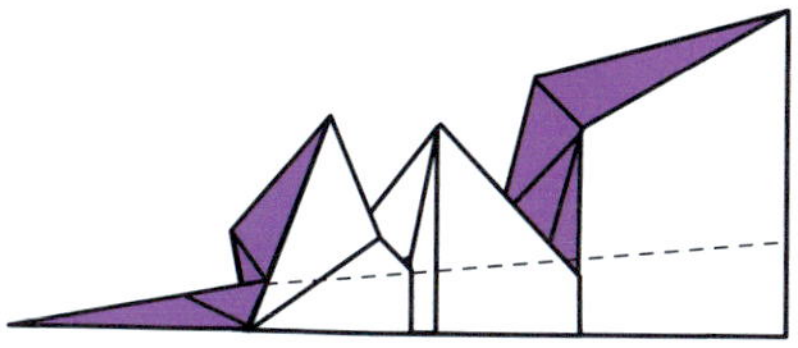

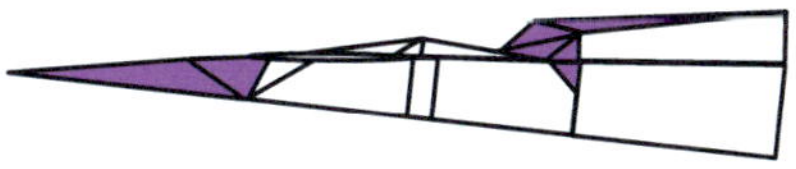

Side View

Front View

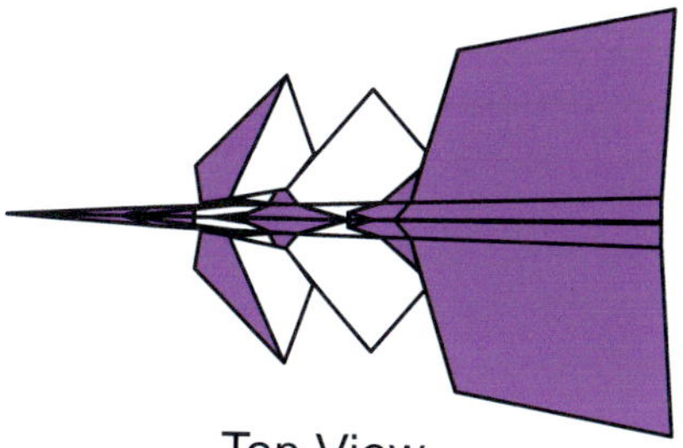

Top View

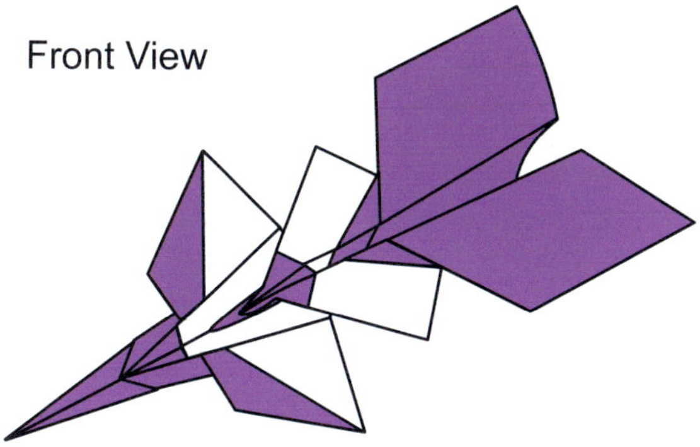

Alpha Zeta Seven

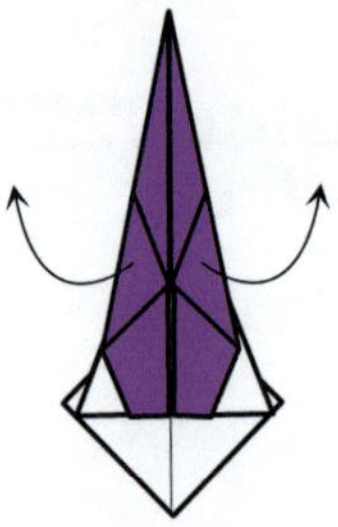

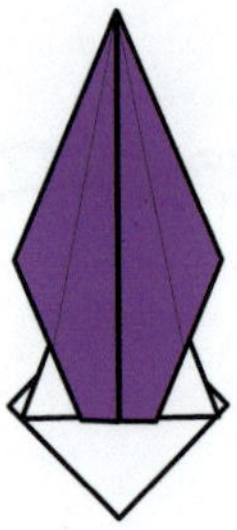

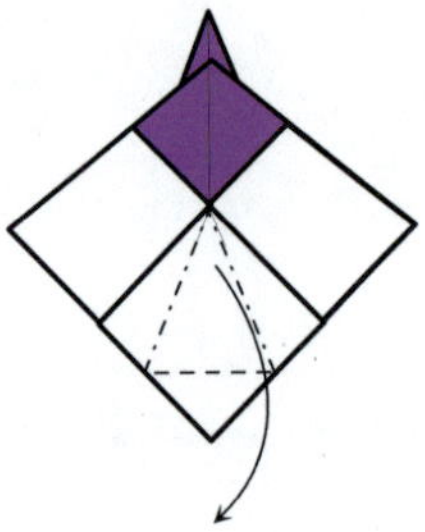

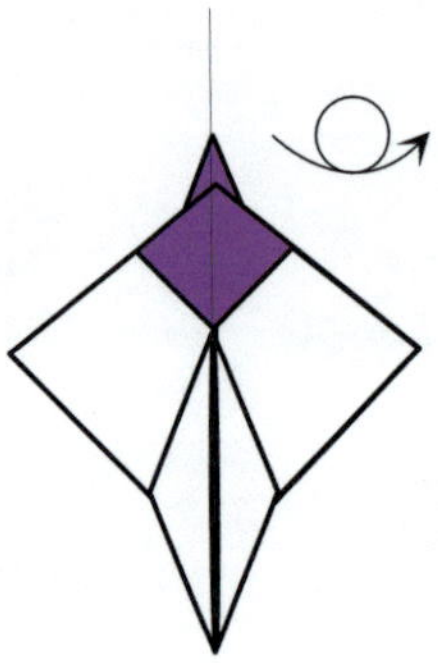

1. To insert into Front Piece Zeta, open up the front flaps.

2. Like so.

3. Petal fold the join of Front Piece Alpha.

4. Turn over.

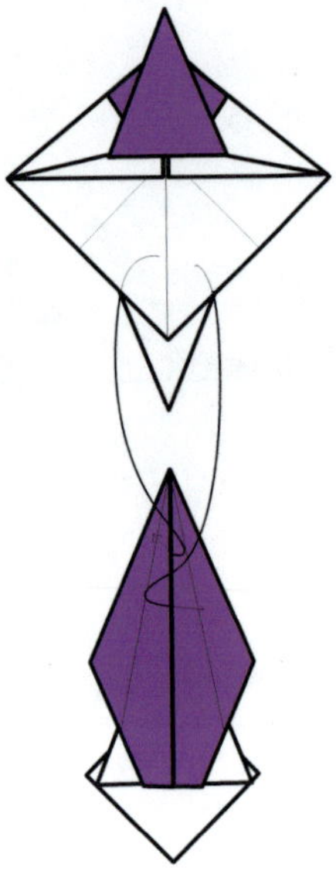

5. Insert the flaps of the join inside the flaps of Front Piece Zeta.

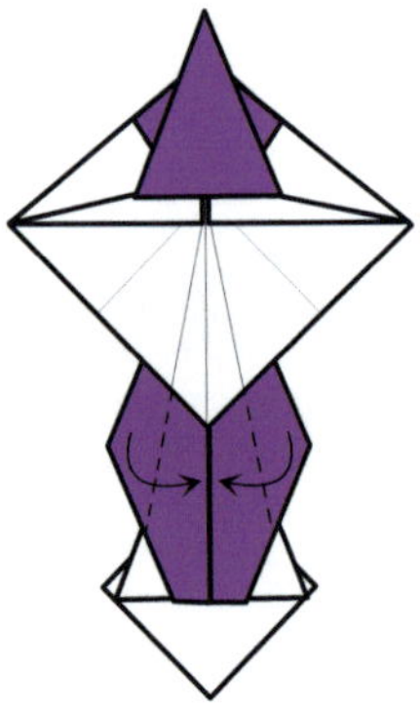

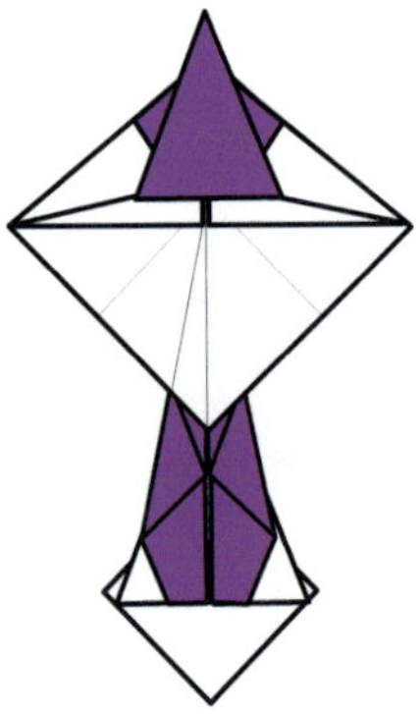

6. Refold the front of Front Piece Zeta along preexisting creases.

7. Insert the join of Front Piece Zeta into Back Piece Seven.

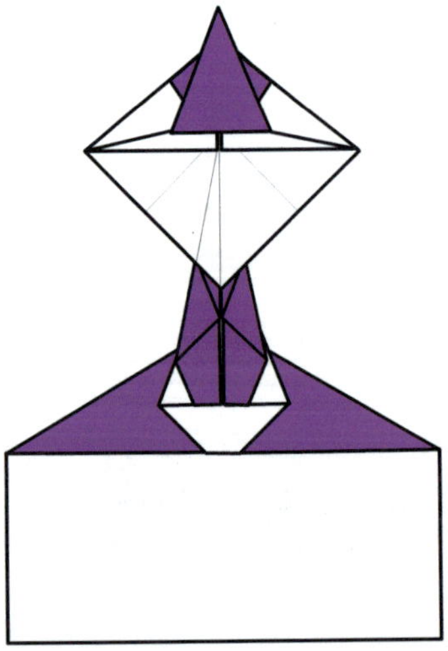

Alpha Zeta Seven Page 2

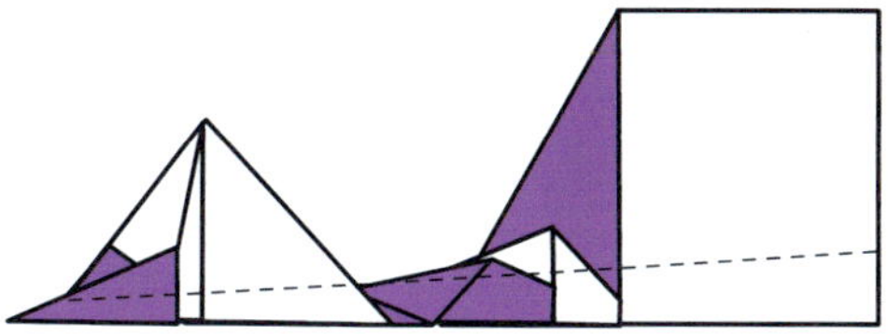

Folding the wings is a bit tricky. Fold to the intersection of Alpha and Zeta. You'll need a straight edge to continue the fold on the back piece.

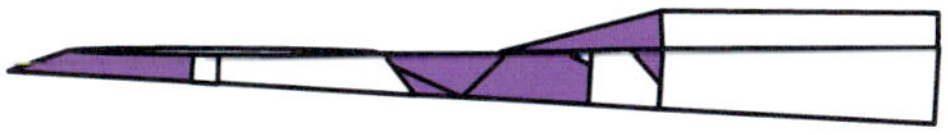

Side View

Front View

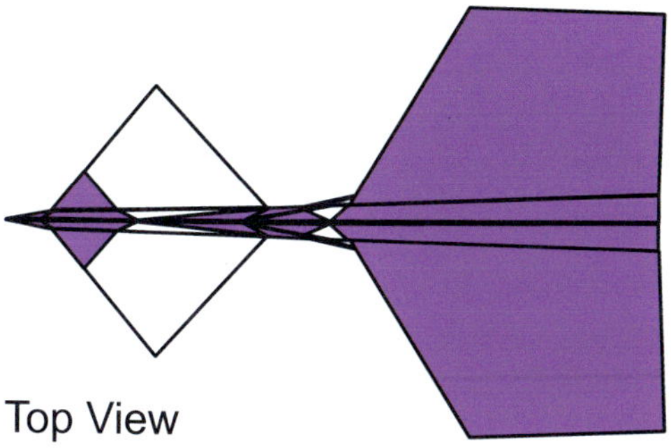

Top View

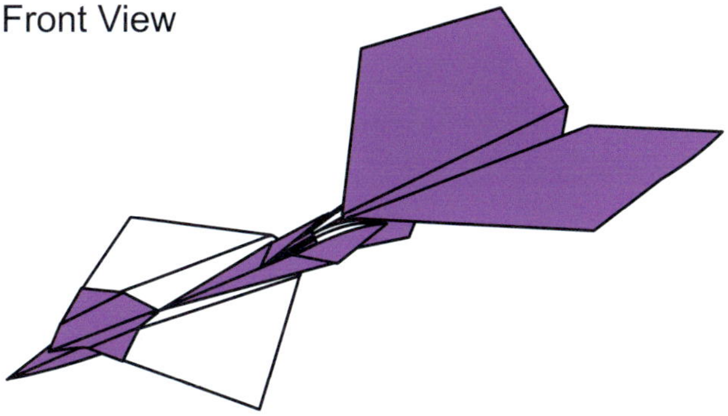

Alpha Squared Eight

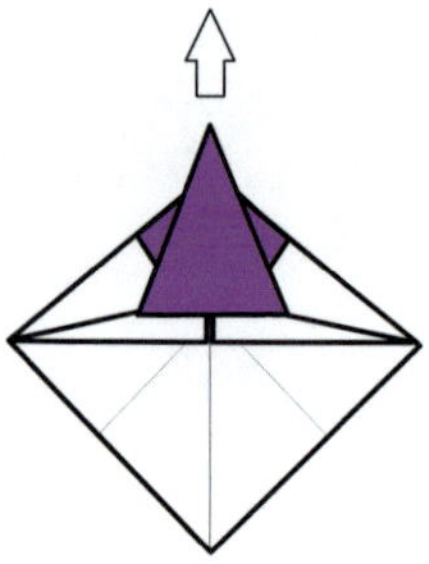

1. Undo the front of Front Piece Alpha.

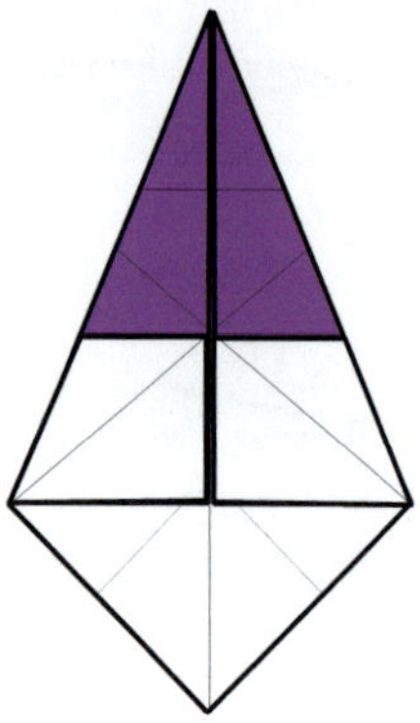

2. Like so.

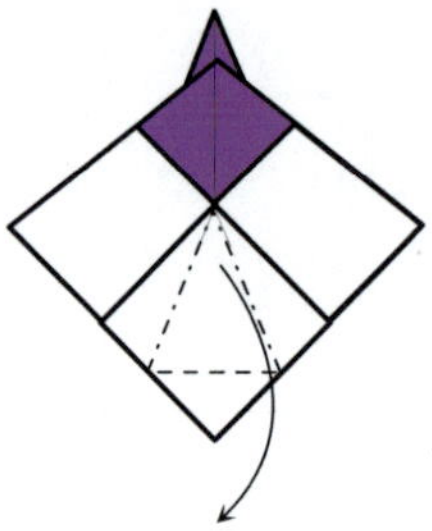

3. Petal fold the join of Front Piece Alpha.

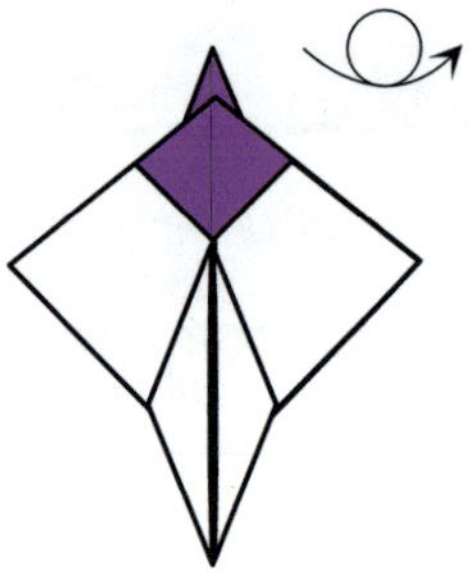

4. Turn over.

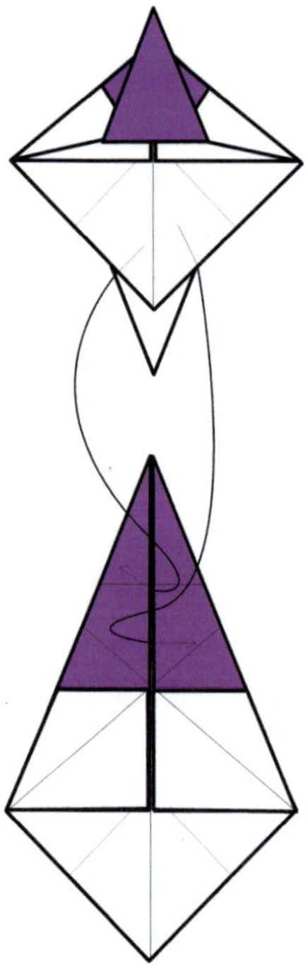

5. Insert the flaps of the join inside the flaps of Front Piece Alpha.

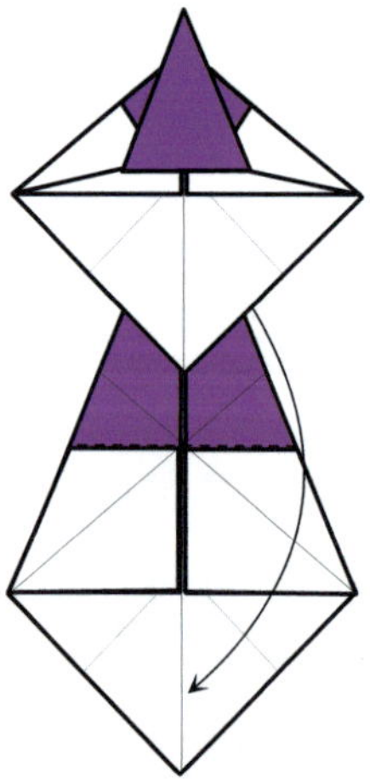

6. Valley fold down along a preexisting crease.

7. Refold the diagonals for Front Piece Alpha.

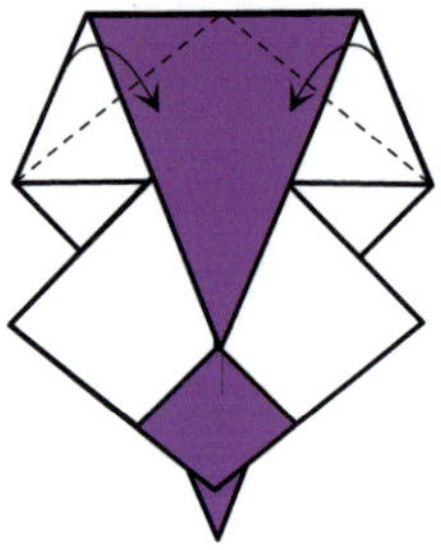

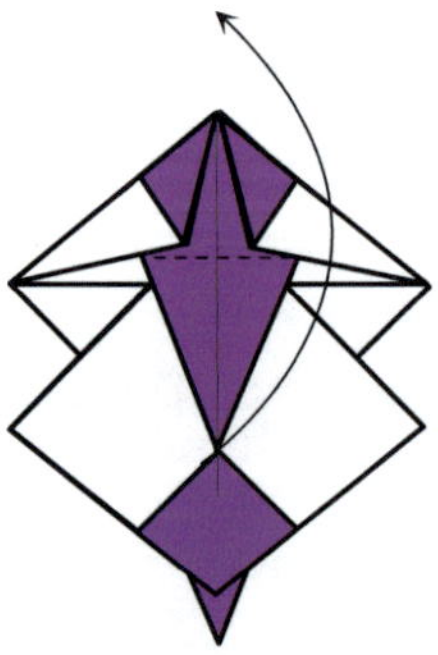

8. Valley fold upwards to complete Front Piece Alpha.

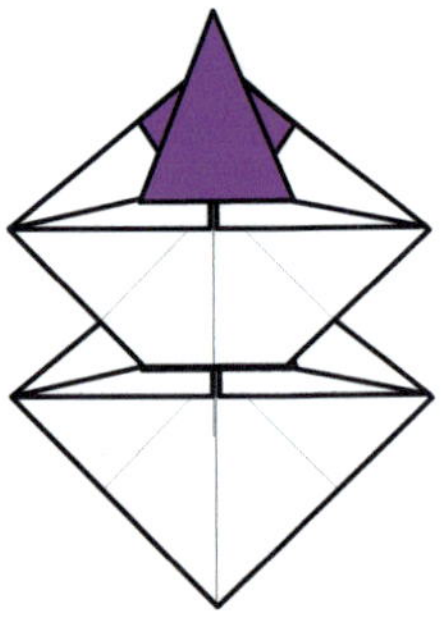

9. Insert into Back Piece 8.

Alpha Squared Eight Page 2

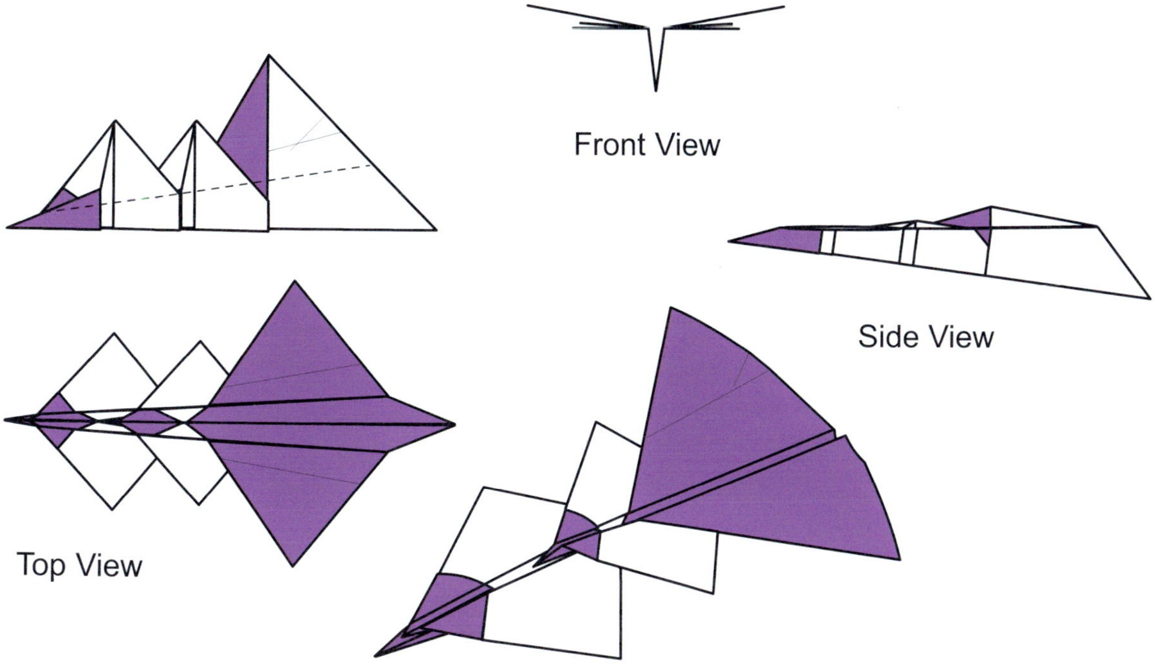

Alpha Gamma Nine

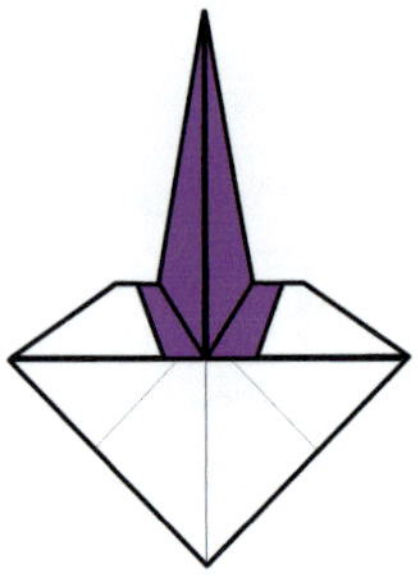

1. Unfold the front of Front Piece Gamma

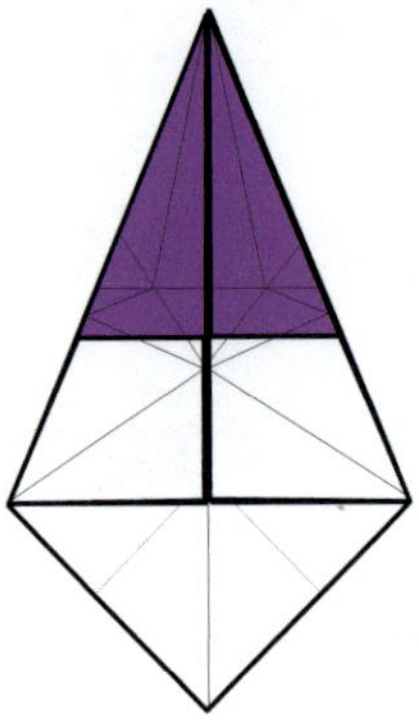

2. Ready for assembly.

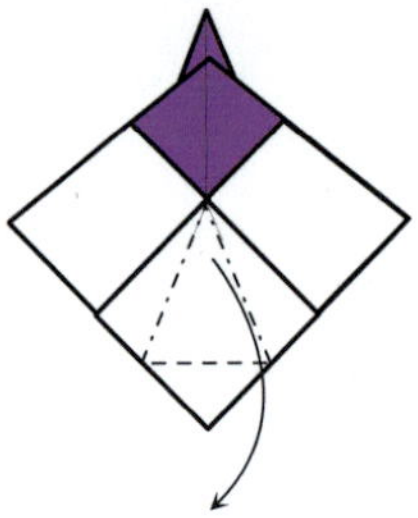

3. Petal Fold.

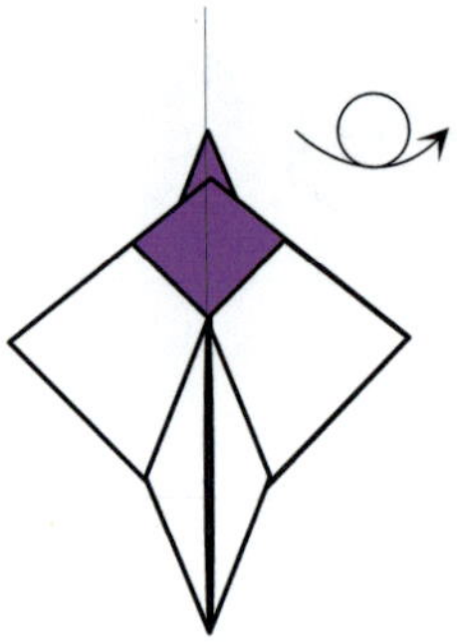

4. Turn over.

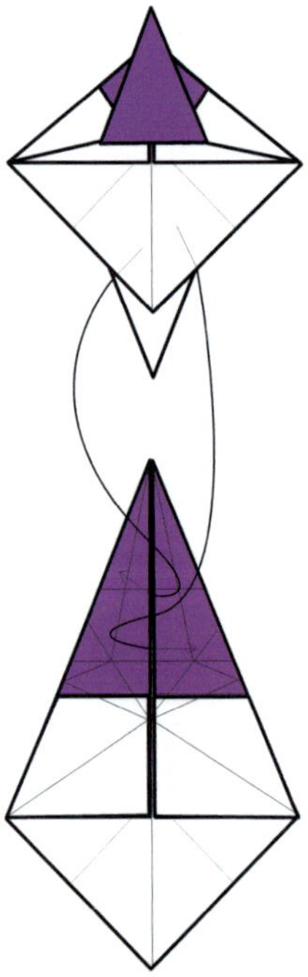

5. Insert the flaps of the join inside the flaps of Front Piece Alpha.

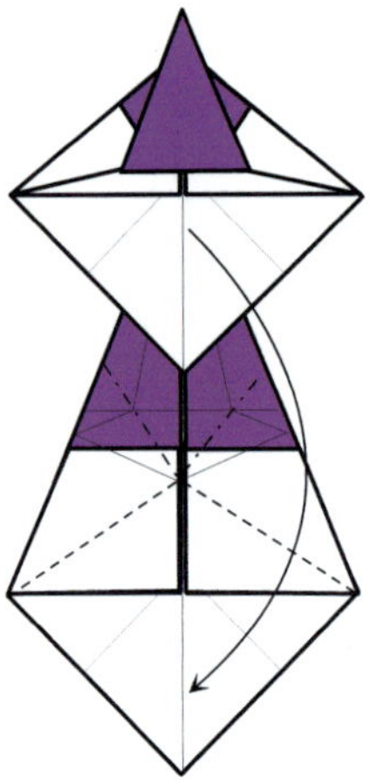

6. Begin reforming Front Piece Gamma along existing creases.

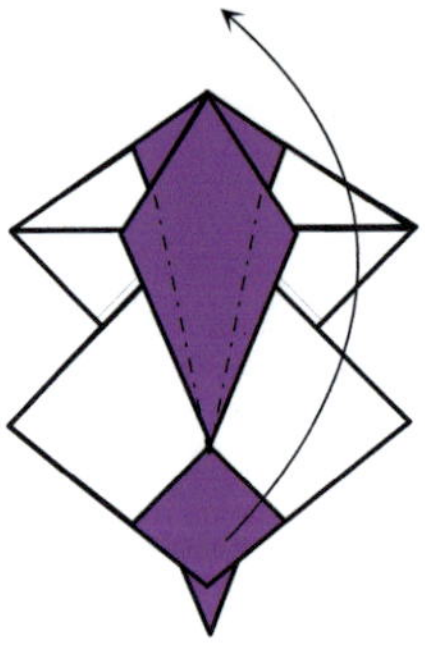

7. Petal fold upwards.

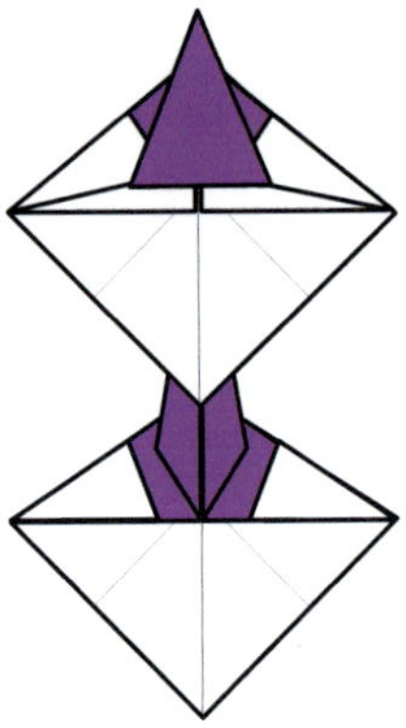

8. Attach to Back Piece Nine.

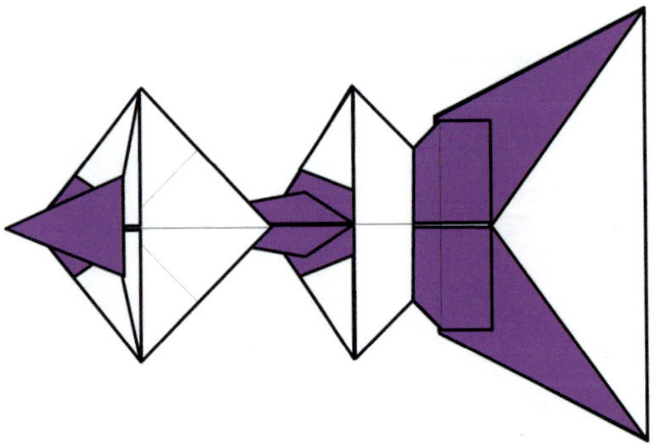

Alpha Gamma Nine Page 2

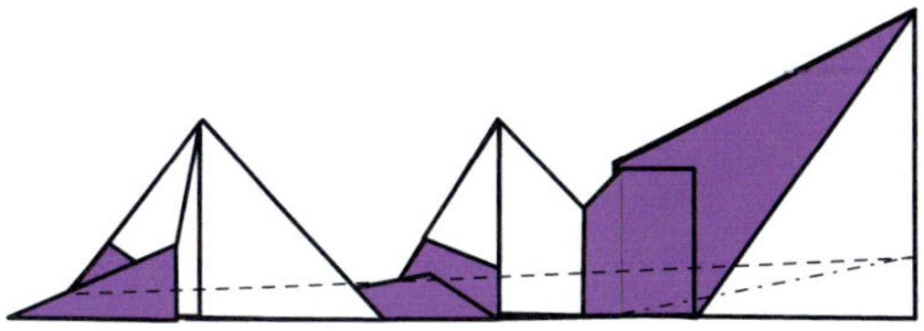

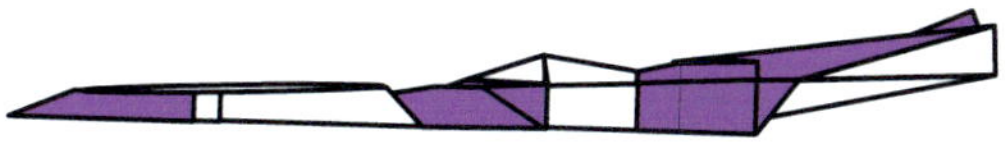

Side View

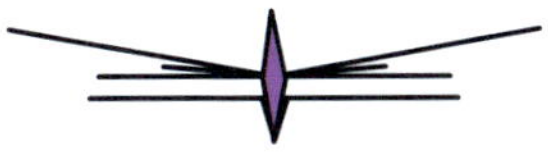

Front View

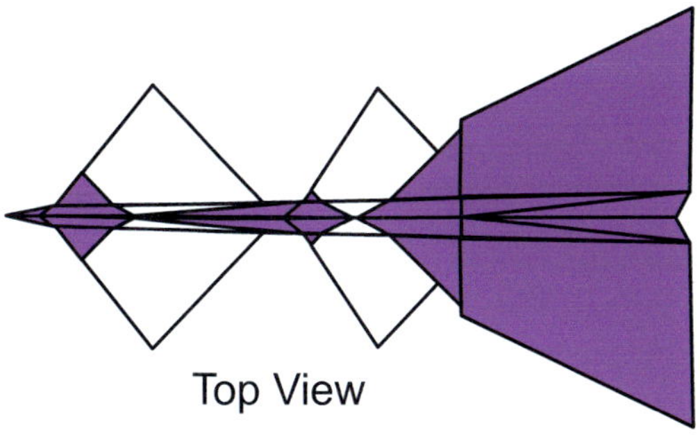

Top View

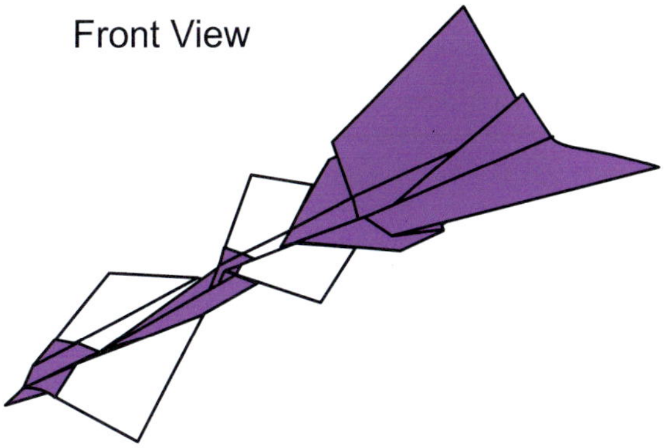

Zeta One Ten

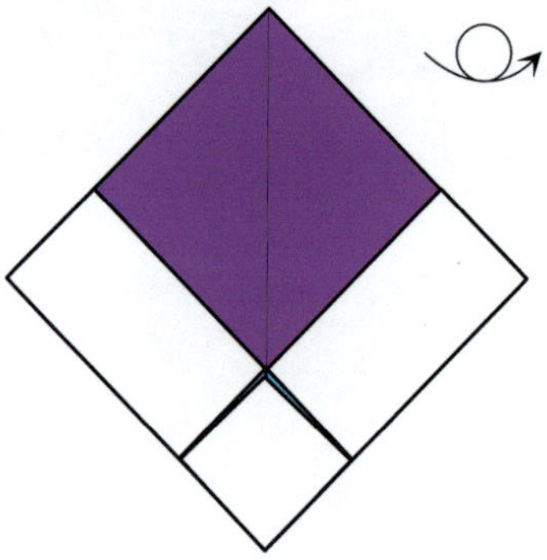

1. To use Back Piece One as a middle piece, begin with the front piece base. Turn over.

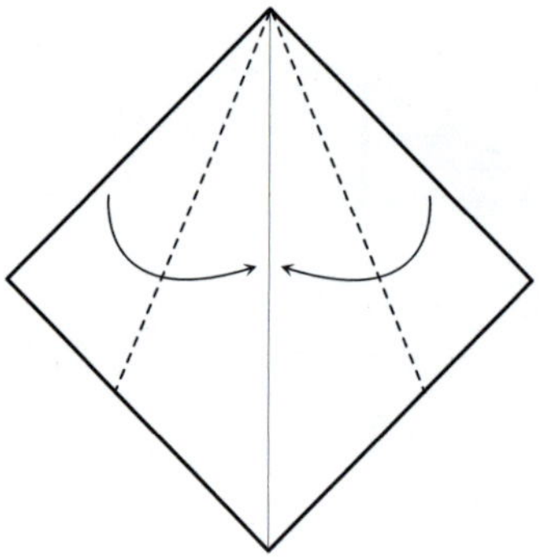

2. Valley fold so the raw edges lie in the middle.

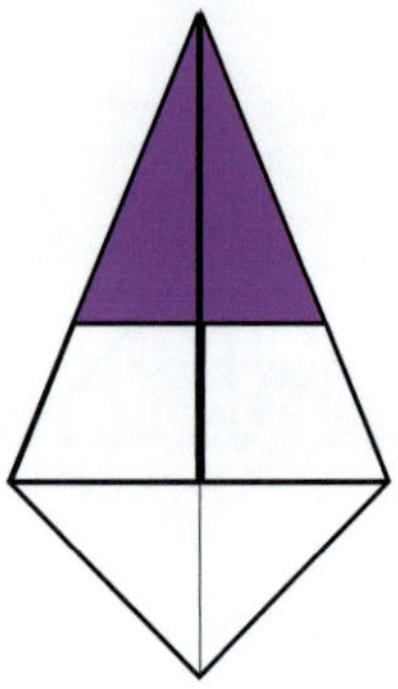

3. Continue to assemble Back Piece One.

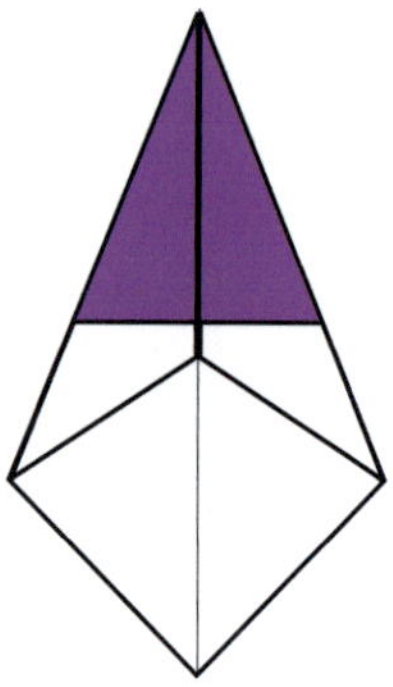

4. Finished for now.

5. Petal fold the join of Front Piece Zeta.

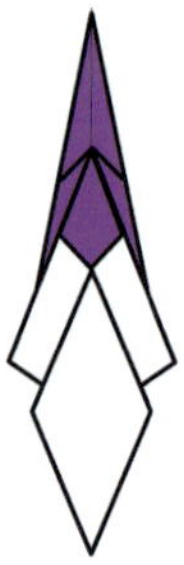

6. Insert Front Piece Zeta into Back Piece One.

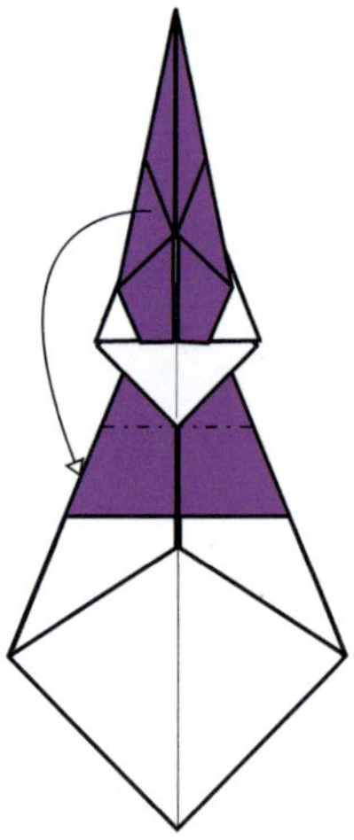

7. Mountain fold Front Piece Zeta behind where it meets Back Piece One.

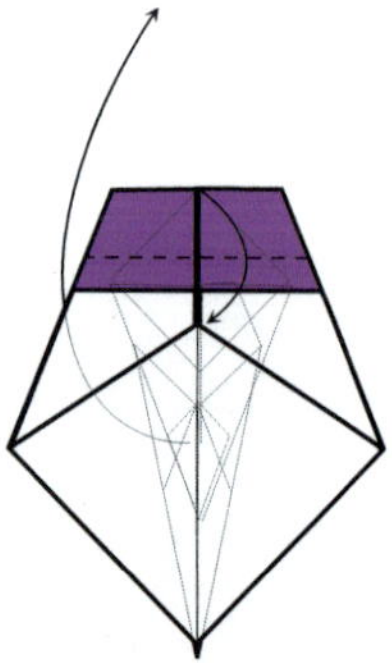

8. Valley fold, bringing the top edge to the intersection shown.

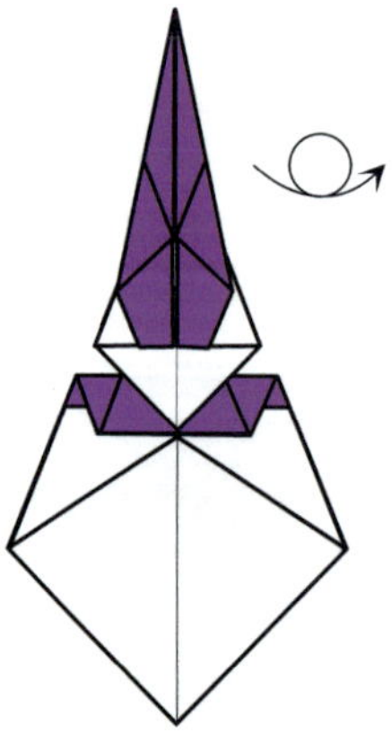

9. Turn over.

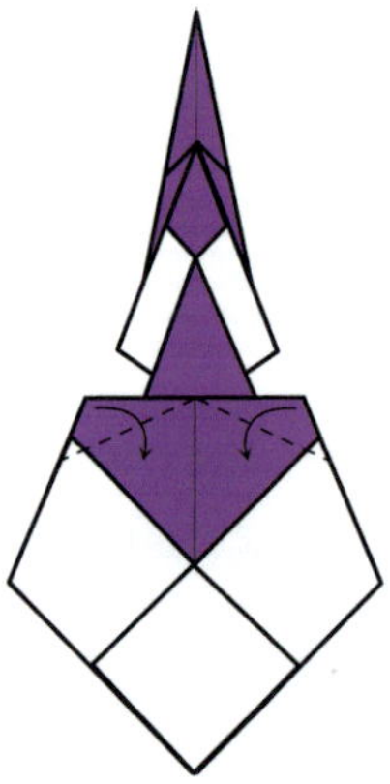

10. Valley fold the corners as far as they'll go.

Zeta One Ten Page 2

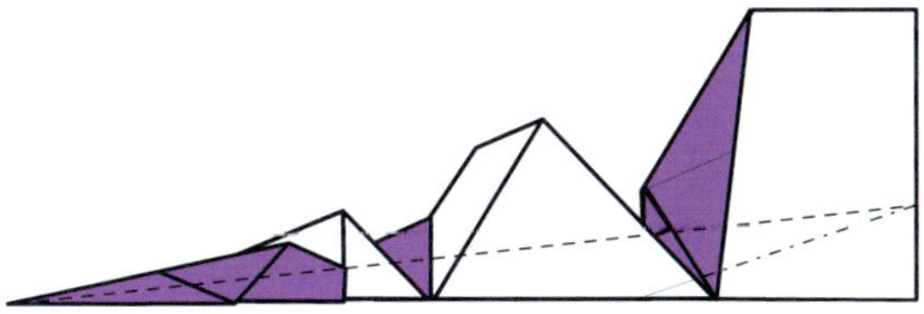

The wings are created by folding the front angle in half and following the angle back. This aircraft also gets a tail.

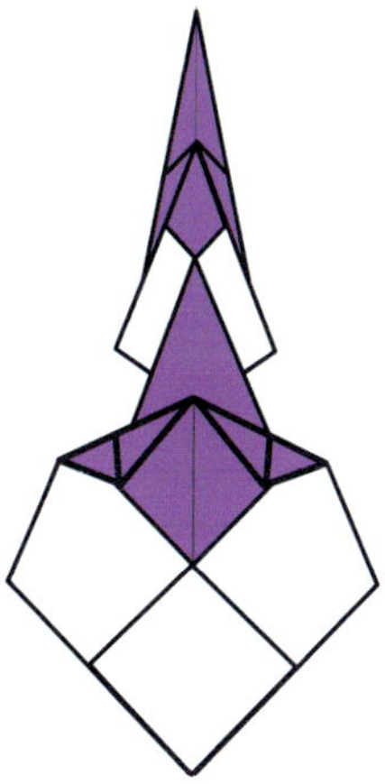

11. Insert into Back Piece Ten.

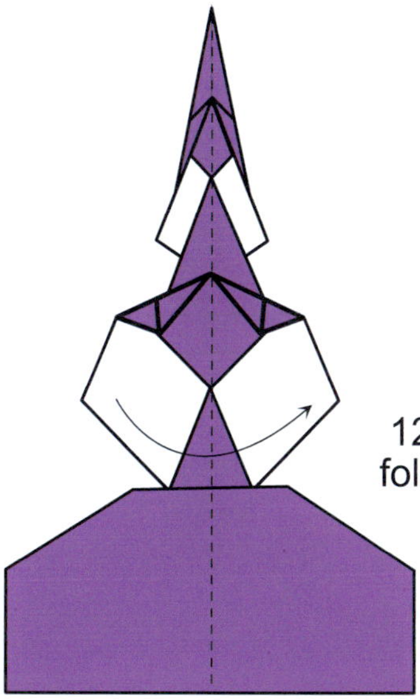

12. Valley fold in half.

Front View

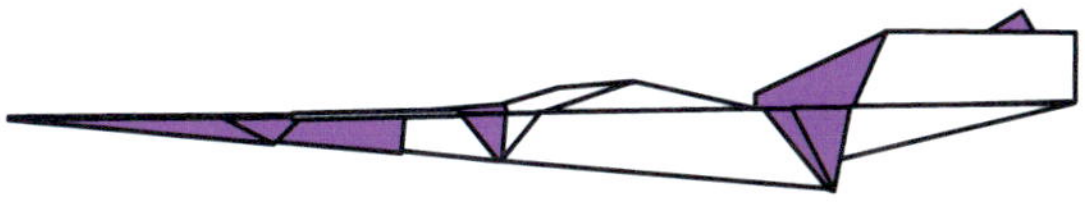

Side View

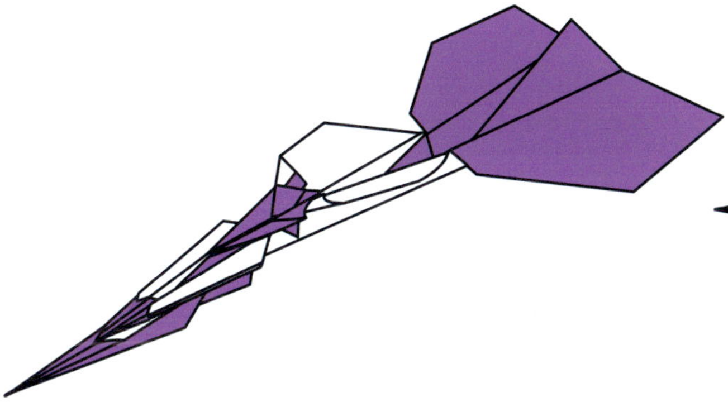

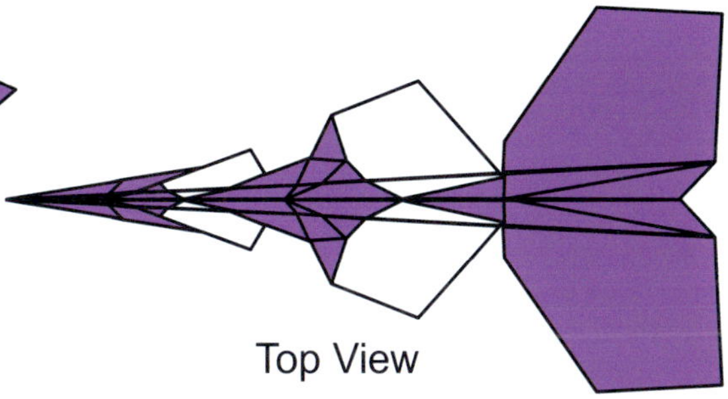

Top View

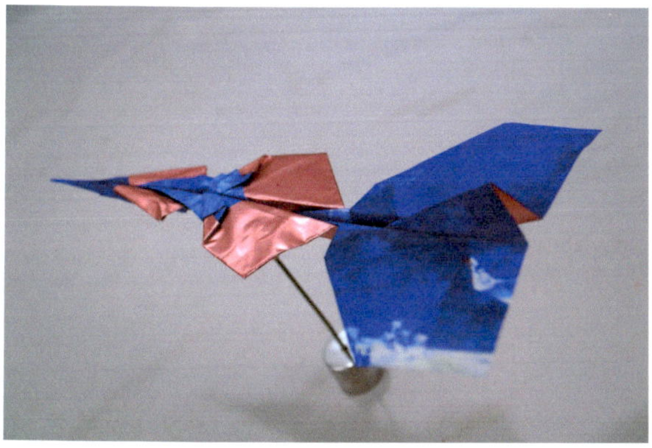

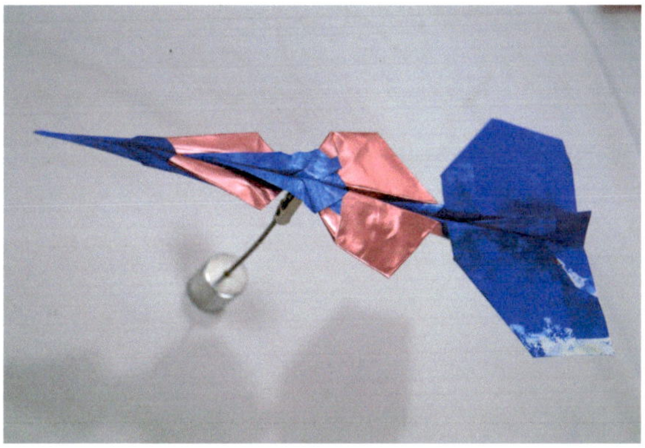

Made in the USA
Monee, IL
05 November 2026

67387504R00079